AF573891

Clinical Research Methods for Surgeons

Clinical Research Methods for Surgeons

Edited by

David F. Penson, MD, MPH

Departments of Urology and Preventive Medicine
Keck School of Medicine, University of Southern California
Los Angeles, CA

John T. Wei, MD, MS

Department of Urology, University of Michigan Medical School
Ann Arbor, MI

Foreword by

Lazar J. Greenfield, MD

Professor of Surgery and Chair Emeritus
University of Michigan Medical School
Ann Arbor, MI

HUMANA PRESS ✱ TOTOWA, NEW JERSEY

999 Riverview Drive, Suite 208
Totowa, New Jersey 07512

www.humanapress.com

This publication is printed on acid-free paper. ∞

ANSI Z39.48-1984 (American Standards Institute) Permanence of Paper for Printed Library Materials.

Cover design by Patricia F. Cleary

For additional copies, pricing for bulk purchases, and/or information about other Humana titles, contact Humana at the above address or at any of the following numbers: Tel: 973-256-1699; Fax: 973-256-8341; E-mail: humana@humanapr.com, or visit our Website: http://humanapress.com

Printed in the United States of America. 10 9 8 7 6 5 4 3 2 1

eISBN 1-59745-230-0

Library of Congress Cataloging-in-Publication Data

Clinical research methods for surgeons / edited by David F. Penson, John T. Wei.
p. ; cm.
Includes bibliographical references and index.
ISBN 1-58829-326-2 (alk. paper)
1. Medicine--Research--Methodology. 2. Surgery. [DNLM: 1. Biomedical Research--methods. 2. Surgery--methods. 3. Research Design. WO 20 C641 2006] I. Penson, David F. II. Wei, John T.
R850.C56 2006
610.72--dc22

2006002639

To our mentors, colleagues, and trainees,
whose collective experiences have been essential
to the conceptualization and compilation of this text

Foreword

With his keen analytical mind and penchant for organization, Charles Darwin would have made an excellent clinical investigator. Unfortunately for surgery, his early exposure at Edinburgh to the brutality of operations in 1825 convinced him to reject his father's plan for his career and pursue his interest in nature. His subsequent observations of how environmental pressures shaped the development of new species provided the essential mechanism to explain evolution and the disappearance of those species that failed to adapt. Today, surgeons face the same reality as new technology, progressive regulation by government and payers, medico-legal risks, and public demands for proof of performance force changes in behavior that our predecessors never imagined.

We know that surgeons have always prided themselves on accurate documentation of their results, including their complications and deaths, but observational studies involving a single surgeon or institution have given way to demands for controlled interventional trials despite the inherent difficulty of studying surgical patients by randomized, blinded techniques. That is why this book is so timely and important. In a logical and comprehensive approach, the authors have assembled a group of experienced clinical scientists who can demonstrate the rich variety of techniques in epidemiology and statistics for reviewing existing publications, structuring a clinical study, and analyzing the resulting data. As these techniques become incorporated as standards into the curriculum of medical, public health, and nursing schools, the surgical professions must include them in their graduate training programs, professional meetings, and reporting practices. To ignore these new standards is to risk failing to continue to attract the best and brightest students into the field and becoming labeled as more technologically than scientifically advanced.

Recent evidence suggests that even the most rigorously designed randomized clinical trial can be corrupted by biased reporting or data withheld on adverse events. The potential threat of industry control of such information must be a part of the training and review process as clinical research becomes more dependent on industry funding. Full disclosure of business relationships between industry and clinician-investigators has been a good start in defining ethical limitations, but it is essential that full disclosure include the registry of all clinical trials in a national database as recommended by the Consolidated Standards of Reporting Trials statement (*Ann. Intern. Med.* 2004; 141:781–788) and adopted by the International Committee of Medical Journal Editors in 2004. These editors declared that its members would not publish the results of trials that had not been publicly registered, and most surgical journals have followed this lead. Currently there are several registries in existence and the World Health Organization is working on an online portal that would bind these databases into a single source.

Darwin taught us that change in response to environmental pressures is essential to survival of the species, and leads not only to successful adaptation, but also to new directions for potential development. Surgeons have always been leaders in exploring new fields and this book will be a useful guide to better methods of clinical research. We should be grateful to the authors for pointing the way; the rest is up to us.

Lazar J. Greenfield, MD
Professor of Surgery and Chair Emeritus
University of Michigan School of Medicine
Ann Arbor, MI

Preface

Clinical research is the branch of scientific endeavor devoted to the evaluation of patients and the analysis of associated health outcomes. These analyses serve to identify potential areas for change in physician or patient behavior or in clinical processes. Implicit in the concept of clinical research is the notion that the findings will be used to modify clinical practice to achieve better outcomes. As such, clinical research has always been a necessary prerequisite for the advancement of surgery as a practice.

In the past decade, basic science research in the surgical disciplines has advanced at a dizzying pace. Clinical research in surgery, however, has lagged far behind surgical basic science research. For example, the selected case series from a single academic center still remains one of the most common study designs employed by surgeons who address clinical research questions, despite the known limitations of this design. Although such clinical research techniques were appropriate 50 years ago—when the primary focus was on advances in surgical technique—they are inadequate for addressing the broader policy issues and clinical management questions faced by the surgeon today. The clinical research questions facing surgeons in the 21st century require sophisticated research techniques that most surgeons are, at best, only vaguely familiar with and, at worst, completely unfamiliar with.

Evidence-based medicine is the foundation on which clinical research is built and is the explicit use of scientific data in decision making for clinical care. It is as critical to surgical practice as to any other medical discipline. A requisite for evidence-based practice is the availability of high levels of evidence. Our colleagues in internal medicine have successfully adopted clinical research methods and have disseminated this information to trainees and practicing physicians through textbooks, educational series, and fellowship programs, such as the Robert Wood Johnson Clinical Scholars Program. Although it may be tempting to use existing resources to educate surgeons in clinical research methods, one must remember that diseases requiring surgical treatment are often unique, and that many of the methods used for looking at research questions in internal medicine are not easily applied to the surgical fields. Patients faced with the prospect of a major surgical procedure must often deal with physical and psychological challenges as a result of treatment that are quite distinct from those facing patients undergoing medical therapy for chronic illness. Existing methodologies used in internal medicine, pediatrics, or other nonsurgical fields will fail to capture some of the distinct aspects in surgical diseases. It is incumbent on the surgeon-scientist to understand clinical research methodology and to develop new techniques for addressing important research questions. This need for new well-trained clinical researchers, and original clinical research in the surgical fields, is so great that funding agencies such as the National Institutes of Health, the American College of Surgeons, the American Academy of Head and Neck Surgeons, and the American Urological Association Foundation have specifically allocated research funding to assist in the development of physicians with formal training in clinical epidemiology and health services research to improve clinical research in the surgical disciplines. This trend will no doubt increase in the coming years as the complexity of clinical research questions further increases.

The purpose of *Clinical Research Methods for Surgeons* is to provide the surgeon with an easy-to-use guide for interpreting published clinical research. With so many articles published even in the most arcane of surgical journals, the need to separate the wheat from the chaff requires one to be critical when reviewing the study design and methods for each article. This book is also intended to serve as a reference guide for the surgeon who wishes to conduct clinical research either to answer scientific, hypothesis-driven questions or simply to evaluate his or her outcomes. The book itself is divided into four parts. The first provides a general overview of the infrastructure of clinical research. It describes the thought process required for undertaking new studies and discusses both the ethical and financial issues involved in running a clinical research unit. It should be of particular interest to young surgeons who are about to undertake new studies. The second part describes specific study designs and statistical techniques used in clinical research, whereas the third part describes methods for assessing clinical outcomes. These two sections will be of interest to both investigators interested in performing clinical research and those who are just reviewing the literature and applying it to their practice. Finally, the last section addresses special research techniques and topics that will certainly be of interest to the active investigator. As a whole, *Clinical Research Methods for Surgeons* provides insights to the clinical investigator and clinician reading the literature.

It is our sincere hope that the text will allow the reader to have a clear understanding of clinical research methods. However, it is still highly recommended that the surgeon develop collaborations with an experienced analyst or a biostatistician if the surgeon himself or herself has no such expertise. This is no different than in basic science research, where the mantra has been to train surgeons as "translational scientists" to bridge the gap between the laboratory and the bedside. So the same applies to clinical research where we need to bridge the gap between epidemiological/statistical science and the bedside.

Although it is always tempting for a surgeon to report his or her "experience" on a topic, it is important to recognize that this does not provide high levels of evidence and will undoubtedly not change practice. High-quality surgical research takes time, involves planning, and, most importantly, requires an appreciation of methods and the clinical setting. If a clinical research project is worth doing, then it is worth doing right. Surgeons must be open minded about learning new clinical research methods so that horizons can be expanded and patient care improved. Failure to do so may lead surgery to become stifled and allow other parties to dictate the care of surgical patients. In the end, use of high-quality clinical research methods is a necessity for the surgical discipline as our practices expand with new basic science discoveries and new surgical techniques, and as other scientific discoveries abound. The reader is encouraged to become familiar with these methods and to incorporate them into his or her surgical practice. In this new millennium, the enlightened clinical researcher/surgeon must command a thorough understanding of the latest methodologies for analyzing clinical data. It is our sincere hope that this text will be the first step in that direction.

David F. Penson, *MD, MPH*
John T. Wei, *MD, MS*

Contents

Contributors

PETER C. ALBERTSEN, MD, MS • *Professor and Chief, Division of Urology, University of Connecticut Health Center, Farmington, CT*

DAVID A. AXELROD, MD, MBA • *Assistant Professor of Surgery, Surgical Director of Kidney and Pancreas Transplant, Dartmouth Medical School, Hanover, NH*

DONNA L. BERRY, PhD, RN, AOCN, FAAN • *Professor and Vice Chair of Research, Biobehavioral Nursing and Health Systems, University of Washington; Adjunct Associate Professor of Urology, University of Washington School of Medicine, Seattle, WA*

EUGENE H. BLACKSTONE, MD • *Head, Clinical Research, Department of Thoracic and Cardiovascular Surgery; Staff, Department of Quantitative Health Sciences, Cleveland Clinic, Cleveland, OH*

JUSTIN B. DIMICK, MD • *Postdoctoral Fellow, Department of Surgery, University of Michigan Medical School, Ann Arbor, MI*

RODNEY L. DUNN, MS • *Biostatistician, Department of Urology, Comprehensive Cancer Center, University of Michigan, Ann Arbor, MI*

WILLIAM J. ELLIS, MD • *Professor, Department of Urology, University of Washington School of Medicine, Seattle, WA*

DAVID ETZIONI, MD, MPH • *Public Health Sciences Division, Fred Hutchinson Cancer Research Center, Seattle, WA*

RUTH ETZIONI, PhD • *Public Health Sciences Division, Fred Hutchinson Cancer Research Center, Seattle WA*

JUDITH FINE • *Research Coordinator, Division of Urology, University of Connecticut Health Center, Farmington, CT*

HOWARD A. FINK, MD, MPH • *Assistant Professor of Medicine, Geriatric Research Education and Clinical Center; Minneapolis VA Center for Chronic Disease Outcomes Research, Minneapolis, MN*

ARVIN KORUTHU GEORGE, BA • *Division of Urology, Beth Israel-Deaconess Medical Center, Boston, MA*

LAZAR J. GREENFIELD, MD • *Professor of Surgery and Chair Emeritus, University of Michigan Medical School, Ann Arbor, MI*

MARYLOU V. H. GREENFIELD, MPH, MS • *Department of Anesthesia, University of Michigan Medical School, Ann Arbor, MI*

RODNEY HAYWARD, MD • *Professor of Medicine, Associate Director, Robert Wood Johnson Clinical Scholars Program, Department of Internal Medicine, University of Michigan Medical School, Ann Arbor, MI*

WILLIAM G. HENDERSON, MPH, PhD • *Professor, University of Colorado Health Outcomes Program and Senior Biostatistician, NSQIP, University of Colorado School of Medicine, Denver, CO*

NADIA HOWLADER, MS • *Department of Biostatistics, University of Washington School of Medicine, Seattle, WA*

SHUKRI F. KHURI, MD • *Chief, Cardiothoracic Surgery, Boston VA Healthcare System, and Professor of Surgery, Harvard Medical School, Boston, MA*

CLIFFORD Y. KO, MD, MS, MSHS • *Associate Professor, David Geffen School of Medicine, University of California, Los Angeles; Greater Los Angeles Veteran Affairs Health Care System, Los Angeles, CA*

MARK S. LITWIN, MD, MPH • *Professor of Urology and Health Service, Associate Chairman of Urology, David Geffen School of Medicine, University of California, Los Angeles, Los Angeles, CA*

DEBORAH P. LUBECK, PhD • *Associate Director, Health Economics and Outcomes Research, Amgen, Thousand Oaks, CA*

SALLY L. MALISKI, PhD, RN • *Assistant Researcher, Jonsson Comprehensive Cancer Center, Department of Urology, David Geffen School of Medicine, University of California at Los Angeles, Los Angeles, CA*

JULIE C. MCLAUGHLIN, MPH, MS • *Research Associate, Department of Urology, University of Michigan Medical School, Ann Arbor, MI*

DAVID C. MILLER, MD, MPH • *Lecturer and Research Fellow, Department of Urology, University of Michigan, Ann Arbor, MI*

JESSICA B. O'CONNELL, MD • *Department of Surgery, David Geffen School of Medicine, University of California, Los Angeles; Greater Los Angeles Veteran Affairs Heath Care System, Los Angeles, CA*

DAVID F. PENSON, MD, MPH • *Associate Professor, Departments of Urology and Preventive Medicine, Keck School of Medicine, University of Southern California, Los Angeles, CA*

MICHAEL P. PORTER, MD, MS • *Assistant Professor, Department of Urology, Adjunct Professor, Department of Epidemiology, University of Washington School of Medicine, Seattle, WA*

ANDREW L. ROSENBERG, MD • *Assistant Professor, Department of Anesthesia; Director, Critical Care Research, University of Michigan Medical School, Ann Arbor, MI*

MARTIN G. SANDA, MD • *Associate Professor, Division of Urology, Beth Israel-Deaconess Medical Center, Harvard School of Medicine, Boston, MA*

ARUNA V. SARMA, PhD, MHA • *Assistant Research Professor, Departments of Urology and Epidemiology, University of Michigan, Ann Arbor, MI*

LYNN STOTHERS, MD, MHSc, FRCSC • *Director, Bladder Care Centre, Associate Professor of Urology, Associate Member, Department of Health Care and Epidemiology, University of British Columbia, Vancouver, British Columbia, Canada*

JOHN T. WEI, MD, MS • *Associate Professor, Associate Chairman for Clinical Research, Department of Urology, University of Michigan School of Medicine, Ann Arbor, MI*

TIMOTHY J. WILT, MD, MPH • *Professor of Medicine, Section of General Internal Medicine, University of Minnesota School of Medicine; Minneapolis VA Center for Chronic Disease Outcomes Research, Minneapolis, MN*

I

Building a Foundation for Clinical Research

1 Planning the Research

Eugene H. Blackstone, MD

CONTENTS

1. WHY PLAN FOR RESEARCH?

It would be unthinkable for a surgeon to begin an operation without first formulating a plan. The plan has *specific objectives,* whether it is to remove a tumor, repair a traumatized visceral organ, correct a congenital anomaly, or achieve a cosmetic result. It embodies a *temporal sequence of implementation steps* from surgical incision to closing that are intended to achieve the objective. It should include at least immediate, if not long-term, *assessment of success* of the procedure in reaching its objectives. Even before operation, however, *alternative therapies* must be considered. As Kirklin and Barratt-Boyes originally pointed out, it is only after considering the data for alternatives that the indications for surgical therapy can be formulated *(1–3)*.

Clinical research is no different: its most important key is a *research plan*. Thus the format of this chapter is based on a template research plan (*see* Appendix, p. 28) that has a proven track record in (1) achieving research objectives, (2) identifying the sequence of purposeful steps that facilitates generating and disseminating new knowledge, and (3) avoiding pitfalls that result in unsuccessful and frustrating research experiences. As with an operative plan, however, the role of clinical research in advancing the knowledge and practice of surgery must be appreciated, as must alternatives to pursuing that objective.

1.1. WHY UNDERTAKE RESEARCH?

It is clear that advancing the knowledge and practice of surgery is needed to:

- Better understand the underlying disease process
- Appreciate superimposition of anesthesia and surgical trauma on accompanying acute, chronic, and genetically driven disease processes
- Generate new concepts, more effective operations, and less-invasive equivalent treatments

From: *Clinical Research for Surgeons*
Edited by: D. F. Penson and J. T. Wei © Humana Press Inc., Totowa, NJ

- Make evidence-based individual patient care decisions *(4)*
- Obtain objective informed consent from patients for operations *(5)*
- Improve short- and long-term surgical outcomes
- Assess quality and appropriateness of care *(6)*
- Contribute to developing a rational basis for regulatory decisions, including addressing ways in which surgical treatment can be made more efficient, more cost effective, and more accessible

Research is but one of several means of advancing the knowledge and practice of surgery and achieving these goals. Specifically, patient-centered research (also known as clinical research) is particularly effective because it provides the opportunity to take the next direct step beyond research—namely, putting into clinical practice the inferences derived from research, a process termed "development" in the business world.

1.2. What Is Clinical Research?

Although the definition of clinical research may seem intuitive to most clinicians, it is still important to specifically define the broad spectrum of activities that make up this important discipline. In response to this need, an American Medical Association Clinical Research Summit and subsequent ongoing Clinical Research Roundtable (established by the Institute of Medicine) endeavored to define the broad spectrum of activities that make up clinical research (Box 1) *(7)*. Although by no means a succinct definition, its important aspects are that (1) clinical research is but one component of medical and health research aimed at producing new knowledge, (2) knowledge gained should aid in understanding the nature of disease, its treatment, and prevention, and (3) clinical research embraces a wide spectrum of categories of research.

1.3. What Is the Value of Research?

From a patient-centered perspective, the value of research lies in its positive impact on disease. To realize that value, research results must be disseminated in presentations and publications. Research that is conducted but never culminates in a peer-reviewed publication has limited impact on patient care. The full value of published research is achieved when investigators take a step beyond summarizing results and draw clinical inferences translated into new patient management strategies aimed at reducing therapeutic failures. From a socioeconomic perspective, there is little doubt that the economic value of clinical research outweighs its costs, as documented in the scholarly and sophisticated set of articles collected by Murphy and Topel *(8)*.

2. BASIS FOR A RESEARCH PLAN

Our current understanding of a viable clinical research plan in the 21st century is based on three separate principles. The first is embodied in the statement, "Let the data speak for themselves." This thought hardly seems radical today, but it turned the direction of scientific investigation upside down in the 18th century. Originally proposed by Sir Isaac Newton, this then-novel idea led to a new method of investigating the nature of natural phenomena *(9)*.

Newton's method had two strictly ordered phases. The first phase was *data analyses*, whereby observations of some small portion of a natural phenomenon were examined and dissected. The second phase was *synthesis*, whereby possible causes for the observations

Box 1: Definition of Clinical Research

Clinical research is a component of medical and health research intended to produce knowledge valuable for understanding human disease, preventing and treating illness, and promoting health. Clinical research embraces a continuum of studies involving interaction with patients, diagnostic clinical materials or data, or populations, in any of these categories:

- disease mechanisms;
- translational research;
- clinical knowledge, detection, diagnosis, and natural history of disease;
- therapeutic interventions, including clinical trials;
- prevention and health promotion;
- behavioral research;
- health services research;
- epidemiology; and
- community-based and managed care-based research.

Source: Association of American Medical Colleges Task Force on Clinical Research. For the health of the public: ensuring the future of clinical research. Washington, DC: AAM, 1999, p. 16.

were inferred, revealing some small aspect of Nature *(10)*. He thus formalized the *inductive method* in science: valuing first and foremost the observations made about a phenomenon, then "letting the data speak for themselves" in suggesting possible natural mechanisms.

The antithesis of Newton's inductive method was the *deductive method* of investigation that was successful in developing mathematics and logic, but less successful in advancing natural sciences. The deductive method began with what was believed to be the nature of the universe ("principles" or "fundamentals," but referred to by Newton as "hypotheses"), from which logical predictions were deduced and tested against observations. If observations deviated from logic, the data were suspect, not the principles behind the deductions; the data did not speak for themselves. Newton realized that it was impossible to possess complete knowledge of the universe. Therefore, a new inductive methodology was needed to permit use of observations of just portions of Nature, with less emphasis on synthesizing the whole. The clinical research plan is based directly on inductive Newtonian ideas for proceeding from observation to inference.

The second basis for the research plan is the structure of a scientific paper as formalized by Louis Pasteur. He established the "IMRD" format for reporting scientific information: Introduction, Methods, Results, Discussion *(11–13)*. Although it provides a valuable structure for thinking about, and expressing clearly, the findings of one's research, this format is also useful for formulating the plan for research from its inception so that it fits neatly into this reporting structure. The introduction sets forth the background of the research and its purpose. The methods identify the "material" (patients), the observations and how they were made, and the methods of analysis. The results present some portion of the findings of various investigations. The discussion synthesizes the results into the context of what is known about a phenomenon and draws inferences about what the research means.

The third basis for the research plan is the observation of both successful and unsuccessful research endeavors. Thus, as with the research study itself, the research plan arises from inferences based on observation! The research plan template presented here is one that has evolved over a 40-yr involvement in clinical research, mostly as a full-time physician-investigator, and is pragmatically focused on those aspects that generally lead to success. Nevertheless, it is likely somewhat incomplete, flawed in certain respects, more generalized than may be warranted, and reflects observer bias.

3. PLANNING THE RESEARCH

The vehicle selected to convey the ideas of planning for successful research is the template clinical research proposal format used routinely with our surgical residents, fellows, and faculty, and for studies involving a multidisciplinary collaborative research group. It's the basis for setting priorities for research, for assigning or obtaining resources for conducting the study, and for initiating the study. Importantly, it serves as a valuable tool for communication within the group and as a checklist and reality check during conduct of the research.

The research plan is not a linear workflow document. Rather, it represents the end result of a number of steps that include dependencies and iterative and collaborative reevaluation in the process of its development. It cannot be put together hurriedly to get a study under way; rather, success of a study depends greatly on the care given to the plan.

Nevertheless, the research plan is not a static proposal, but one that is refined during the course of discovery *(14)*. Writing the initial plan takes place over the course of a few days to weeks, but the process is not protracted unduly to contemplate every conceivable contingency. The research plan ultimately forms the basis for internal reporting of results of various phases of the research as well as for the content of the resulting manuscript. The clinical research protocol template is presented in annotated outline form in the Appendix. It is a reorganized and refined version of the one presented in Chapter 6 of *Cardiac Surgery,* 3rd edition *(3)*. The text that follows discusses each of its elements in turn.

3.1. Title

Another way to approach the title of the research is to consider the question: "Are you just working, or are you working on something?" Even it its formative stage, a research study needs a title that succinctly and accurately answers the question: "What are you working on?" The title should be as short as possible, but must clearly reflect the question (topic) being addressed. If the title is broad, such as "Results of Coronary Artery Bypass Grafting" or "Esophagectomy," the topic is likely unwieldy and the research unfocused. In contrast, "Postoperative Bleeding after Cardiac Surgery: Risk Factors, Impact on Hospital Outcomes, and Clinical Implications" is still a large project, but focuses clearly on: (1) the topic (postoperative bleeding), (2) target population (cardiac surgical patients), (3) approach (risk factor analysis), (4) end points (hospital outcomes), and (5) inferences (clinical implications). It is improbable that the title of a proposal will be that of the finished manuscript. Rather, it will likely evolve in the process of performing the research and then distilling its essence *(15)*.

3.2. Contributors

Successful clinical research is rarely accomplished without collaborators. Too often, names are attached to manuscripts as a courtesy after the research is completed, without meaningful input from these individuals. Indeed, Rennie and colleagues propose dispensing with the notion of authorship of publications altogether and adopting the concept of contributors and guarantors of integrity of the research *(16)*.

Collaborators (contributors) should include fellow surgeons, and one should not shy away from those who may hold different opinions! Because surgeons often work within the context of some disease-related discipline, medical colleagues and other clinicians or health care providers in that discipline likely will add valuable insight as well and should be included as key collaborators.

Quantitative scientists—in this context, those with broad expertise in analyzing the data, including biostatisticians, epidemiologists, outcomes researchers, and possibly those involved in bioinformatics, quantitative genetics, and other quantitative disciplines—should not be considered simply service providers. Rather, these contributors, professionally trained in research design, should be brought in early in the planning process. They can ascertain sample size and other aspects of feasibility, an essential task in determining whether a study should go forward. Investigators frequently expend enormous effort gathering data but no effort constructing the plan for analysis. Sometimes this results in such a mismatch that no appropriate analysis can be performed to address the original research question.

Data management experts are a vital link between a proposal and actually having data to work with for determining results of the study. They should provide insight into what variables are available from electronic sources and what will need to be collected *de novo*; they can construct controlled-vocabulary databases for entry of additional data; and they will eventually be involved in cleaning and formatting the data in such a way as to make them analyzable *(17)*.

Think about each end point and whether a person expert with respect to it should be consulted. This may mean imaging experts, pathologists, psychologists, immunologists, and others. Often a single meeting with a number of these persons, with individual follow-up, can importantly enhance a proposal. On the other hand, multiplicity of input can have the effect of unfocusing the project or directing it along an undesirable tangent; advice has to be sifted! This is best done at the end of such a meeting, so that there is general agreement on "next steps" and who will be responsible for taking them. In this way, the roles, responsibilities, and expectations of each contributor are established.

3.3. Research Question

The most important ingredient for successful research is a well-framed research question. This is sometimes called the study objective, aim, purpose, or hypothesis. The research question must be well formulated, clearly stated, and focused. Examples of poorly focused questions are:

- What is our experience with tracheostomy?
- Why don't you look up our experience with postinfarct ventricular septal defects (VSD)?
- Is positron emission tomography (PET) any good?

In each of these cases, several interesting, focused questions could have been asked that would lead to successful research. For example:

- Tracheostomy
  a. What is the time-related mortality of patients who receive tracheostomies after a specified surgical procedure?
  b. What factors are associated with tracheostomy after a specified surgical procedure?

c. Can subgroups of patients be identified for whom tracheostomy is a futile intervention and others for whom results are favorable?
d. Has changing technique of tracheostomy from rigid to flexible tubes reduced tracheal stenosis?

- Postinfarct VSD

a. What factors influence the interval from infarction to VSD development?
b. What are the patient and procedural risk factors for time-related mortality after postinfarct VSD repair?
c. What is the predicted survival of patients who have been turned down for repair of postinfarct VSD compared to actual?

- PET

a. What is the sensitivity and specificity of PET in detecting clinical stage of lung cancer?
b. What is the work-up bias-adjusted diagnostic value of PET in various stages of working up lung cancer *(16,18,19)*?
c. Does PET add diagnostic information over and above that from spiral computed tomography of the chest?

The research question should be revisited, revised, restated, and kept uppermost in mind throughout the study and its eventual presentation. The study cannot be initiated without this step, because the study group, end points, variables, analyses, and feasibility all depend on it.

Research questions arise from clinical observations, often from clinical failures the surgeon feels helpless to prevent in future patients. Counsell suggests that some of the most relevant questions are those asked directly or indirectly by patients *(20)*. Their questions often relate to their disease (diagnosis), expected survival (prognosis), treatment options (treatment evaluation, preventive measures), and whether the latest "cure" they have heard about is any good (therapeutic comparisons).

3.3.1. What Types of Research Questions Should Be Asked?

There are various types of research questions. For statisticians, the type of question often suggests an appropriate "experimental design." What follows is a general taxonomy of the types of research questions that can be addressed.

3.3.1.1. "Is It True?"

One of the pillars of science is testability. Thus some research is aimed at confirming observations of other investigators. When such a study is focused on confirmation and a case is made that confirmation is important, this is a worthy type of study to pursue. An important class of confirmatory study is meta-analysis of randomized or nonrandomized studies or of individual patient data *(21–24)*. However, too often, confirmatory studies are redundant, reflecting inadequate preparation of the research plan in identifying key literature citations or failure to consult knowledgeable colleagues.

3.3.1.2. "What Is the Truth?"

Sometimes research is pursued to question "conventional wisdom" or what has become established knowledge. If, as some of us suspect, a major failing of published studies is underpowering (insufficient data), then there is plenty of room for studies that focus on factors that were not found to be associated with outcomes, those not accounted for in previous studies that may make a difference in inferences, or those that reflect what is believed to be flawed thinking. An important class of such studies is elucidation of confounding variables and surrogate end points *(25)*. A confounding variable is one that is associated with both the primary variable being studied and one or more end point of interest in such a way as to distort relationships, falsely magnifying or suppressing real relationships and biasing inferences *(26)*. A typical example is the apparent association of alcohol intake with laryngeal cancer that may spuriously represent a causal relation to cigarette smoking. This comes about because patients who drink often smoke. A legitimate avenue of research may be to raise the question of confounding with new information that attempts to measure and account for it.

3.3.1.3. "Is It Better (Different)?"

Clinical trials generally are comparative studies of two or more treatments, including surgical vs medical strategies. Increasingly, special statistical techniques are being used for making causal inferences based on observed data *(27–30)*. The claim of developers of these techniques is that they bridge the gap between identifying mere *associations* with outcome and discovering *causation (30)*. These techniques can be useful when it is not feasible to perform a randomized clinical trial because of patient resistance to randomization (which is often the case with surgical trials), high costs of performing a randomized trial, or other barriers. Although these techniques can eliminate some of the selection bias commonly seen in observational studies, it is quite difficult to eliminate all confounding from surgical studies given the unique nature of the intervention.

3.3.1.4. "What Don't We Know?"

There are huge gaps in medical knowledge, and a fundamental purpose of research is to fill one or more of these gaps. Most variables associated with disease, its treatment, and its outcome represent weak associations or surrogates for causes and mechanisms *(31)*. Thus there are many gaps in knowledge to fill. It is not surprising, then, that reviewers of manuscripts are generally asked whether the research is "new." The goal for most clinical research studies should be discovery of something *novel*. The introduction of the 50-word (maximum) ultra-miniabstract into *The Journal of Thoracic and Cardiovascular Surgery (15)* was to encourage authors to identify for readers (and reviewers before them) the essence of their study—what is new, what has been discovered, what question has been answered, and what gap in knowledge has been filled?

3.3.2. Significance and Relevance of the Research Question

This raises the important issue of the significance and relevance of the research question. Knowledge for knowledge's sake is valuable. However, clinical research is generally focused on knowledge that is hoped to improve outcomes for patients in the future. Thus a research question may be well focused, but if no one cares what the answer is, the research will likely fail, either from lack of support or rejection of the ultimate manuscript. Thus one test of a good research question is *clinical relevance*, which translates into the importance attached to the answer among those experienced in the field.

3.3.3. What Is the Next Step in the Research?

After a research question has been identified and answered, the researcher may ask, "What is the next logical step?" Many studies turn into a sequence of studies that examine different aspects of a phenomenon, with each research question arising from the previous study. For example, in a protracted series of studies, we examined the natural history of tetralogy of Fallot *(32)*, primary vs staged repair *(33)*, quantitative relations of somatic growth and pulmonary valve and artery growth *(2)*, prediction of postrepair pressures on the basis of angiography for surgical decision making *(34)*, the status of pulmonary artery arborization in tetralogy with pulmonary atresia *(35)*, and so forth. In no case did we go back and repeat a study; rather, one study naturally led to another in a sequence of "logical next steps."

It is difficult to provide any general advice about how to recognize the "next logical step or steps" other than the following observation: it is not uncommon in the course of pursuing a research study that the assumptions or knowledge base on which the study was thought to be grounded is found to be faulty or incomplete. A decision then must be made as to whether the next logical step is to take a step backward and fill in those gaps before pursuing a study further, or to continue the study and come back later to fill in details, or to pursue multiple leads simultaneously. Another observation is that a study often raises questions that naturally lead to new research. For example, in the study of repair of tetralogy of Fallot mentioned previously, differences were found in outcome between those with and without pulmonary atresia, leading to questions about the subgroup with pulmonary atresia that became the next logical step. A study of an interesting echocardiographic marker of outcome may lead to a quantitative 3D image reconstruction study to further clarify the findings.

3.3.4. Is the Research Well Aligned?

Even when a research question is good, the research may be doomed if it is pursued in an environment that is not well aligned with the clinical interests of the investigator. A vascular surgeon asking important research questions about the thoracic aorta in collaboration with cardiovascular surgeons has well-aligned research and clinical interests. In contrast, an obstetrician studying myocardial protection during heart surgery has misaligned research and clinical interests. The ideas may be good (isn't a slowly contracting uterus just a slow-motion analog of the preejection phase of heart contraction?), but credibility of the investigator to those in heart surgery may preclude serious consideration of the work. Similarly, colleagues may question why he or she is not addressing pressing questions in obstetrics! On the other hand, a collaborative study of fetal cardiac surgery by this obstetrician and a cardiac surgeon represents the interface of disciplines, and it is such interfaces that generate the most exciting research ideas and discoveries.

3.4. Background

No matter how novel the research, it does not take place in a knowledge vacuum. A thorough review of what is known about the research topic has several purposes. First, from textbooks, literature, or discussions with collaborators, it provides an overview of the context of the research. For example, if one is proposing a reclassification scheme for esophageal cancer, it would be important to acquire an overview of esophageal cancer and its treatment as well as current staging classification and its perceived difficulties.

Second, review of what is already known establishes the status of the specific area of research. Some areas are so overresearched that it may not be possible to propose anything novel. An example is the status of the field of myocardial protection in cardiac surgery, for which there are tens of thousands of publications. In such a case, a recent textbook or in-depth review article may be most helpful in identifying remaining gaps in knowledge, followed by a pinpoint search of literature relevant to the proposed research.

Third, a thorough review of the existing knowledge is needed to ascertain whether research mirroring what you have proposed has already been done. If so, and the conclusions have been confirmed, life is too short and the literature too replete to undertake a redundant study. On the other hand, you may think there is valid ground to question what has been done or the approach taken, in which case the research is no longer redundant.

Fourth, a review of the existing knowledge reveals how others have approached this or similar questions. What data have been gathered? Have important correlates or associations with outcomes been identified? What data analysis methods have been employed? At the same time, be cautious about insisting on using substandard research methods simply because they have been used in the past. For example, in ascertaining clinical status of patients after surgery, it was once common practice to record and analyze just the status at last follow-up. Preoperative and postoperative status was then simply compared. Such an approach to data is flawed because (1) it ignores possible temporal changes in status; (2) if temporal changes are considered, it assumes overoptimistically that a single assessment of each patient will give an accurate representation of individual temporal changes; (3) it does not use all the available data from previous follow-ups; and (4) it likely ignores death as a censoring mechanism (no longer able to ascertain status). There are important new analytical methods that permit great insight into the pattern of temporal outcomes *(36,37)*.

Other egregious examples not to emulate include dichotomization of continuous variables and failure to recognize skewed distributions of continuous variables (such as many laboratory values, length of stay, and financial data). Helpful information on how to assess such matters in detail has been provided by Ferraris and Ferraris *(38)*, who refer to a growing literature on misuse of statistical methods in even the best medical journals.

3.4.1. How to Review the Existing Knowledge on a Research Topic

As with performing a difficult operation, the surgeon-researcher must have a strategy for reviewing the existing knowledge about a specific research question. There are a number of resources researchers should consider when developing their strategy.

3.4.1.1. Contributors

In discussing a project seriously with collaborators and colleagues (one of the keys to successful research), you should poll them for what they believe to be the key references relevant to the proposed research. If one or more of these contributors is a senior person, he or she may also be able to bridge the gap to older literature that may otherwise be overlooked from the false assumption that the most important work is recent work. A few relevant references may lead you to other specific relevant references.

Contributors can also identify individuals outside your institution who are active in the particular field of research. Performing a search of those individuals' papers and of some of their references may quickly lead to a set of key references without the necessity of a general literature search.

3.4.1.2. Reviews and Textbooks

Although it may be argued that review articles and textbooks are out of date the moment they are published, these should be treated as a primary resource for an overview of the research context. They also are a source of individual citations that may deal with a more focused aspect of the picture that is relevant to your proposed research.

3.4.1.3. Medical Librarians

Particularly if your proposal is an observational study of the literature (a common form of meta-analysis) *(24)* or a thorough review of the literature, a medical librarian is indispensable. In this type of research, a comprehensive literature search is necessary. Medical librarians are trained in eliciting the many key words that result in a sophisticated and complete literature search. It is rare that a physician is sufficiently familiar with this linguistic search process to perform an adequate search *(20)*.

3.4.1.4. Online Search

The National Library of Medicine (NLM) in the United States hosts a web-based publicly accessible system to search its many online databases. "Gateway," accessed at http://gateway.nlm.nih.gov/gw/Cmd, is more comprehensive than its NLM companion PubMed®, found at http://www.ncbi.nlm.nih.gov/entrez/query.fcgi, but has a less intuitive syntax for searching. Sources of information used by Gateway in searching are listed in Table 1. Reference managers such as EndNote®, which are essential tools for preparing manuscripts, can access these and several other online bibliographic sources, accurately download citations and abstracts, and format them as required by various medical journals. Another online resource is standard Web search engines. These are sometimes valuable, but filter far less specifically than NLM databases.

3.4.2. How to Synthesize the Collected Information

After obtaining an overview from review articles and textbooks, studying the literature references provided by collaborators and colleagues, and reviewing the references of those known to be active in the field, you are ready to tackle the sometimes daunting task of sifting through the long list of references provided by the library or generated by online searches. You need a strategy to expedite this task. I suggest a multiple-pass approach.

Pass 1: Title scan. Quickly scan article titles, discarding those clearly not relevant to the specific research question.

Pass 2: Abstract scan. For those articles not discarded in Pass 1, view the abstracts. You need to read only the first line or two and the conclusions.

Pass 3: Detailed abstract review. Read the entire abstract of those deemed relevant in Pass 2. This will further narrow the list.

Pass 4: Synthesis of the literature. This pass is not yet a careful dissection of individual references. Depending on the nature of the research, you are looking for:

- Knowledge that has been generated and seems true
- Gaps in that knowledge
- Inferences that you question
- Methods used in formulating the study group (*see* Chapter 8), gathering values for variables, and analyzing the data
- Duplication of the study you have proposed (in which case you may abandon the study, recognize that there has been no independent corroboration of the study, or question the study)

You are also interested in quickly ascertaining the quality of the article. The higher the quality, the more valuable the background information it provides. This can be determined from the following:

- Credibility of the reference (which may be based on institution or authors and their known contributions to the field, as well as more objective information such as

Table 1
Sources of Information Searched When Using the National Library of Medicine's Gateway Web-Based System

Category	*Collection*	*Data Type*
Journal citations	MEDLINE/PubMed	Journal citations, 1966 to present
	OLDMEDLINE	Journal citations, 1953–1965
Books/serials/AVs	LOCATORplus	Catalog records for books, serials, AV materials
Consumer health	MEDLINEplus Health Topics	Health information from NIH and other sources
	MEDLINEplus Drug Information	Generic and brand name drug information
	MEDLINEplus Medical Encyclopedia	Articles about diseases, tests, symptoms, injuries, and surgeries
	ClinicalTrials.gov	Information about clinical research studies
	DIRLINE	Directory of health organizations
Meeting	Meeting abstracts	Meeting abstracts on selected subjects
Other	HSRProj	Health services research projects in progress

Abbreviation: AV, audiovisual.

MEDLINE® (Medical Literature, Analysis, and Retrieval System Online) is the U.S. National Library of Medicine's (NLM) bibliographic database, containing more than 12 million references to journal articles in life sciences with a concentration on biomedicine.

OLDMEDLINE contains citations published in the *Cumulated Index Medicus* from 1960 through 1965 and in the *Current List of Medical Literature* from 1953 through 1959. OLDMEDLINE covers the fields of medicine, preclinical sciences, and allied health sciences.

LOCATORplus is the NLM's online catalog, including more than 800,000 records for books, audiovisuals, journals, computer files, and other materials in the library's collections.

MEDLINEplus is the NLM's web site for consumer health information.

ClinicalTrials.gov is a registry of clinical trials for both federally and privately funded trials "of experimental treatments for serious or life-threatening diseases or conditions."

DIRLINE (Directory of Information Resources Online) is the NLM's online database containing location and descriptive information about a wide variety of organizations, research resources, projects, and databases concerned with health and biomedicine.

Meeting Abstracts contains meeting abstracts from the former AIDSLINE, HealthSTAR, and SPACELINE databases. It also includes new meeting abstracts on AIDS/HIV, and meeting abstracts from the Academy for Health Services Research and Health Policy (formerly the Association for Health Services Research), the International Society of Technology Assessment in Health Care, and the Cochrane Colloquium annual conferences.

HSRProj provides project records for health services research, including health technology assessment and development and use of clinical practice guidelines.

number of patients, duration and thoroughness of follow-up, comprehensiveness, and sophistication of statistical analysis)

- Age of the reference (although one must be cautious about assuming that more recent references are more valuable)

- Source of the citation (prestigious general medical journal such as *The New England Journal of Medicine* or *Lancet,* specialty journal with a high Science Citation Index®, journal with known lesser standards of review, non–peer-reviewed citations, and letters to the editor)
- Editorials accompanying the article (or printed discussion)

This fourth pass through the references should provide candidate articles for selecting "key" references you deem most relevant to your proposal, of highest quality, and seminal in the field. These are the articles from which you will select those that need to be cited in the Discussion section of your eventual article. It is rare for there to be more than about a dozen or two such references, and sometimes there are only one or two.

Pass 5: Thorough study of key references. Key references should be read carefully, using the techniques cited by, for example, Ferraris and Ferraris *(38)*, even though they have slanted their recommendations toward the thoracic surgery literature. They suggest asking three questions:

- Are positive results really positive? They note that there is a bias toward publication of positive results, and that most errors in the medical literature occur in articles that contain positive results. Your task is to ascertain as best you can what is true. They give some hints as to how to make that assessment: (1) looking at the way end points have been measured, (2) looking at the explanatory variables collected, (3) looking at the statistical analyses used, and (4) looking at interpretation of apparently small *p* values.
- Are negative results really negative? Underpowered studies leading to false-negative results identify a course for novel discoveries.
- Is there any evidence of bias? This is a particular challenge in analyzing clinical experience rather than randomized trials.

Their admonition is to be certain of positive results, be skeptical of negative results, and assume bias exists. It is probable that some key references will be set aside. Others will be cited in the eventual manuscript, with comments on why their inferences may be different from yours.

Having digested the relevant background information, the research question should be revisited. Sharpen it, change it, or abandon it!

3.5. Study Group Definition

What is the appropriate study (patient) group pertinent to answering the research question? To help answer this question, consider the characteristics of future patients for whom your study's inferences will likely be relevant; that is, how you hope the study will be generalized. Those characteristics should be used to define inclusion and exclusion criteria.

Although there are infinite gradations, it is useful to discuss polar approaches to defining the study group as "lumping" and "splitting." A *lumper* seeks to encompass all possibly relevant patient subsets. Thus this person may ask the research question: "What are the incidences of time-related events after heart valve replacement and what are their modulating (risk) factors?" The lumper would include in the study all patients undergoing heart valve replacement in any position and with prostheses of any type. The matters of position and type would be considered potential modulators. A *splitter* seeks to narrow the focus of the study to homogeneous groups of patients. Thus this person would focus on cryopreserved allograft aortic valve prostheses inserted as a root in patients 75 yr of age or older.

There are advantages and disadvantages to each of these approaches. The lumper will produce a manuscript that is more of an overview (the danger being information content overload and superficiality), makes the important contribution of finding either that certain groups cannot be combined or that they should be combined, and has the advantage of larger numbers. The splitter will produce a more specific manuscript that will likely have greater depth in a circumscribed area. However, the numbers will be smaller, producing underpowered negative findings. The limitation may also not allow trends to be identified (such as with age), because the more restricted the range, the more difficult it is to resolve trends. These two approaches represent, in a sense, the clinical research dilemma of the Heisenberg Uncertainty Principle—depth and breadth are likely not achievable simultaneously.

3.5.1. Inclusion Criteria

If you are identifying patients from electronic sources, formal search logic must be generated. Note that at this stage you are not identifying the variables you wish to extract, but the patients you wish to study. For most surgical studies, you will need to define accurately the surgical procedure performed (statisticians might call this the *exposure*) and often the type of disease for which it was performed. For example, search criteria for esophagectomy may be for (1) cancer or some specific stage of cancer, such as superficial carcinoma of the esophagus, or (2) benign disease, such as achalasia. Inclusion criteria may also include nonsurgical, nondisease criteria, such as patients 18 and older, diabetics, patients in heart failure, or elective operations.

3.5.2. Exclusion Criteria

Inclusion and exclusion are two sides of the same coin, although for ease of communication, specific exclusion criteria may be established. These could be characteristics of the operation (e.g., excluding patients undergoing any concomitant procedure), disease (excluding those with metastatic tumors), or patient (excluding those on renal dialysis).

3.5.3. Inclusive Time Frame

Definition of a study group must include the time frame within which all patients meeting inclusion and exclusion criteria will be identified. For surgical studies, generally this will be the interval during which patients were operated on. Avoid "strange" time frames such as January 1, 1990, to April 23, 2003. Taking the study to the nearest year or half year avoids suspicion on the part of a reader of the eventual manuscript that something bad happened after that date and you are hiding it.

Having defined the ending time limit, the question arises as to the earliest operation that will be accepted. If you want to study long-term persisting or recurring hypertension after neonatal coarctation repair, you cannot limit your study group to the most recent patients operated on. You may object that "everything has changed" in the approach to operation. If you really believe that there is no relation between what has transpired in the past and what is being done today, then you will have to wait a few years until sufficient time has passed to obtain meaningful data and hope in the meantime that the technique of operation has not "completely" changed once again. Of course, the obvious alternative is to have some faith in the continuity of nature *(39)* and take the changes of technique (or the surrogate of date of operation) into account in the analyses.

3.5.4. Comparison Group

The most common failing in defining the study group is to forget about a comparison group. Any time you wish to infer that something is better or changed or different, the question to ask is "than what?" This is called "numerators in search of denominators" *(39)*.

There has been a limited repertoire of techniques for making meaningful comparisons on the basis of clinical experience other than formal randomized clinical trials and multivariable adjustment. Development of balancing score techniques (such as the propensity score) with exquisite patient matching has changed this *(27–30,40,41)*. Now, as long as operations are not absolutely systematically applied, at least some degree of compensation for confounding from patient selection is possible. Indeed, these tools are particularly suited to discovery of the nature of selection biases when criteria are not explicit or are heterogeneous. What cannot be adjusted for is selection criteria based on clinical variables not recorded or extracted from the medical record. It is this bias protection that is provided by randomized clinical trials *(42)*.

It is worth noting that thoughtful consideration of inclusion and exclusion criteria and comparison groups provides another opportunity to refine the research question. On the other hand, if you find that essential criteria to identify a group of patients are not available because they are not in electronic format or have not been recorded systematically, a different research question should be pursued. For example, you may have been interested in the sensitivity and specificity of PET scans for lung cancer, but if your institution does not do such scans, you need to redefine your research question, perhaps directing it to spiral computed tomography. You may be interested in mitral valve repair in Marfan syndrome, but if for some reason the syndrome has not been systematically coded or recorded, you need to refine your research question to some aspect of mitral valve repair that does not depend on knowledge of Marfan syndrome.

3.6. End Points

End points is a synonym for outcomes or results of a surgical procedure or in the group of patients that you propose studying. In some statistical analysis settings, a synonym is *dependent variable*. Each end point must relate and contribute to answering the study question. This statement may seem obvious, but a common failing is to decide that "while I am there," I might as well collect this end point and that end point. John Kirklin called this the "Christmas tree effect." It is deadly to clinical research success for several reasons. First, gathering data consumes as much as 80% of study resources. If an end point is not needed for answering a question, bypass it for economy's sake. Second, every additional variable collected increases the risk of introducing errors unless the same amount of vigilance against data errors is expended on each item. Third, in assembling datasets for statistical analysis, preparation and code is written for every item; if some end points are not used, this wastes data preparation time. What you want are high-quality, believable, reproducible end points that are clearly relevant to answering the research question.

Define each end point exactly and reproducibly. Make no assumptions, but be ready to compromise. For example, if you want to study some particular mode of death, be advised that worldwide there are fewer and fewer autopsies being done, and death certificate causes or modes of death are often inaccurate *(43)*. Instead, all-cause mortality, although possibly less specific, is less subjective and may serve well in answering the research question.

It is important to plan how each end point will be assessed in every patient in the study. If time-related survival is the relevant end point, follow-up questionnaires may need to be devised for *active* follow-up, or death registries consulted for *passive* follow-up. In our experience, the most difficult end points to assemble accurately for appropriate analyses are time-related events. Although such end points seem straightforward, they are not. They generally require active follow-up of patients, which is time consuming. They require accurate capture of the date of occurrence. Thus it is worth pausing to consider what is needed for such an analysis. For any time-related end point, three questions must be answered for each patient: (1) What is the event? (2) When is time zero? (3) Who is at risk?

3.6.1. What Is the Event?

Defining the event for an analysis may be straightforward, such as death from any cause. However, some "events" are processes. For example, time-related renal failure after an operation may reflect a sudden event or a gradual process. In such cases, a surrogate event may be created, such as the date of first institution of renal dialysis. Importantly, for those experiencing the event, its date of occurrence must be recorded; for those not as yet experiencing the event, the last date of active follow-up must be recorded (more details of the data requirements for time-related analyses are given in Chapter 6 of *Cardiac Surgery*) *(3)*.

Processes that can be measured at multiple times are best studied by longitudinal data analyses rather than time-to-event analyses, and all observations of the process should be recorded, including every date of observation. Some events may be ephemeral, such as paroxysmal atrial fibrillation. Analysis of such events, which are difficult to date, may require longitudinal analysis of prevalence in a group of patients from a series of electrocardiograms rather than a time-to-event analysis *(44)*.

3.6.2. When Is Time Zero?

The moment a patient becomes at risk of experiencing the event of interest is called *time zero*. For patients who undergo interventions such as a surgical procedure, time zero is often the time of the procedure. Under many circumstances, however, defining time zero is not so simple. For example, it is not easy to date the onset of a cancer or peripheral vascular disease, although it may be easy to identify the date symptoms developed or a diagnosis was made. Because these occur at various stages of the disease, the state of the disease at "time zero" needs to be recorded. Techniques to work backward from diagnosis to onset have been developed, and are important, for example, in the field of HIV infections *(45)*.

3.6.3. Who Is at Risk?

Patients remain at risk of experiencing the event from time zero to either the occurrence of the event or the time at which they no longer can experience the event; the latter are called censored observations for historical reasons. Defining who is at risk demands thought. For example, if the event is reoperation for bioprosthetic structural valve deterioration, then patients receiving a mechanical prosthesis are never at risk. This distinction may not be obvious to a statistician asked to analyze structural valve deterioration as a time-related event, unless the surgeon-investigator explains it in detail. In this example, patients receiving a bioprosthesis also become no longer at risk of this event the moment the bioprosthesis is explanted for other indications. They are permanently censored at that point. Note that if a repeating morbid event is being analyzed, such as transplant rejection or stroke, patients continue to remain at risk after each occurrence of the event until they are censored by death, end of follow-up, or, for these examples, retransplantation or removal of the valve prosthesis.

3.7. Variables

Somewhat artificially, end points have been discussed separately from variables that may either influence end points or be confounded with them. This artificiality is useful, however, and the text that follows focuses on variables that may be associated with end points.

No study is without heterogeneity and potential sources of selection and other forms of bias. Thus study of end points unadjusted for variables representing this heterogeneity is often not sufficient. This may be necessary for randomized clinical trials, although even for them, it is important to examine possible modulating variables that may render treatment effective in one subgroup and harmful in another *(46)*.

Ability to account for heterogeneity is limited, however. If one is studying binary (yes/no) end points, the effective sample size is number of events, not number of patients. If one follows the general rule that one can only reliably identify one risk factor (variable associated with outcome) per 10 events *(47,48)*, and there are few events, it may not be possible to achieve good risk adjustment. If two treatments are being compared, however, it is possible to achieve considerable risk adjustment with just a single factor—the propensity score *(27,40,41,49–52)*.

The words *variable* and *parameter* are often used interchangeably, although they are actually antonyms. An attribute of a thing that can take on different values from one thing to another is called a variable. For example, systolic blood pressure is a variable whose value differs from patient to patient. In contrast, a constant used to characterize some attribute of a population, such as a mean value, is called a parameter. One generally uses a sample of patients to estimate such constants. These constants are commonly (but not always) designated by letters or symbols in mathematical equations called *models*. Perhaps the most familiar parameter is mean value; another is standard deviation. Each of these is a parameter in the equation (or model) that describes the Gaussian distribution.

Just as accurate and reproducible (precise) definitions must be established for end points, the same is true for all variables in a study. Often, but not always, variables existing in electronic sources will have a companion data dictionary that defines each variable both clinically and electronically. This is called metadata *(53)*. However, definitions may evolve over time. So it is also important to know what a variable meant at one time and how that definition has changed. Variables may be classified according to (1) their role in a study, (2) temporal occurrence, and (3) the nature of their values.

3.7.1. Classification by Role

Variables can be grouped with respect to their role in a study. A common grouping is that used in this chapter: (1) dependent variables and (2) explanatory variables. *Dependent variables* are the study end points, also called the result or the response or outcome variables. In simple multiple regression, such as size of children as a function of size of parents, the dependent variable is size of children, which mathematically appears on the left side of the equals sign, and the explanatory variable is size of parents, which is on the right side of the equals sign. In analysis of dichotomous non–time-related end points, the dependent variable is synonymous with an indicator or response variable for occurrence of the event. In time-related analysis, the dependent variable is the distribution of times to an event, although the indicator variable (such as death) is often cited inaccurately as the dependent variable.

Explanatory variables are characteristics examined in relation to an outcome. Alternative names include *independent variables*, *correlates*, *risk factors*, *incremental risk factors*, *covariables*, *confounders*, and *predictors*. No important statistical properties are implied by any of these alternative names. The least understood is *independent variable* (or *independent risk factor*). Some mistakenly believe it means the variable is uncorrelated with any other variable. All it describes is a variable that by some criterion has been found to be associated with outcome and to contribute information about outcome in addition to that provided by other variables considered simultaneously. The least desirable of these terms is *predictor*, because it implies causality rather than association.

3.7.2. Classification by Temporal Occurrence

Explanatory variables may be usefully classified as (1) those available before time zero, (2) those available during treatment, and (3) those that occur after time zero. This classification is particularly useful for studies involving interventions, in which time of intervention serves to differentiate the temporal availability of data.

Variables available before time zero are patient characteristics and diagnostic testing results available at the time of "entry" into a study (for example, preoperative factors). They are called *baseline* values for variables. Because they are available at the time of decision-making, they can be incorporated into analyses whose purpose is to provide strategic decision support. Perhaps controversial is the claim that baseline variables can include a number of features of the intended procedure; I include this category of variables as "baseline" because it is known (or is being discussed with the patient) at the time of decision making before time zero.

Variables available *during treatment* include those related to details of the immediate operation being performed. If one is evaluating the operation, then surely these, too, occur before time zero and, thus, any end point; however, they are unavailable at the time of patient decision making.

Variables that become available *after time zero* are often called time-varying covariables. They may be end points themselves, but often they represent changes in patient condition that one wants to evaluate with respect to outcomes. Examples include surgical complications, interim events that happen during follow-up, such as occurrence of a myocardial infarct or stroke, institution or withdrawal of medication, or further surgery. Although it would seem obvious to take such variables into account in analyzing end points, they are analytic nightmares! For example, one of the assumptions underlying most analyses of time-related events is that times of occurrence of all other events are uncorrelated with one another; these time-varying covariables tend to be confounded with the outcomes—may even be surrogates for the outcomes—and this confounding is not easily accounted for by readily available analytic methods.

3.7.3. Classification by Value

Variables are usefully classified according to the nature of the values they can assume. Each class of variable implies different ways in which they are expressed and analyzed. The various classes of variables are discussed later in this book in the chapter on biostatistics (Chapter 8).

3.7.4. Organization of Variables

Medical organization of variables is key both to communication with those analyzing the data and to meaningful data analysis. Following is an example organization scheme for variables if one were engaged in clinical research related to cardiac surgery. The list is easily translated to other surgical specialties. Under each of the following heads, one would specify the variables available for the study.

Demography: e.g., age, height, weight

Clinical condition: e.g., New York Heart Association functional class, Canadian Angina Class, presence of cardiogenic shock, evolving myocardial infarction

Morphology: e.g., segmental anatomy of congenital heart disease, echocardiographic findings, coronary angiographic estimates of stenosis

Cardiac-related comorbidity: e.g., left ventricular ejection fraction, left ventricular dimensions and mass, associated dextrocardia

Noncardiac comorbidity: e.g., diabetes, peripheral vascular disease, hypertension, history of stroke, degree of carotid artery occlusion, chronic obstructive pulmonary disease

Surgical details: e.g., number of distal coronary anastomoses, use of internal thoracic artery to bypass stenosis of the left anterior descending coronary artery, stapling of left atrial appendage

Concomitant procedures: e.g., simultaneous carotid endarterectomy, mitral valve repair, pulmonary vein isolation for atrial fibrillation

Support mechanisms: e.g., on pump, off pump, duration of cardiopulmonary bypass, use of retrograde cardioplegia

Experience: e.g., date of operation, surgeon, institution

3.7.5. Values for Variables

To analyze data, values must be obtained for variables. These come from two main sources: electronic and non-electronic.

3.7.5.1. Electronic Data Sources

Unless you have a small number of patients, there must be some way to identify the proposed study group, and these days this generally means an electronic source of at least some patient data. The most fundamental system is a simple registry consisting of a small amount of information about all operations (a full discussion of the various types of databases available for research is included in Chapter 11 of this book) *(54,55)*. For many years, our group at the University of Alabama at Birmingham was productive using a system in which basic demographic information, disease, and type of surgery were kept in a retrievable fashion along with accessible key documents that included copies of the patient's demographic profile ("admission slip"), diagnostic test reports, operative note, discharge summary, and follow-up information. Detailed values for variables had to be extracted "by hand," but the system was inexpensive and effective.

A step up from a simple registry is a disease-oriented registry database based on a set of core variables identified by national or international specialties or governmental agencies *(56)*. In thoracic and cardiovascular surgery *(57)* as well as interventional cardiology *(58)*, such databases, and even the software for them, have been established both in North America and Europe.

Rarely, institutions (e.g., Duke University, Cleveland Clinic) have maintained comprehensive disease-oriented combined registries and research data repositories that are prospective and nearly concurrent with patient care. These are expensive to maintain. They have the advantage, however, of providing readily available data about demographics, disease, operations performed, and patient comorbid conditions that are a useful starting point for many types of research. Generally, one must supplement these variables with ones focused on the purposes of the study.

If variables come from additional electronic sources, one must assess the reliability of the values for the study's key variables. Complete audit may be necessary. It is also possible that from registry audits, sampling of medical records, or degree of missing data, one may be able to identify a subset of variables needing comprehensive verification. For this, range checks, attention to units of measure, and correlation (such as height with weight) can assist in identifying incorrect data. Further discussion of databases in research is included in Chapters 6 and 11.

3.7.5.2. Non-Electronic Sources

For many studies, at least some values for variables will be unavailable from electronic data sources, or available electronic sources will not contain up-to-date information, such as current cross-sectional follow-up. This requires developing a database for their acquisition or extending existing databases. Principles described in the preceding text should govern this process:

- Avoid the Christmas tree effect: stipulate only those variables that are clearly relevant to the proposed study
- Define each variable explicitly using (when possible) definitions that have general agreement within the specialty (e.g., STS definitions)
- Stipulate exhaustively all possible values for variables from which pick lists are derived (controlled vocabulary)
- Specify default values (preferably an answer indicating that no one has yet looked for values for the variable) and what value will indicate a truly unknown value
- For any medical encounter variable, consider a time stamp for its values
- For values that could come from multiple sources (diagnostic procedures, patient history, operative reports, pathology reports, and so forth), state source of data for purposes of both verification and accounting for different methods of gathering information
- Avoid free text that is used for anything other than nonanalyzed comments
- Never put multiple values into a "value set"; for example, blood pressure should have two (or three) columns specifying systolic blood pressure, diastolic blood pressure, and probably anatomic source of the pressure (e.g., brachial artery).

It is best to use a "real" database product for this activity rather than spreadsheet technology that is less controlled, less geared to many-to-one relations (such as values for multiple echocardiograms or reoperations), and more easily damaged (such as sorting on only one column, which leaves the remaining columns unsorted and results in misregistration of these values with the patient for whom they were collected).

3.8. Data Analysis

A research proposal needs a plan for data analysis. Such a plan includes determining sample size so the study will not be underpowered. It is useful in refining the list of end points and explanatory variables. Details of proposed analytic methodology should be formulated in collaboration with a statistician or other quantitative analyst. This collaboration should reveal appropriate methodology and whether the proposed manner in which data are to be collected will meet requirements of that methodology. The surgeon-investigator often does not know the most appropriate methodology to use.

For example, if you propose to study the type of structural deterioration of a heart valve prosthesis, you may elect to use as a surrogate end point the date the prosthesis was explanted. A statistician will help you understand that you must ascertain this end point on every patient. Thus, you will need to identify every prosthesis explant for whatever reason (necessitating a systematic follow-up that, for example, cannot rely on passive information about vital status because you must know about an event that transpired during life), record information about every explant, categorize in particular those deemed to be from structural valve failure, and then document the date of follow-up assessment for patients not yet experiencing valve failure. Alternatively, you may decide to look at the temporal process of valve failure using echocardiograms. This will involve different statistical methods than used for the time-related event of explant. You will need to gather every echocardiogram made and, ideally, supplement that information with a systematic assessment of each valve by cross-sectional echocardiographic follow-up.

Perusal of the surgical literature for what other groups have "gotten away with in the past" in even top-tier journals is not a good way to determine appropriate methods *(38)*. Some inappropriate methods are used because the statistician has not been brought into the formative stages of the research plan to recommend how data should be gathered. Sometimes no truly appropriate method is thought to be available to answer the research question directly. By involving the statistician in the planning stage of the research, sufficient time is given to investigate methodologic issues. If the most appropriate analytic method is not available, then methodologic research can be commenced.

3.8.1. Elements of the Analytic Plan

The analytic plan has two primary components: research objectives and analytic objectives. In most cases, the research question leads to a series of specific objectives that must be addressed individually by data analysis to answer the primary research question. For example, a research question might be, "Is preoperative atrial fibrillation a risk factor or a marker for long-term mortality after coronary artery bypass grafting?" Specific objectives for data analysis might be to (1) characterize factors that distinguish patients having and not having preoperative atrial fibrillation (perhaps refining the variable to chronic atrial fibrillation or characterizing the type); (2) identify patients well matched with respect to these differences; (3) compare survival; or (4) identify subgroups of patients within the entire dataset who may be most vulnerable to atrial fibrillation.

Some statisticians will frame research objectives or questions in the format of either an informal hypothesis (the idea the investigation is designed to demonstrate, such as that preoperative atrial fibrillation leads to reduced long-term survival after operation) or a formal statistical hypothesis. The latter is often framed in an archaic manner that is uncomfortable to the investigator, such as "presence of preoperative atrial fibrillation does not impact long-term survival." It is helpful to know that ordinary statistical hypothesis testing has its roots in ancient Roman ideas of law *(59)*, one of whose tenets is that an individual is innocent until proven guilty beyond doubt. Statistical testing can be thought of as the evidence against "innocence" and the *p* value in particular as a measure of doubt or surprise with respect to the matter of innocence. Box 2 explains some terms statisticians use in transforming research questions into formal hypotheses that can be tested. Basic statistical analyses for use in clinical research are explained in greater detail in Chapter 8 of this book.

Box 2: Hypothesis (Significance) Testing

Statistical Hypothesis

A statistical hypothesis is a claim about the value of one or more parameters. For example, the claim may be that the mean for some variable, such as creatinine, is greater than some fixed value or than some value obtained under different conditions or in a different sample of patients. It can be calculated only if the distribution of the data is known.

Null Hypothesis

The null hypothesis is a claim that the difference between one or more parameters is zero or no change (written H_0). It is the claim the investigator generally is arguing against. When a statistician infers that there is a "statistical significance," it means that by some criterion (generally a p value) this null hypothesis has been "rejected." Some argue that the null hypothesis can never be true and that sample size is just insufficient to demonstrate this fact. They emphasize that the magnitude of p values is highly dependent on n, so other "measures of surprise" need to be sought.

Alternative Hypothesis

An alternative hypothesis is the "investigator's claim" and is sometimes called the *study hypothesis* or *informal hypothesis*. It is generally not the same as a medical hypothesis about mechanisms. Generally, the investigator would like for the data to support the alternative hypothesis, although the statistician will be testing the null hypothesis in most instances.

Test Statistic

A test statistic is a number computed from the distribution of the variable to be tested in the sample of data that is used to test the merit of the null hypothesis.

Type I Error

Rejecting the null hypothesis when it is true (false negative) is called a type I error. The probability of a type I error is designated by the Greek letter alpha (α).

Type II Error

Not rejecting the null hypothesis when it is false (false positive) is called a type II error. The probability of type II error is designated by the Greek letter beta (β).

The research objectives then lead to detailed analytic objectives that allow for estimation of the resources needed and provide some idea of the time required to complete the analysis. The analysis objectives, then, should address:

- Whether the study group is appropriate to answer the research question
- Whether a control arm has been forgotten
- How missing values for variables will be managed
- Specific analytic objectives that will lead to answering the research question
- Statistical methods that will be used to obtain answers to each specific analytic objective

As the analysis plan evolves, some of these items will become incorporated into various aspects of the study proposal. Other items will evolve into an analysis report consisting of a general description of study group characteristics, and, for each specific research objective, the rational for the analytic objective, methods used to answer the specific question, results of analyses, relevant tables and figures, comments on the findings and limitations, and implications of the results of each objective. The combined research proposal and analysis report constitute the bulk of a manuscript, enormously facilitating its preparation.

3.9. Feasibility

Successful studies are built on ascertaining that:

- The study population can be identified reliably
- Values for variables are either already in electronic format or can be obtained reliably by review of medical documents
- Sample size is sufficient to answer the question
- Clinical practice is not completely confounded with the question being asked *(27)*
- Institutional resources are available
- The anticipated timetable to complete the study is tolerable
- Study limitations do not present fatal flaws

If these criteria are not met, the study is not feasible and should be abandoned or a long-range plan devised for prospectively obtaining and recording the needed data. Ascertaining feasibility as early as possible is as central to successful research as the question being asked.

For any study, a minimum sample size (number of patients) is needed to detect an effect reliably. If a comparison group is used, sample size calculations similar to those used for randomized clinical trials provide guidance to the size needed. What is required is ascertaining from either experience or a preliminary look at the data the magnitude of clinically meaningful difference in end point value. If the sample size is too small to identify a clinically meaningful difference reliably, the study is underpowered and should be abandoned or other end points sought, surrogate end points used, the study group enlarged by less restrictive inclusion or exclusion criteria or a wider study time frame, or a multi-institutional study mounted. Note that for studies of events, effective sample size is proportional to number of events rather than number of patients *(25)*.

When considering feasibility, it is important to develop a timetable for data abstraction, dataset generation, data analysis, and reporting. If the timetable is intolerable, either abandon the study or narrow its scope. Physicians' most scarce commodity is time. All too often, however, an investigator is willing to devote many hours to collecting and verifying data, and then does not allow those analyzing the data sufficient time to do a good job. It is difficult to complete a study and submit a manuscript in a year from start to finish. This emphasizes both the bottlenecks of research and the need for lifelong commitment.

Study timetables are often driven by abstract deadlines. Although they should provide some incentive, it is incorrect for studies to be driven by such deadlines. What matters is publication of the research. It is preferable to develop a disciplined approach to writing manuscripts. During the writing process, one is often driven back to the data or more analyses. In the end, one must ask, "Have I answered the research question?" Beyond that, the conclusions should support clinical implications (inferences). These should be thought through and discussed with collaborators. After this is done, and the first major draft of a manuscript is completed, the study is a candidate for constructing a meeting abstract.

Finally, when assessing feasibility, it is important to consider the limitations of the study. These can be identified by a brief but serious investigation of the state of all the considerations above. If any appear insurmountable or present fatal flaws that preclude later publication, the study should be abandoned before it is started.

3.10. Institutional Review Board

Any research proposal that does not simply use existing data that have already been approved for use in research by an Institutional Review Board requires study-specific Institutional Review Board approval before any research is commenced *(60)*. Similarly, at least in the United States, each investigator needs to be certified for performing studies on human subjects. The US Health Insurance Portability and Accountability Act (HIPAA) places important restrictions on interactions of investigators with both patients and data; ensuring that your research complies with these restrictions is an essential feature of successful and ethical research *(61)*.

In the research plan, compliance issues with respect to patient privacy and confidentiality, informed patient consent, use of patient data, and certification of all investigators must be addressed. Issues surrounding informed consent, the Institutional Review Board and HIPAA are discussed in greater derail in Chapter 2.

REFERENCES

1. Kirklin JW, Barratt-Boyes B. Cardiac surgery. New York: Churchill Livingstone, 1986.
2. Kirklin JW, Barratt-Boyes BG. Cardiac surgery. New York: Churchill Livingstone, 1993.
3. Kouchoukos NT, Blackstone EH, Doty DB, Hanley FL, Karp RB. Cardiac surgery. Philadelphia: Churchill Livingstone, 2003.
4. Bigby M. Evidence-based medicine in a nutshell. A guide to finding and using the best evidence in caring for patients. Arch Dermatol 1998;134:1609–1618.
5. Guidelines and indications for coronary artery bypass graft surgery. A report of the American College of Cardiology/American Heart Association Task Force on Assessment of Diagnostic and Therapeutic Cardiovascular Procedures (Subcommittee on Coronary Artery Bypass Graft Surgery). J Am Coll Cardiol 1991;17:543–589.
6. Kirklin JW, Blackstone EH, Naftel DC, Turner ME. Influence of study goals on study design and execution. Control Clin Trials 1997;18:488–493.
7. Association of American Medical Colleges Task Force on Clinical Research. For the health of the public: ensuring the future of clinical research. Vol. 1. Washington, DC: AAM, 1999.
8. Murphy KM, Topel RH. Measuring the gains from medical research: an economic approach. Chicago: University of Chicago Press, 2003.
9. Guerlac H. Theological voluntarism and biological analogies in Newton's physical thought. J Histol Ideas 1983;44:219–229.
10. Newton I. Philosophiae naturalis principia mathematica, 1687.
11. Hamilton CW. How to write and publish scientific papers: scribing information for pharmacists. Am J Hosp Pharm 1992;49:2477–2484.
12. Day RA. How to write and publish a scientific paper. Westport, CT: Greenwood, 1998.
13. Uniform requirements for manuscripts submitted to biomedical journals. International Committee of Medical Journal Editors. N Engl J Med 1997;336:309.
14. Blackstone EH, Rice TW. Clinical-pathologic conference: use and choice of statistical methods for the clinical study, "superficial adenocarcinoma of the esophagus." J Thorac Cardiovasc Surg 2001;122: 1063–1076.
15. Kirklin JW, Blackstone EH. Notes from the editors: ultramini-abstracts and abstracts. J Thorac Cardiovasc Surg 1994;107:326.

16. Rennie D, Yank V, Emanuel L. When authorship fails. A proposal to make contributors accountable. JAMA 1997;278:579–585.
17. Kouchoukos NT, Blackstone EH, Doty DB, Hanley FL, Karp RB. Cardiac surgery. Philadelphia: Churchill Livingstone, 2003, p. 254–350.
18. Begg CB, Greenes RA. Assessment of diagnostic tests when disease verification is subject to selection bias. Biometrics 1983;39:207–215.
19. Kosinski AS, Barnhart HX. Accounting for nonignorable verification bias in assessment of diagnostic tests. Biometrics 2003;59:163–171.
20. Counsell C. Formulating questions and locating primary studies for inclusion in systematic reviews. Ann Intern Med 1997;127:380–387.
21. Glass G. Primary, secondary and meta-analysis of research. Education Res J 1976;5:3.
22. Bero L, Rennie D. The Cochrane Collaboration. Preparing, maintaining, and disseminating systematic reviews of the effects of health care. JAMA 1995;274:1935–1938.
23. Chalmers I, Dickersin K, Chalmers TC. Getting to grips with Archie Cochrane's agenda. Br Med J 1992;305:786–788.
24. Berry DA, Stangl DK. Meta-analysis in medicine and health policy. Biostatistics. New York: Marcel Dekker, 2000.
25. Fleming TR, DeMets DL. Surrogate end points in clinical trials: are we being misled? Ann Intern Med 1996;125:605–613.
26. Vineis P, McMichael AJ. Bias and confounding in molecular epidemiological studies: special considerations. Carcinogenesis 1998;19:2063–2067.
27. Blackstone EH. Comparing apples and oranges. J Thorac Cardiovasc Surg 2002;123:8–15.
28. Rosenbaum PR, Rubin DB. The central role of the propensity score in observational studies for causal effects. Biometrika 1983;70:41–55.
29. Rosenbaum PR, Rubin DB. Reducing bias in observational studies using subclassification on the propensity score. J Am Stat Assoc 1984;79:516–524.
30. Rubin DB. Estimating causal effects from large data sets using propensity scores. Ann Intern Med 1997;127:757–763.
31. Gordon T, Kannel WB. Multiple risk functions for predicting coronary heart disease: the concept, accuracy, and application. Am Heart J 1982;103:1031–1039.
32. Bertranou EG, Blackstone EH, Hazelrig JB, Turner ME, Kirklin JW. Life expectancy without surgery in tetralogy of Fallot. Am J Cardiol 1978;42:458–466.
33. Kirklin JW, Blackstone EH, Pacifico AD, Brown RN, Bargeron LM, Jr. Routine primary repair vs two-stage repair of tetralogy of Fallot. Circulation 1979;60:373–386.
34. Blackstone EH, Kirklin JW, Bertranou EG, et al. Preoperative prediction from cineangiograms of postrepair right ventricular pressure in tetralogy of Fallot. J Thorac Cardiovasc Surg 1979;78:542–552.
35. Shimazaki Y, Maehara T, Blackstone EH, Kirklin JW, Bargeron LM, Jr. The structure of the pulmonary circulation in tetralogy of Fallot with pulmonary atresia. A quantitative cineangiographic study. J Thorac Cardiovasc Surg 1988;95:1048–1058.
36. Blackstone EH. Breaking down barriers: helpful breakthrough statistical methods you need to understand better. J Thorac Cardiovasc Surg 2001;122:430–439.
37. Diggle PJ, Heagerty PJ, Liang KY, Zeger SL. Analysis of longitudinal data. New York: Oxford University Press, 2002.
38. Ferraris VA, Ferraris SP. Assessing the medical literature: let the buyer beware. Ann Thorac Surg 2003;76:4–11.
39. Bockner S. Continuity and discontinuity in nature and knowledge. In Wiener PP, ed. Dictionary of the history of ideas: studies of selected pivotal ideas. Vol. 1. New York: Charles Scribner's Sons, 1968: 492–504.
40. Rosenbaum PR. Optimal matching for observational studies. J Am Stat Assoc 1985;84:1024–1032.
41. Rosenbaum PR. Obervational studies. New York: Springer-Verlag, 2002.
42. Piantadosi S. Clinical trials: a methodologic perspective. New York: John Wiley & Sons, 1997.
43. Lauer MS, Blackstone EH, Young JB, Topol EJ. Cause of death in clinical research: time for a reassessment? J Am Coll Cardiol 1999;34:618–620.
44. Gillinov AM, Blackstone EH, McCarthy PM. Atrial fibrillation: current surgical options and their assessment. Ann Thorac Surg 2002;74:2210–2217.
45. Brookmeyer R. AIDS, epidemics, and statistics. Biometrics 1996;52:781–796.
46. Singh BN. Do antiarrhythmic drugs work? Some reflections on the implications of the Cardiac Arrhythmia Suppression Trial. Clin Cardiol 1990;13:725–728.

47. Harrell FE, Jr., Lee KL, Califf RM, Pryor DB, Rosati RA. Regression modelling strategies for improved prognostic prediction. Stat Med 1984;3:143–152.
48. Marshall G, Grover FL, Henderson WG, Hammermeister KE. Assessment of predictive models for binary outcomes: an empirical approach using operative death from cardiac surgery. Stat Med 1994;13:1501–1511.
49. Drake C, Fisher L. Prognostic models and the propensity score. Int J Epidemiol 1995;24:183–187.
50. Parsons LS. Reducing bias in a propensity score matched-pair sample using greedy matching techniques. Proceedings of the Twenty-Sixth Annual SAS Users Group International Conference. Cary, NC: SAS Institute, Inc, 2001.
51. Rubin DB. Bias reduction using Mahalanobis' metric matching. Biometrics 1980;36:295–298.
52. Rosenbaum PR, Rubin DB. The bias due to incomplete matching. Biometrics 1985;41:103–116.
53. Kirklin JW, Vicinanza SS. Metadata and computer-based patient records. Ann Thorac Surg 1999;68:S23–S24.
54. Williams WG, McCrindle BW. Practical experience with databases for congenital heart disease: A registry versus an academic database. Semin Thorac Cardiovasc Surg Pediatr Card Surg Annu 2002;5:132–142.
55. Lauer MS, Blackstone EH. Databases in cardiology. Topol EJ, ed. Textbook of cardiovascular medicine. Philadelphia: Lippincott Williams & Wilkins, 2002:981.
56. Hannan EL, Kilburn H Jr, O'Donnell JF, Lukacik G, Shields EP. Adult open heart surgery in New York State. An analysis of risk factors and hospital mortality rates. JAMA 1990;264:2768–2774.
57. Ferguson TB, Jr., Dziuban SW, Jr., Edwards FH, et al. The STS National Database: current changes and challenges for the new millennium. Committee to Establish a National Database in Cardiothoracic Surgery. The Society of Thoracic Surgeons. Ann Thorac Surg 2000;69:680–691.
58. Mullin SM, Passamani ER, Mock MB. Historical background of the National Heart, Lung, and Blood Institute Registry for Percutaneous Transluminal Coronary Angioplasty. Am J Cardiol 1984;53:3C–6C.
59. Post G. Ancient Roman ideas of law. In Wiener PP, ed. Dictionary of the history of ideas. vol. 2. New York: Charles Scribner's Sons, 1968:492–504.
60. Agich GJ. Human experimentation and clinical consent. In: Monagle JF, Thomasma DC, eds. Health care ethics: critical issues for the 21st century. Gaithersburg, MD: Jones & Bartlett, 1998.
61. Muhlbaier LH. HIPAA training handbook for researchers. Marblehead, MA: HCPro, 2003.

APPENDIX: CLINICAL RESEARCH PROPOSAL TEMPLATE

1. Title

The title of a research proposal should reflect the question being addressed. It should indicate the (1) topic, (2) target population (including control group), (3) analytic approach, (4) end point, and (5) clinical importance.

2. Contributors

Name the principal investigator (guarantor) and contributors. The role, responsibilities, and expectation of each should be agreed upon. As a minimum, contributors include colleagues, statisticians, data managers, and experts with respect to end points of the proposed research.

3. Research Question

The purposes, aims, or informal hypotheses of the study must be clearly stated; often, this is best accomplished in the form of one or more specific research questions. The research question must be well formulated and focused, because it is the single most important ingredient for success. It should be revisited, revised, restated, and kept uppermost in mind throughout the study and its eventual presentation. The study cannot be initiated without this step, because the study group, end points, variables, and analyses all depend on it. A well-framed question may deliberately confirm others' observations, may question conventional wisdom or what has been thought to be established knowledge, or may attempt to fill in a gap in knowledge.

4. Background

Synthesize the state of knowledge and identify what is unknown or controversial to indicate why a study is needed. The clinical motivation and biologic rationale for proposing the study should be established. Background information comes from colleagues, textbooks and review articles, and literature searches, the result of which is a list of key references.

5. Study Group Definition

What is the appropriate study (patient) group pertinent to answering the research question? Define both inclusion and exclusion criteria and justify them. Define well-justified inclusive dates. If one proposes that outcome is improved or different, a comparison group is needed. If one proposes to study an *event*, this is a numerator; *both numerator and denominator are needed.*

6. End Points

End points are the study outcomes. Each must relate and contribute to answering the study question. State them specifically—their exact, reproducible, and unequivocal definitions—determine how they can be assessed in *each* individual in the study, and show how each relates to the study. One temptation is to specify many end points that are not clearly linked to answering the study question while spending too little time thinking about what end points are critical to the study. Time-related and longitudinal end points require definition of the event, time zero, and the patients at risk.

7. Variables

No study is without heterogeneity and potential sources of selection and other forms of bias. Thus, study of end points alone is rarely sufficient. Other variables need to be available to characterize the study group and to consider when interpreting each end point. Variables are usefully categorized by their role in the study, their temporal occurrence, and their values. Organizing variables medically is key to meaningful data analysis.

8. Data Analysis

Details of analytic methodology should be formulated in collaboration with a statistician or other quantitative analyst. The surgeon-investigator often does not recognize or know the most appropriate methodology. Collaboration with a statistician or other quantitative professional should reveal appropriate methodology and whether the proposed manner in which data are to be collected will meet the requirements of the methodology. Sometimes there are no appropriate methods for meaningful and accurate analysis of data, and methodologic research is required in parallel with the clinical research.

9. Feasibility

a. Determine the sample size needed. For any study, a minimum sample size is needed to detect a clinically meaningful effect. For events (e.g., death), sample size depends on number of events, not size of the study group. If a comparison group is used, sample size calculations similar to those used for randomized clinical trials provide guidance to the size needed. If the sample size is sure to result in an underpowered study, abandon the study, seek other end points, or engage in a multi-institutional study.

b. Develop a timetable for data abstraction, dataset generation, data analysis, and reporting. If the timetable is intolerable, either abandon the study or narrow its scope. It is rare for a study to be completed in a year from start to finished manuscript. This emphasizes both the bottlenecks of research and the need for lifelong commitment. Although abstract deadlines often drive the timetable, this is a poor milepost.

c. Identify limitations and anticipated problems. These can be identified by a brief but serious investigation of the state of all the considerations above. If any appear insurmountable or present fatal flaws that preclude later publication, the study should be abandoned before it is started.

10. Institutional Review Board

Any proposal that does not use existing data already approved for use in research by an Institutional Review Board requires study-specific Institutional Review Board approval *before* any research is started.

2 Ethical Issues in Clinical Research

David F. Penson, MD, MPH

CONTENTS

As both clinicians and researchers, surgeons are expected to behave in an ethical manner and put the interests of their patients above all else. This was originally codified in the Hippocratic Oath and is included in most medical professional societies' mission statements. This includes the American College of Surgeons Fellowship Pledge that contains the following text:

> *I pledge to pursue the practice of surgery with honesty and to place the welfare and the rights of my patient above all else. I promise to deal with each patient as I would wish to be dealt with if I was in the patient's position and I will respect the patient's autonomy and individuality (1).*

Medical ethics provide the foundation for the modern practice of surgery.

Patient-based research, however, may present ethical and moral dilemmas for the surgeon. After all, many of the interventions under study are investigational and may place the patient at significant risk. Often, we justify this risk by reminding ourselves that, if the treatment works, the patient's condition will improve and, even if the intervention does not work, the knowledge gained will benefit all patients with the disease. Although this rationalization may ultimately prove true, the surgeon-scientist must carefully consider each research situation and determine if the benefits of the study outweigh the risks, not just for society in general, but for the individual patient.

Unfortunately, although we would all like to believe that clinicians and scientists always act in the best interests of their patients, there are numerous historical examples (many of which have been well publicized) of unethical research practices that have negatively affected patients' health. These incidents have led to the establishment of numerous regulations and a fairly impressive infrastructure aimed at ensuring safe and

From: *Clinical Research for Surgeons*
Edited by: D. F. Penson and J. T. Wei © Humana Press Inc., Totowa, NJ

ethical research practices. The goal of this chapter is to review these entities and help the surgeon researcher to design and perform research in an ethical manner. We will begin by briefly reviewing the historical events that have lead to the current regulations and practices surrounding the ethical practice of clinical research. We will then specifically discuss the role of the Institutional Review Board (IRB), the local body which is responsible for research oversight at most institutions. We will then address specific issues surrounding ethics and surgical studies. Finally, we will discuss the Health Insurance Portability and Accountability Act of 1996 (HIPAA), as it relates to the conduct of clinical research. With this broad overview, the surgeon-scientist should be able to interact more amicably and efficiently with his or her local IRB and should be fully aware of the ethics surrounding clinical research in surgery.

1. ETHICAL RESEARCH: AN HISTORICAL PERSPECTIVE

Although there are a number of historical examples of inappropriate and/or unethical research, there are three specific events which have had the greatest impact on federal regulations surrounding the protection of human research subjects and, therefore, have contributed the most to the development of the current infrastructure to ensure ethical research. These three events are: (1) the 1946 Nuremberg Doctors Trial; (2) the rash of birth defects associated with Thalidomide use in the 1960s; and (3) the 1972 exposé on the Tuskegee Syphilis Study. Each of these historical events led to the enactment of new codes and regulations specifically designed to protect human subjects in research.

1.1. The 1946 Nuremberg Doctors Trial

This case first brought the issue of unethical research to public attention and underscored the need for regulation in this area. During World War II, Nazi physicians in Germany performed numerous horrible experiments on concentration camp internees in an effort to aid the German war machine. For example, the German Air Force was concerned about the effect of low atmospheric pressure on pilots who might bail out of their aircraft at high altitudes. Therefore, they performed a series of experiments on prisoners that included placing healthy subjects into vacuum chambers and lowering atmospheric pressure and oxygen levels. Approximately 40% of the subjects died of various causes, including anoxia and ruptured lungs from the low pressure in the chambers. In other experiments, traumatic wounds, such as stabbings or gun shots, were inflicted on subjects. Resulting wounds were stuffed with contaminants, such as glass, dirt, and various bacteria, to simulate battlefield conditions. Various experimental antibiotics were then administered. In another experiment, numerous limb amputations were performed followed by attempts at various forms of transplantation. Although it is easy to state that this could never happen today, the reader should bear in mind that these barbaric experiments occurred in highly civilized Western Europe only 60 yr ago *(2)*.

In August 1945, the Allied governments created a military tribunal in Nuremberg, Germany, to place the Nazi leadership on trial. After the trial of the military leadership, a number of trials were held to judge the Nazi physicians involved in the human studies. The defendants were charged with murder, torture, and other atrocities, of which the majority were ultimately found guilty. The primary defense of the accused was that they were simply following the orders of their superiors. This motivated the inclusion of what has come to be known as the "Nuremberg Code" in the final trial judgment *(3)*. The full text of the code is available elsewhere *(4)*, but, in summary it states that:

- Subjects participating in research should give informed consent without coercion.
- The anticipated benefits of the study should justify the research and the risks associated with it.
- Human studies should be based on prior animal studies and knowledge of the natural history of the condition under study.
- Physical and mental suffering and injury should be avoided.
- During the study, the subject should be able to withdraw at anytime he or she sees fit.
- The study should be performed by qualified scientific personnel and these individuals should be prepared to terminate the study at any stage if they believe that the continuation of the study will result in injury, disability or death of the subject.

The World Medical Association then applied the principles elucidated in the Nuremberg Code to the practice of medical research. Development of these "rules" for medical research started in 1953, culminating with adoption of a formal declaration of ethical principles for medical research involving human subjects by the World Medical Association in Helsinki in 1964. The Declaration of Helsinki, as it is commonly known, has been updated and reendorsed by the World Medical Association numerous times since, with the last being in 2000 in Edinburgh, Scotland. The full-text of the Declaration of Helsinki is available at http://www.nihtraining.com/ohsrsite/guidelines/helsinki.html. In summary, the Nuremberg Trial first brought the need for regulations to protect human research subjects to the attention of the public and resulted in the development of the Nuremberg Code and the Declaration of Helsinki, two important documents regarding the conduct of ethical research.

1.2. The Thalidomide Tragedy of the 1960s

Although not truly the result of "unethical research," the rash of birth defects associated with thalidomide use in the early 1960s documented the somewhat unethical business practices of certain pharmaceutical companies at that time and the need for stronger regulations regarding "experimental" drugs in the United States. Thalidomide had been approved in Europe as a sedative in the late 1950s. Although it did not have approval in the United States, the drug's manufacturer provided samples to American physicians who received payment to assess the its efficacy and safety. This form of "research" was not uncommon at the time; however, it quickly became apparent that thalidomide was extremely teratogenic, resulting in limb deformities in newborn children whose mothers had used the agent during the first trimester of pregnancy *(5)*. This led to a worldwide ban of the drug *(6)* and ultimately to the 1962 passage of the Kefauver-Harris Amendments to the Food, Drug, and Cosmetic Act. These amendments, with additional legislation in 1963 and 1966, required that subjects be informed that they were receiving experimental agents that had not be approved by the US Food and Drug Administration and that explicit consent be obtained before administration of the experimental agent. They also specifically stated that the subject must be informed if he or she might receive a placebo. These regulations laid the foundation for the current new drug approval process in the United States and provided the legal basis for the protection of human subjects in research in this country.

1.3. The Tuskegee Syphilis Study

Although the Nuremberg Trial and the Thalidomide tragedy provide a historical perspective for the protection of human subjects in research, it is the Study of Untreated

Syphilis in the Negro Male, initiated in 1932 by an agency within the United States Public Health Service (a forerunner of the Centers for Disease Prevention and Control), that truly motivated the current set of rules and regulations regarding ethical research in the United States. That the study focused on an ethnic minority, continued into the 1970s, and was funded by the federal government underscored the pressing need for regulation in this area. The Public Health Service initially identified Macon County, Alabama, as an area with an extremely high prevalence of syphilis. They then designed a study to assess the health effects of syphilis on untreated African-American men. At first, the study was to end after the initial enrollment and assessment of the infected patients' disease status with a nontherapeutic spinal tap. However, researchers never informed the subjects that they had syphilis or that the primary goal of the study was to assess the effects of this infection on health. Rather, the subjects were told they were part of a study that would provide free examinations and medical care. In 1933, the researchers added a surveillance phase during which the subjects were followed and received additional testing. At this point, penicillin was not yet recognized as curative treatment for syphilis; therefore, in theory, one might argue that, although the study was of questionable value, it was not completely unethical, because no alternative treatments were available. Although this line of reasoning is likely flawed (after all, weren't the researchers morally obligated to inform the men they had syphilis and to offer some sort of palliative treatment?), it may have been acceptable until 1943, when penicillin became widely available and accepted as curative treatment for syphilis *(7)*.

At this point, researchers conspired with the local draft board to make study participants exempt from the military to prevent them from receiving treatment for their condition. Furthermore, researchers continued to withhold penicillin from the subjects, without informing the patients that they had syphilis, to prevent contamination of the study population and allow them to continue to study the long-term natural history of untreated syphilis. The study continued until 1973, when it was closed in response to the 1972 publication of an exposé in the *Washington Star*. The Tuskegee Syphilis Study led to significant new legislation to protect human subjects in research and effectively led to the current infrastructure surrounding the ethical performance of research *(7)*. In addition, President Bill Clinton formally apologized to the study's participants in 1997.

The first piece of legislation that was passed in response to the Tuskegee Study was the National Research Act of 1974. This legislation included requirements for informed consent and mandated the establishment of local IRBs to oversee the ethical practice of research. It also established the National Commission for the Protection of Human Subjects of Biomedical and Behavioral Research. This commission published a document that has come to be known as "The Belmont Report," which provided the foundation on which the federal regulations for the protection of human subjects in research are based. The full text of the Belmont Report can be found at http://www.nihtraining.com/ohsrsite/guidelines/belmont.html.

After the release of the Belmont Report, the Department of Health and Human Services and the US Food and Drug Administration published convergent regulations regarding informed consent and IRBs that were based on the principles in the Belmont Report. Fifteen additional government departments and agencies reviewed these regulations and, after 10 yr of negotiation, these agencies agreed to adopt a set of basic human subjects protections that have come to be known as the "Common Rule." The regulations established under the common rule follow.

- 45 CFR46 Protection of Human Subjects
- 21 CFR50 Protection of Human Subjects
- 21 CFR56 Institutional Review Board
- 21 CFR312 Investigational New Drug Application
- 21 CFR812 Investigational Device Exemptions

The complete text of these regulations and a great deal of additional information regarding the federal oversight of human subjects in research can be found at http://www.hhs.gov/ohrp.

Since the adoption of the Common Rule, there has been continued federal activity in the area of bioethics. Specifically, President Clinton established the National Bioethics Advisory Commission in 1995 with the specific purpose of continually reviewing the federal regulations on ethical human experimentation and making recommendations to the government on research topics and legislation as these issues arise. When the commission's charter expired in 2001, President George W. Bush appointed the President's Commission on Bioethics with a similar mission to the National Bioethics Advisory Commission.

One of the developing issues that face bioethicists today is involves conflicts of interest among clinical researchers. Specifically, many translational researchers who develop new agents for use at the bedside also hold financial interests in the companies who develop the agents *(8)*. An example of this ethical issue comes from a recent gene transfer experiment. In 1999, an 18-yr-old man with a mild form of ornithine transcarbamylase deficiency that had been controlled with diet and medications volunteered to participate in a gene-transfer study. The study itself had been reviewed and approved by the US Food and Drug Administration, National Institutes of Health, and the local IRB. The subject understood that he would not directly benefit from the treatment, but felt that it would ultimately help children born with a more severe form of the deficiency. After giving informed consent, the subject received an injection of an adenovirus transfected with the ornithine transcarbamylase gene. Unfortunately, he rapidly developed liver failure and died as a result of the treatment. At first, the patient's father defended the scientists at the University of Pennsylvania, acknowledging that his son knew that this was a novel agent and that there were risks associated with it *(9)*. Ultimately, it came to light that the investigators had not been forthcoming regarding prior adverse events and risks at the time of informed consent, that they had "loosened" the protocol inclusion criteria to improve enrollment, and had not provided adequate safeguards for the patients' well being *(10)*. The National Institutes of Health ultimately halted all gene transfer experiments at that institution and sought to disqualify the investigators from receiving future federal funding or performing further clinical research *(11)*. One of the primary reasons the government took these drastic steps was the perceived conflict of interest that the investigators had in running the Phase I study. One of the lead investigators specifically had formed a biotechnology company that provided resources to the Institute of Human Gene Therapy at his institution and held numerous patents on the viral technology. In other words, the investigator stood to benefit financially if the treatment was demonstrated to be effective in clinical trials *(12)*.

This type of conflict of interest is becoming more common as clinician-investigators serve as consultants or major investors to biotechnology and pharmaceutical firms. Furthermore, many investigator-initiated studies are directly or indirectly funded by industry, which places the investigator into a real or perceived conflict of interest. Many institutions have developed internal regulations regarding the declaration of financial and other interests by investigators. Some institutions have put limitations in place as to how much outside income an investigator can earn and still participate in related research. Although these regulations will likely prove helpful, it is important for the surgeon-scientist to consider the potential for conflicts of interest before initiating research. If you are participating in an industry-sponsored pharmaceutical study for which you or your research program is receiving compensation, it is probably wise to disclose this to the subject in the informed consent. If you stand to personally materially benefit from the study (e.g., you hold stock in the company or a patent on the technology), it is best that you disassociate yourself from the study and allow one of your colleagues to administer the trials. Even the slightest hint of a financial or material conflict of interest will jeopardize your credibility as a researcher and could result in punitive actions on the part of your local institution or state or federal agencies. As a surgeon-scientist, it is best to avoid any true or perceived conflicts of interest when conducting clinical research. If such is unavoidable, it is advisable that a conflict of interest management plan be developed for the study facilitated either by the IRB, the HIPAA board or another regulatory body within the institution.

2. THE INSTITUTIONAL REVIEW BOARD

Current public opinion regarding health care and research has led many institutions to adopt a defensive posture with regard to the ethics of research. This, in turn, has prompted local IRBs to more closely scrutinize each protocol. Many researchers have interpreted this as "obstructionism" on the part of the IRB committee and have, unfortunately, developed an adversarial relationship with committee. In truth, the IRB is a valuable resource for researchers that only helps to protect the investigators from ethical problems and ensures high-quality research that advances science and brings credibility to the institution.

The role of the IRB is specifically spelled out in the federal regulations regarding human research (Code of Federal Regulations, vol. 21). By law, the IRB must review all research and ensure that the following requirements are met before approving any study:

- Minimization of risks to subjects
- Assessment of risks to ensure that the risk/benefit ratio is reasonable. It is important to bear in mind that this assessment is limited to the risks and benefits for the subjects in the study, as opposed to society in general
- Equitable selection of subjects
- Overview of the informed consent process (including documentation) and assurance that consent will be obtained from all subjects, as appropriate
- Data monitoring, if appropriate, to ensure the subject's safety
- Protection of subject's privacy
- Protection of special populations who may be vulnerable to coercion (e.g., children, pregnant women, prisoners, handicapped, mentally disabled or educationally or economically disadvantaged people), if appropriate.

The IRB's review should be focused on these issues primarily and comments should be related to concerns with these issues. At times, the IRB will comment on the scientific merit of the study, which most consider outside the purview of the IRB. However, some scientific review most also be undertaken as part of the review process, because it is the

IRB's responsibility to assess the risk and benefits of the study. If the IRB does not believe that the science is reasonable, how can they justify putting the subject at any risk? If it believes that the scientific hypotheses proposed are not reasonable or do not provide adequate benefit (to outweigh the risk) to the subject, the IRB must, by law, question the proposal. It is helpful to keep this in mind when one considers comments from the IRB.

It is also relevant to consider the composition, operation, and responsibilities of the IRB. The composition is mandated in the Code of Federal Regulations, Title 21 Food and Drugs, Part 56. The IRB must, by law, review most clinical research. There are studies that are exempt or may be eligible for expedited review, but, even in these cases, the IRB should be made aware of the study and should agree that the research is exempt or appropriate for expedited review. The IRB itself must have at least five members with varying backgrounds. There is no requirement that the committee members be experts in your particular clinical specialty or area of research. The IRB must include one member whose primary concerns are in the general scientific area (in this case, biomedical research of any sort) and one whose primary concerns are in nonscientific areas. This second requirement is generally interpreted as the inclusion of a "lay person" or "community representative" on the committee. In addition, there is a requirement that at least one member have no affiliation with the institution and have no immediate family member affiliated with the institution. Again, this requirement is usually met by having a "lay person" or "community representative" from outside the institution. For complex issues, the IRB may invite ad hoc reviewers, although these individuals are not allowed to vote on the proposal. There are no requirements regarding how often the committee must meet, but, when it does meet, written records of the meeting must be maintained. In this respect, an IRB meeting is not unlike a grant review panel. There are usually a number of reviewers assigned to a project who summarize the study and voice any concerns they may have regarding the project. After it has presented their reviews, the committee will discuss the project and will vote on whether or not the study should be approved. There is no scoring system involved, however. The IRB must maintain compliance with the federal regulations and is subject to administrative actions on the part of the government if they are noncompliant.

Most IRBs will spend a significant amount of time reviewing the informed consent document to ensure that it is easy to read and is understandable to a lay person. Because it is also a legal document, the IRB reviews the informed consent to ensure that both the institution and the investigator are disclosing all the necessary information for the ethical performance of the study and that there are adequate protections in place for both the subject, the institution and the investigator. The informed consent is essentially a contract between the subject and the investigator. By signing the informed consent, the subject is agreeing to potentially expose himself or herself to risk in return for any potential benefits he or she might gain. This benefit may include the understanding that he or she may help other patients as a result of participation. The goal of the IRB review of the informed consent is to prevent the subject from stating at a later date that he or she was not made aware of all the risks or did not understand them. In this respect, the IRB is working as much for the investigator as for the subject.

A full discussion of how to best interact with your local IRB is beyond the scope of this book. Simply put, there are too many differences between institutions and specific research proposals to cover all the possibilities. However, there are several basic rules investigators should follow when dealing with an IRB. First and foremost, the IRB is not your enemy. The committee serves a purpose dictated by law that should not be at odds with your goals in most cases. If there are conflicts, your research may expose the subject to greater risk than benefit and may not be entirely ethical. It is clearly better to deal with this *before* you undertake your study. Remember that the IRB protects you, the investigator, as well.

Second, if there are ever any questions regarding ethical issues or concerns with your research, it is important to contact the IRB directly and immediately. If you are performing a study in which there are adverse events, there are usually strict reporting criteria in the protocol that include notification of the IRB. If you are unsure if you should alert the IRB, you are better off erring on the side of caution and informing the IRB. To this end, you should not be afraid to call the IRB and ask questions, particularly during protocol development. Remember that a verbal discussion is not binding and, therefore, oral reporting of adverse events or formal complaints is not acceptable. However, if you are concerned about wording in an informed consent or whether or not a formal informed consent is needed for a study, it is better to call and ask before anything is put in writing, so that you can edit the protocol accordingly and submit an improved document, expediting the approval process. Remember that after you put it in writing, it is difficult to change; therefore, it is better to discuss issues in advance and avoid confrontation later.

Third, remember that you must update the IRB on your study's progress and renew the research with the IRB on a regular basis. Most studies that require full IRB review will require annual re-review and renewal. The reporting requirements usually are not overly burdensome and consist of a short progress report, information regarding the number of participants enrolled in the study, and any adverse events. It is your responsibility to ensure that your study gets renewed by the IRB. If you forget and enroll a patient after the study's approval has expired, you will be held responsible. Most IRBs will alert you that you need to renew the study, but you should not count on this mechanism, because letters can get lost or e-mails get accidentally deleted. It is wise to keep a personal record of your dealings with the IRB and maintain a list of IRB-approval expiration dates for all studies you are undertaking.

Finally, and most important, it is pointless to argue with the IRB. Simply put, you cannot win. Assuming the IRB is acting in what it believes is the best interests of the research subjects (which it probably is) and it is in compliance with the federal regulations regarding human research, there is little you can do to reverse the decision of the IRB. There is no appeals process. Effectively, you must address the IRB's concerns if you wish your research to proceed. The overwhelming majority of IRBs has good intentions and wants you to do good research. If your study is returned by the IRB with requested changes, you can either make the changes or provide justification as to why you feel the changes are inappropriate or unnecessary. Although the IRB may be reasonable and drop the requested changes, often it will not. It is always easier and quicker to make the requested changes and expeditiously receive IRB approval. Therefore, a final piece of advice is: simply make the changes, acknowledge that they may be more objective than you are on the topic, and move on.

3. SPECIFIC ETHICAL ISSUES REGARDING CLINICAL RESEARCH IN SURGERY

Clinical research in the surgical disciplines is subject to many of the same ethical concerns that research in other biomedical disciplines raises. However, although there are some common themes across all specialties, research in the surgical fields does present some unique ethical challenges. Many conditions that are treated with surgical procedures are acute and critical in nature and require quick decision making on the part of the patient and the provider. This, in turn, can affect the informed consent process, because patients often are quite sick and may not be stable enough to give consent or to participate in research. In addition, the intervention itself, if surgical in nature, can involve significant risks to the patient, which must be considered in the course of the research. Finally, although studies of medical interventions can sometimes include a placebo arm, the inclusion of a placebo arm in surgical trial is considerably more difficult, because sham surgery will always carry some risks with minimal benefit to the subject. To this end, we will review a number of ethical issues in research that specifically related to the surgical disciplines.

3.1. Informed Consent

The informed consent process itself is not unique to surgical research. As discussed in the Belmont Report, the informed consent process must have three qualities to be valid: information, understanding, and voluntary agreement. In the case of surgical patients, there may be situations in which these three qualities cannot be easily attained. For example, consider the investigator who is studying the use of a new antibiotic in the treatment of patients who have experienced traumatic closed head injury. Many of those patients will be unresponsive and will not be able to assimilate the information, understand what they are agreeing to, or to give voluntary agreement. Given that these patients are often acutely ill, there can be little delay in administering potentially life-saving antibiotics, so how is the clinician-researcher to proceed? Obviously, in this setting, the investigator might discuss the study with the patient's next of kin and try to obtain consent from this individual, but is this acceptable?

The informed consent process should include the subject and the investigator (or the investigator's designated and IRB-approved representative). In some cases, the subject may not be able to participate in the process. In the example presented here, the subject is too sick to participate. In other cases, he or she may not be legally competent to participate. In this setting, federal regulations direct the investigator to see consent from the subject's legally authorized representative (45CFR46 and 21CFR50). Most investigators, therefore, will seek out the next of kin, or, in the case of pediatric patients who are not of legal age to give consent, seek consent from the parents. The process, however, is not complete after the legally authorized representative gives consent. Whenever possible, the investigator should obtain affirmative assent from the subject as well. This obviously will not be possible in the case of the nonresponsive trauma patient, but is possible in children older than age 7 and in some adults with limited mental capacity. If possible, assent should also be obtained in written form and it is standard for children older than age 12 to complete a written assent form. Finally, in these special cases, there may be a need for continuing evaluations of mental capacity and consent understanding. In the example of the trauma patient, assume that the patient becomes responsive and is now able to understand the study and give informed consent. It is now the responsibility of the investigator to discuss the study with the subject and ensure that he or she wishes to participate. It would probably be wise to have the subject now complete the formal informed consent process. These types of situations are often discussed at IRB meetings during the approval process for any given study and it is wise for the investigator to

Table 1
Required Elements of the Informed Consent Process, as Dictated by the Federal "Common Rule" Regulations

- Statement that the activity or intervention is considered research
- Purpose of the research
- Description of the study procedures
- Potential risks of the study
- Potential benefits of participation
- Alternative treatments
- Methods used to maintain confidentiality
- Compensation for injury statement (if study greater than minimal risk)
- Investigator contact information
- Statement that participation is voluntary
- Statement that there may be unforeseen risks
- Reasons that a subject may be involuntarily removed from the study
- Any additional costs for participation
- Adverse health consequences for early withdrawal (if any)
- Information regarding notification of findings or data from study, if relevant
- Number of subjects in the study

contact the IRB if he or she has any questions regarding the best way to handle these situations.

The consent document itself has a number of required elements that are presented in Table 1 (21CFR50). Many institutions use a common template for their informed consent, and it is wise for the investigator to try to stay as close to this template as possible. The template will include all of the required elements in the table and will also be worded in a way that is already fairly acceptable to the local IRB. In summary, the informed consent is one of the most important parts of the clinical research process. It ensures that both the subject and the investigator are aware of the risks and benefits of the study and that they each know their rights and responsibilities. For the surgical investigator, there may be special cases where consent must be obtained from other individuals, but these situations can be handled in an ethical manner if the investigator uses common sense and seeks the counsel of the local IRB.

3.2. Ethical Considerations Related to the Unique Characteristics of the Surgical Intervention

The study of surgical interventions presents many unique problems for the clinical researcher. Specifically, surgery itself has a very powerful placebo effect that can often lead to improvement in patient symptoms. A recent study in the *New England Journal of Medicine* randomized patients with knee pain to receive either arthroscopic knee surgery for osteoarthritis or sham surgery. Both groups reported improvement in pain 2 yr after surgery and no differences in pain were seen between the active surgery groups and the sham surgery group *(13)*. This unique study underscores the strong placebo effect of a surgical intervention on a patient-centered outcome and also introduces one of the unique problems in trials of surgical intervention: the difficulty with blinding the subject to the intervention. In this study described, the surgeons actually made a skin incision in the sham surgery group, but this is often not a reasonable option for many surgical diseases and their treatments. After all, many patients have life-threatening conditions, and a placebo intervention simply is not an option. In addition, sham surgery itself carries some risks, such as anesthesia and scarring. Many IRBs would not have approved the knee surgery study (the IRB at the Houston VA Medical Center did approve the study). In the case of other surgical interventions, it is likely that an IRB would only approve a sham procedure if the potential benefit far outweighs the risks of the sham surgery. Pharmaceutical trials are much easier to blind because placebo pills only carry the risk of no active therapy. Sham surgery is obviously much more invasive.

How should the surgeon deal with for this problem? Often he or she simply cannot maintain blinding nor have a true placebo group, which will affect patient-centered outcomes. Therefore, one way around this problem is simply to choose a more objective outcome. Whether this is survival, disease recurrence, or some radiologic or laboratory marker, an objective, quantifiable, reliably measured outcome serves to minimize the placebo effect of surgery. In studies in which the primary outcome is a patient-centered outcome (such as quality of life), the investigator should consider a placebo arm. For example, imagine you were to do a trial aimed at determining the effectiveness of laser therapy in reducing lower urinary tract symptoms in men with benign prostate hyperplasia, you would have to account for the placebo effect of any treatment (which in some studies has been shown to be up to 40%) *(14)*. One option would be to randomize the patient to laser therapy vs sham surgery. The patient would receive a light anesthetic and undergo either the laser procedure or cystoscopy. Postoperatively, all interventions would be the same and the patient would never be told which intervention he received. At the conclusion of the study, all patients who required further treatment would be offered the laser therapy. Given that lower urinary tract symptoms are not life-threatening and the patients ultimately would receive treatment if they wanted it, one could make a compelling argument that the risks of anesthesia in this setting are relatively small and that the study is ethical.

In summary, there may be situations were sham surgery could be ethically performed assuming: (1) the disease is not life-threatening; (2) the patient will be offered salvage treatment at the end of the study; (3) the sham surgery presents minimal risk and does not leave permanent scars and is disfiguring in anyway; and (4) there is true clinical equipoise around the question and the study would provide significant scientific information that would advance the field. Of course, each local IRB will have its own opinion on the issue of sham surgery, so it would again be best to discuss the study with the individual IRB.

Finally, the surgeon has to realize that, although surgical trials may be needed for a particular issue, they may not be feasible when the course of treatment for a particular condition is in rapid evolution. To some degree, this is the case in prostate cancer, in which investigators have failed to complete trials comparing the various primary therapies for localized disease *(15)*. This may be more of a feasibility issue than an ethical issue, but, simply put, patients often are not likely to participate in randomized clinical trials if they have strong feelings about a particular treatment or after if a particular treatment has gained popularity in the media or general public. The American College of Surgeons' Oncology Group recently undertook a randomized clinical trial comparing brachytherapy with surgery in localized prostate cancer. The study accrued poorly and closed 1 yr after it opened. Too many patients had preconceived notions about both therapies and would not submit to a computer randomization of treatment *(15)*. The study demonstrated how difficult it can be to get patients to agree to enroll in a randomized surgical trial. Although this may be more a feasibility issue than an ethical issue, some researchers have tried to provide potential subjects with various incentives to enhance recruitment. Although this may seem reasonable on the surface, it raises a number of additional ethical issues that ultimately may prevent the investigator from using this strategy. Despite the trials and tribulations of conducting a randomized surgical trial, significant work has been accomplished, particularly within the context of the clinical trial networks funded by the National Institutes of Health.

4. HIPAA: THE 800-POUND GORILLA OF CLINICAL RESEARCH

HIPAA has arguably had as great an impact on the ethical conduct of clinical research as all of the prior federal legislation combined. The legislation (contained in 45 CFR, parts 160 and 164), originally designed to assist individuals in getting health insurance if they had a preexisting condition, established strict regulations regarding the use of protected health information (PHI). Importantly, it established severe fines and punishments for institutions or individuals that violated the law. Although this legislation was developed for the clinical practice of medicine, it also applies to clinical research. As a surgical investigator considering clinical studies, you *must* be aware of your obligations under HIPAA and *must* act in a manner consistent with the law. If you violate HIPAA, even unknowingly, you run the risk of harsh penalties. It is important to note that ignorance of the law is not an acceptable defense.

The HIPAA legislation establishes the "privacy rule," which basically protects the use and disclosure by "covered entities" of identifiable health information, referred to as PHI. The rule regulates the transmission of information related to health care. In these settings, permission must be obtained from the patient to transmit or use his or her information. A list of the various definitions of "covered entities" is presented in Table 2. For the clinical researcher, it is almost certain that you are working within a covered entity of some type and, therefore, must comply with the HIPAA privacy rule. The rule itself applies to PHI, which is individual health information that is individually identifiable and is created or received by a covered entity. A list of HIPAA PHI identifiers is presented in Table 3. As can be seen from the table, any variable that could be used, alone or in combination, to identify the patient is considered PHI and is covered by the privacy rule. Bear in mind that, by including a category of PHI that consists of any other unique identifying number, characteristic, or code, the use of any information (even if it is not specifically mentioned in the list) that could reidentify the individual is considered protected and is subject to HIPAA.

The significance of this list lies in the fact that health information itself is not necessarily covered by HIPAA, but rather only individually identifiable health information is. In other words, a chest X-ray or a serum sample is not covered by HIPAA *unless* there is an identifier associated with it that would allow someone to trace the X-ray or specimen back to the patient. This becomes a key consideration when designing clinical studies. After all, we usually assign a subject a study number to ensure this data remain confidential. Therefore, if we use "coded data," are we now in HIPAA compliance and able to proceed without worry? The answer is "no," if a link exists between the study number and any patient identifiers. Most researchers place their coded data into a large research database, but maintain a separate smaller and protected database that links the study numbers to patient identifiers. If this common strategy is used and the linkage maintained, even under electronic "lock and key," the researcher must act within the confines of HIPAA. However, after the data are deidentified (that is, the linkage between the study

Table 2
Examples of Covered Entities Under HIPAA

- Institutional covered entities: a "covered function" is anything that makes the entity either a health care provider, health plan, or health care clearinghouse.
  - — Outpatient clinics
  - — Community hospitals that only provide medical care and have no other non–health care related functions
  - — Private practice doctors offices
- Hybrid entities: complex institutions that provide health care but also have noncovered functions that are unrelated to health care. If an institution designates itself as a "hybrid" institution, it must isolate its "covered" functions from its "noncovered" functions to prevent the unauthorized exchange of PHI.
  - — Universities (including university medical centers)
  - — VA Medical Centers
  - — Certain health maintenance organizations
  - — Other health plans
- Affiliated covered entities: according to the law, legally separate but affiliated institutions may choose to designate themselves as a single-covered entity under HIPAA if they are under common ownership or control. This creates efficiencies within the system and may facilitate the transfer of data within the institution as a single common notice of privacy practices can be used for the affiliated institutions.
  - — Universities (including university medical centers)
  - — VA Medical Centers
  - — Certain health maintenance organizations
  - — Other health plans

number and all patient identifiers is destroyed or the recipient of the data could not possibly identify the data because he or she has no access to the linkage), the data are no longer covered by the HIPAA Privacy Rule. Therefore, whenever possible, surgical researchers should strive to use deidentified data.

If the researcher does wish to use PHI data, he or she must obtain an authorization from the subject to use the data in research. These authorizations must include: a description of the PHI to be used; who the PHI will be disclosed to; who will be using the PHI; a description of the purpose for its use; whether or not there is an expiration date for the authorization (and if so, when that date is); a notice that the subject may revoke the authorization at any time; a warning that the disclosed information may no longer be covered under HIPAA; a statement that the provision of treatment is not contingent on the authorization; and that subject signature and date. Most IRBs and research institutions have boilerplate text available to researchers from which to draft a HIPAA authorization. It is wise to obtain this authorization even if you are not certain that you will need to use identifying information.

There are situations in which you may be able to use limited PHI without a HIPAA authorization from the patient. HIPAA provides for limited datasets that exclude all PHI identifiers except addresses, dates, and other indirect identifiers. For these datasets, the researcher may apply for a waiver of authorization from the local IRB or HIPAA board. The committee is likely to grant a waiver if the researcher: describes how the dataset will be used, identifies who will have access to the limited dataset, assures the IRB that the

Table 3
HIPAA PHI Identifiers

- Name
- Geographic identifiers (beyond state), which includes city, town, or ZIP code
- All elements of dates (birthdates, date of death, date of admission or discharge), age
- Telephone numbers
- Fax numbers
- E-mail addresses
- Social security numbers
- Medical record numbers
- Health plan beneficiary numbers
- Account numbers
- Death certificate numbers, driver's license numbers, etc.
- Web addresses
- Internet protocol addresses
- Biometric identifiers (e.g., fingerprints, voice prints)
- Full face images
- Any other unique identifying number, characteristic, or code

dataset will not be used to contact individuals and that appropriate safeguards are put in place to prevent uses or disclosures outside the research agreement, and the IRB feels that it would not be practicable to obtain a signed authorization and the research poses no more than minimal risk. The limited dataset strategy is a very reasonable approach for researchers who wish to analyze existing databases. In this setting, the investigators with the identified data could develop a limited dataset and give it to another investigator after IRB approval of a HIPAA waiver of authorization is obtained. Since the inception of HIPAA, many databases researchers have also obtained HIPAA authorizations from the subjects at the time of enrollment, which greatly simplifies the process.

There is one additional situation that the researcher must be aware of when considering the impact of HIPAA on clinical research. HIPAA applies when screening and recruiting subjects in the clinic. In this setting, the researcher often will review the charts of potential subjects to determine eligibility before an office visit. Although this may seem fairly benign, it is not appropriate to screen medical records in this manner under HIPAA. In this setting, the researcher should obtain a partial waiver of authorization from the IRB before proceeding. This waiver will be granted if the screening presents minimal risk to the patient and obtaining prior authorization is not practical. Full authorization can then be obtained if the patient wishes to participate. One might argue that the clinician who is performing the study has to review the medical records as part of his or her routine care. However, this review is part of the clinician's role and is unrelated to the role of a researcher. Therefore, it is wise to obtain a partial waiver of HIPAA authorization as part of the IRB approval process. The same is true for the use of existing databases to identify study cohorts. If no HIPAA authorization was previously obtained, it is wise to obtain a partial waiver of authorization before querying the database and contacting any patients. There is a clause in HIPAA that deals with "reviews preparatory to research" that allows the investigator to review the dataset to assess sample size and determine if there are adequate subjects for the research. In this setting, no HIPAA authorization is required assuming that the PHI used for these reviews is not disclosed or used offsite. In certain settings, individuals within the covered institution may even be able to contact the potential subjects to discuss the research further. However, this should be discussed with the IRB before proceeding to ensure that you are in compliance with HIPAA.

HIPAA may seem quite overwhelming at first glance, but after you obtain a basic understanding of the rules, it is relatively easy to navigate. Most research will require a HIPAA waiver from the patient. This is easily obtained at the time of informed consent and usually remains in force throughout the study. If you are ever unsure about whether HIPAA applies to you or your research, you should contact your local IRB or office of compliance before proceeding. Invariably, they will take a conservative approach to the issue, but this is probably wise, because the penalties for HIPAA violations are severe

5. CONCLUSIONS

The surgeon-scientist must be aware of the ethical issues surrounding clinical research if he or she is to succeed at this endeavor in the 21st century. We have reviewed the history of ethics and clinical research, the role of the institutional review board, and specific issues surrounding surgical studies and HIPAA as they relate to research. There are a few take-home messages that the reader should bear in mind. First, the current landscape surrounding the ethics of clinical research developed in response to a number of isolated incidents that were morally repugnant, but were also more widespread than one might imagine. As evidenced by the recent gene-transfer experiments described previously, ethical dilemmas may arise even with the current safeguards. Second, the IRB serves an important role by protecting researchers from ethical problems and providing guidance as these problems arise. Although many researchers have assumed an adversarial relationship with their IRB, this is a mistake. If you view the IRB as a resource that is there to protect you and give you guidance, you will find your dealings with the IRB more pleasant and productive. It will also improve your research. Finally, the passage of HIPAA changed the way we do clinical research. You must be aware of HIPAA and always ensure that you are in compliance. If you run afoul of the HIPAA regulations, you and your institution will be liable. Because compliance is relatively easily and your local IRB can assist you in maintaining compliance, there is no reason this should ever be a problem for you, assuming you are aware of your responsibilities.

REFERENCES

1. American College of Surgeons. Statements on principles. 2005. www.facs.org/fellows_info/statements/stonprin.html. Accessed June 19, 2006.
2. Caplan A, ed.): When medicine went mad: bioethics and the holocaust. Totowa, NJ: Humana Press, 1992.
3. Annas G, Grodin M, eds. The Nazi doctors and the Nuremberg code: human rights in human experimentation. New York: Oxford University Press, 1992.
4. Trials of war criminals before the Nuremberg military tribunals under Control Council Law No. 10. Washington DC: United States Printing Office, 1949:181–182.
5. McBride WG. Thalidomide and congenital malformations. Lancet 1961:2:1358–1362.
6. Lenz W. A short history of thalidomide embryopathy. Teratology 1988:38:203–208.
7. Jones J. Bad blood: the Tuskegee syphilis experiment. New York: Free Press, 1993.
8. Puttagunta PS, Caulfield TA, Griener G. Conflict of interest in clinical research: direct payment to the investigators for finding human subjects and health information. Health Law Rev 2002;10:30–32.
9. Stolberg SG. The biotech death of Jesse Gelsinger. New York Times Magazine, November 28, 1999; 136.
10. Nelson D, Weiss R. Penn researchers sued in gene therapy death: teen's parents also name ethicist as defendant. Washington Post, September 19, 2000;A3.
11. Stolberg SG. FDA officials fault Penn team in gene therapy death. New York Times December 9, 1999; A22.

12. Couzin J, Kaiser J. Gene therapy. As Gelsinger case ends, gene therapy suffers another blow. Science 2005;307:1028.
13. Moseley JB, O'Malley K, Petersen N J, et al. A controlled trial of arthroscopic surgery for osteoarthritis of the knee. N Engl J Med 2002;347:81–87.
14. Lepor H, Williford WO, Barry MJ, et al. The efficacy of terazosin, finasteride, or both in benign prostatic hyperplasia. Veterans Affairs Cooperative Studies Benign Prostatic Hyperplasia Study Group. N Engl J Med 1996;335:533–540.
15. Penson DF. An update on randomized clinical trials in localized and locoregional prostate cancer. Urol Oncol 2005;23:280–287.

3 Budget Development and Staffing

Judith Fine and Peter C. Albertsen, MD, MS

Contents

1. INTRODUCTION

A research project, whether large or small, and regardless of funding source, requires a budget. A budget is a realistic prediction of what it will cost to complete a study and is constructed by analyzing each of the steps involved in conducting the study. For most grant applications, a budget justification must accompany the budget and usually includes a narrative description of the roles played by personnel requested in the application and how other projected expenses have been determined. The budgeting process requires the investigator to think through the proposed project carefully and identify each step of the research plan. Not only must the investigator estimate the dollar amount of projected expenses, but he or she must also estimate the timing of when expenditures are likely to occur. The budgeting process is an integral part of a grant application and should not be left for the last minute. The process itself can often identify important steps or omissions in the project.

When preparing a budget for a grant application, the investigator should consider three important items: the policies and requirements of the granting agency, the policies and requirements of the investigator's institution, and the projected new costs and existing resources that will be part of the research project. The budget needs to reflect accurately what the investigator plans to do and the proposal itself needs to include the relationship and the need for each of the budget items. Reviewers considering a grant for funding expect to see a carefully prepared budget that is thoroughly justified and documents how calculations were made.

1.1. Types of Research Proposals

Before developing a budget proposal, researchers should learn the types of activities that an agency will fund as part of the grant program. An investigator-initiated project (a

From: *Clinical Research for Surgeons*
Edited by: D. F. Penson and J. T. Wei © Humana Press Inc., Totowa, NJ

project the investigator developed in response to his or her own research interests, as opposed to in response to a request from the granting agency), allows an investigator considerable latitude concerning the scope and extent of a project. These types of projects are typically funded through the R-01 mechanism by the National Institutes of Health. Some projects can be completed within a short period. More commonly, however, an R-01 proposal extends for 5 yr and requires a detailed budget for yr 1 and estimates of expenditures anticipated in yr 2 through 5.

Granting agencies frequently have research interests of their own and ask investigators to submit proposals to address specific research problems. These projects, often referred to as an RFA (Request for Application) when requested by a federal agency, have more restrictions. A federal agency may issue an RFA stating that a certain amount of money is available in a particular fiscal year and that the agency anticipates awarding some number (e.g., six to nine) of grants to successful applicants. The agency may place further restrictions by stating the proposed projects are to last only 2 or 3 yr and may not exceed a certain amount that includes both direct and indirect costs. Direct costs represent the actual costs of performing the research, including staff salaries, laboratory supplies, and analytic costs. Indirect costs represent the administrative and overhead costs to the institution. The indirect costs are usually calculated as a percentage of the direct costs at a rate dictated by the investigator's institution.

The most restrictive type of research proposal is a contract. Under this scenario, the granting agency has a very specific research task to complete. In a process referred to as an RFP (Request for Proposal), the agency asks researchers to bid on the proposal by submitting a research plan and a proposed budget. When awarding a contract, the funding agency may negotiate specific items within the investigator's budget.

Each granting agency has its own rules governing expenses that are allowed and those that are not. These rules differ for each grant competition. Many private foundations, for example, frequently offer modest sized grants of $25,000–$50,000 to encourage researchers to explore novel ideas. Unfortunately, the awards are often restricted to research supplies. They explicitly prohibit grant awards from funding the investigator's salary, graduate student tuition, stipends, professional travel, secretarial help, and some equipment costs. Before preparing a budget, the investigator must carefully evaluate the scope of a proposed project and whether sufficient resources are available.

1.2. Components of a Budget

There are usually three major components of a budget proposal: direct costs, indirect costs, and institutional commitments. This chapter will review each in detail. Direct costs are those associated with actual conduct of the proposed study. Indirect costs are those expenses shouldered by the institution that provides the researcher with a warm, dry, safe place to conduct the research. Most institutions have negotiated an "indirect rate" with the federal government. Private foundations, however, may limit the amount of money that can be allocated to indirect costs.

1.3. Other Factors to Consider

When constructing a research budget, careful attention should be given to the timing of expenses. For example, perhaps not all personnel will need to start immediately. That would affect some other categories of expenses, such as staff travel and consumption of supplies.

Some funding agencies will allow investigators to take unspent funds from one fiscal year and carry them over to the next fiscal year. Others will not. In the latter case, unspent funds must be returned to the granting agency. Such restrictions have important implications for how a researcher plans his or her research activities. It is also important to know whether a funding agency will allow yearly increases for salary and cost of living or requires level funding. Most granting agencies will allow "no-cost extensions." In a "no-cost extension," the investigator is allowed to extend the life of the grant beyond the original funding period by using unspent funds from the final year. In some instances, "no-cost extensions" are unlimited; in others, any remaining unspent funds must be returned after a 1-yr extension.

Institutions also have specific guidelines for grant and contract applications. Before the development of any research proposal, an investigator should contact his or her institution's business office, grants office, and human resources department. Many institutions have very specific hiring policies and salary scales. This is especially true if employees within the institution are members of a union. Most institutions have developed an internal set of procedures to review grant applications and budget proposals to ensure that they comply with their requirements. Multiple signatures will be needed as the proposal moves through the organization. Researchers need to remember that this process takes time and that they should complete their grant proposal several weeks before the submission deadline to avoid last minute crises.

Avoid padding the budget, but also avoid a budget that is too lean. Research, by definition, will encounter unexpected situations. Investigators should anticipate potential problems and build in sufficient resources so that the proposed project can be completed successfully.

2. PARTS OF A BUDGET

2.1. Direct Costs

Using the National Institutes of Health (NIH) "detailed budget for initial budget period – direct costs" (NIH form page 4; *see* Appendix 1) and the companion "budget for entire proposed project period – direct costs" (NIH form page 5; *see* Appendix 2) as an outline, researchers should assemble budgets estimating expenses in each of the categories listed here. The budget should be calculated for the initial budget period, which is usually 1 yr. The budget should then be extended over the entire period of the proposed study, including increases over time, if allowed by the funding agency. Further information on calculating these costs is provided under "Calculating Costs and Budget Justification." The sections of the budget are as follows and each is discussed in turn:

1. Personnel
2. Consultant costs
3. Equipment
4. Supplies
5. Travel
6. Patient care costs, if any
7. Alterations and renovations
8. Other expenses
9. Consortium/contractual costs

2.1.1. Personnel

Start by identifying the number and kinds of tasks needed to be done to conduct the research and the number and kinds of personnel that it will take to perform those tasks. The following questions may be useful.

1. How long will it take to complete the research?
2. How much time will the investigator devote to the project?
3. What tasks will the investigator handle?
4. Will the investigator, or someone else, supervise research staff? If someone else, who?
5. Clinical research usually involves human subjects. Will the investigator correspond with an institution's Institutional Review Board (IRB), or will this task be delegated to a project coordinator?
6. Is a project director required? If so, is this a full- or part-time position?
7. What kinds of knowledge and skills will a project director need?
8. How many of the required tasks can a project director take on?
9. How will data be collected? Who will collect the data?
10. Is a data manager needed? If so, full or part time?
11. What kind of experience will a data manager require?
12. How many of the required tasks can a data manager take on?
13. Does the project require laboratory personnel?
14. What knowledge and skills would be required of laboratory personnel?
15. Is patient care involved? What level of patient care? Inpatient or outpatient?
16. Will specialized medical personnel be involved, such as a registered nurse, or can the project be completed with the assistance of a medical technician?
17. How long will each person be needed on the project?
18. Does the institution have core facilities (sometimes called service centers or recharge centers) that provide specific technical or administrative services for a user fee, or must the researcher perform or contract for all specialized tests or functions?

After a researcher has addressed these questions, he or she must obtain estimates of the salary costs associated with the number and types of personnel that will conduct the project. Frequently, specific salary estimates are available from an institution's human resources department. Calculation of personnel costs will almost always involve fringe benefits. Fringe benefit rates can be substantial and are usually determined by an institution or as part of a collective bargaining agreement. Fringe benefit rates may simply be a percentage of salary or may vary by job classification. Fringe benefits cover such things as FICA, Medicare, retirement plans, insurance, and worker's and unemployment compensation. There are some job categories such as student labor that have no fringe benefit costs. Investigators should remember that fringe benefit rates are usually not under their control and can increase along with salary costs with each year of the project. These increases need to be budgeted in advance.

2.1.2. Consultant Costs

Investigators are not always capable of providing all the necessary professional experience and expertise to conduct a research project. Some skills, such as help with statistical analysis, are best provided by research consultants. Usually, consultants are hired on an hourly or daily rate. These rates should be estimated realistically and should reasonably reflect the consultant's usual salary. An investigator must carefully evaluate the needs of the project to determine how many types of consultants may be required and the amount of effort needed. Large clinical projects, for example, often must include the costs of convening an external data and safety monitoring board or an external advisory committee.

2.1.3. Equipment

Research projects require supplies and equipment. Supplies are items that are consumed on a weekly or monthly basis. Equipment refers to items that frequently have a usable life of at least 2 yr and a purchase value over some amount such as $500 or $1000. Investigators must carefully review a funding agency's regulations to determine what types of equipment may or may not be included in the grant proposal. Researchers must take into account computer and software requirements. Other important items include office equipment such as file cabinets, desks, and chairs. Many institutions have warehouses from which investigators may obtain recycled equipment at steep discounts.

2.1.4. Supplies

Research projects require supplies. At a minimum, researchers must budget for office supplies such as stationary, toner cartridges, copy paper, file folders, pens, and markers. Does the project require letterhead, questionnaires, or other types of preprinted forms? How rapidly will supplies be consumed and need to be reordered?

Investigators conducting laboratory research must also plan for laboratory costs such as chemicals, reagents, plastic ware, and other consumables. Are animals required? If so, how many and what species? In addition, careful consideration needs to be given to where animals will be housed and fed.

2.1.5. Travel

Investigators should carefully consider the travel implications of their project proposal. Does the project call for travel outside of the institution? Does the investigator need to meet with other collaborators or consultants? If so, how often will this occur? Are consultant costs including airfare, hotel accommodations, and mileage carefully estimated? Where will the research results be presented? The investigator may wish to budget for one trip to a national meeting each year.

Do the research personnel need to travel to collect data or meet with study subjects? If so, what is the institution's policy concerning travel reimbursement for use of private car or other modes of travel? Is there a standard mileage reimbursement rate? Do patients need to travel to the investigator's institution? If so, will this travel be reimbursed? Will a fixed per diem be used, or will patients submit receipts for parking, tolls and mileage? Have patient reimbursements been reviewed and approved by an institution's IRB?

2.1.6. Patient Care Costs

Research projects involving patient care can be very costly and complex. Investigators need to review the project proposal carefully to determine what care would be considered routine and what care is part of the investigator's project. Routine health care can be billed to a patient's insurance carrier; tests performed for research purposes cannot. Many institutions have negotiated research patient care rates with the Department of Health and Human Services. If not, researchers should negotiate discounted rates within their institutions for laboratory studies, radiology charges, and inpatient costs. Investigators should not have to pay retail rates for routine hospital services.

When estimating the costs of patient care, careful attention must be given to the rate of patient accrual. This may not be an issue in a small study with relatively few office visits, but it can be a major factor when large numbers of patients requiring multiple office visits are to be enrolled.

Will patient care be conducted in multiple institutions? If so, what part of the care will be done by a central laboratory and what part will be performed locally? Most multi-institutional research projects use a central pathology laboratory and a central laboratory for serum analysis. Investigators should anticipate that the costs of these services will increase on an annual basis.

Who is initiating the research project? Are funds being supplied by a pharmaceutical company to conduct a clinical trial? If so, the investigator needs to estimate the costs of providing appropriate care according to the research protocol outlined in the research proposal. If the investigator has initiated the study, does the institution have a facility such as a General Clinical Research Center to assist in the conduct of the study? If so, the investigator should work with the center director to develop an appropriate budget.

2.1.7. Alterations and Renovations

In some cases, agencies will fund alterations or renovations to physical space required to conduct a research project. Usually these renovations need to be directly related to the conduct of the proposed project and the costs of such alterations carefully documented in the budget justification.

2.1.8. Other Expenses

Expenses that are not specifically detailed in other sections should be categorized here. They can potentially include such items as long-distance telephone calls, charges for telephone lines and voice mail, network computer charges, equipment service contracts, postage, mailings, photocopying charges, medical record charges, or pathology retrieval charges. Cost of rental space may be allowed when a lower off-campus indirect rate is assessed. Other items may include modest incentives to encourage patient enrollment. These items, however, must be approved by an institution's IRB. Finally, some IRBs are now charging a review fee, particularly for studies funded by pharmaceutical companies. These fees should be included as other expenses.

2.1.9. Consortium/Contractual Costs

Some research projects require expertise available only in another institution or an agency of a state government. The two most common arrangements are to construct either a consortium agreement or a contract. The difference dictates how money is dispersed from the granting agency. In a consortium agreement, both institutions have responsibility for the conduct of the entire project and both will receive disbursements from the funding agency. Under a subcontract, the funding agency will distribute the entire grant to the primary institution who will in turn distribute money to the subcontractor. In either of these arrangements, all participating institutions will use their own rules regarding compensation, fringe benefit rates, and indirect costs. Furthermore, each participating consortium/contractual organization must submit a separate detailed budget for both the initial budget period and the entire proposed project period along with their portion of the proposal. Indirect costs appropriate to the respective institutions are included in each contract. Federal funding agencies have specific rules regarding the amount of indirect costs that the primary institution can collect on a subcontract to another institution.

2.2. Facilities and Administrative Costs (Indirect Costs)

A researcher occupies space within an institution. The institution defrays the cost of maintaining that space by leveling a Facilities & Administrative fee on all grants and contracts. This Facilities & Administrative fee is commonly referred to as indirect costs and can represent a sizeable portion of the total budget. Indirect costs are added to reimburse the institution that receives the funding for such things as administrative departments (e.g., payroll, grants & contracts, human resources), depreciation of buildings and equipment, utilities, and libraries. Usually, indirect costs are calculated according to a previously negotiated rate between the research institution and the federal government. Many private foundations accept these rates, but others limit overhead to a specific percentage of the total grant award.

Not all expenses in a research proposal are used to compute indirect costs. Usually only expenses that involve institutional facilities are included in the calculations. These expenses may include personnel, travel, and supplies. Large equipment purchases are usually excluded. Institutions can have multiple indirect rates. A large academic institution may have an on-campus rate, and one or more off-campus rates. In that case, the indirect rate that would be used depends on the location and type of space in which the project would be housed. The indirect rate is applied to all applicable direct costs, and the resulting figure is added to the direct costs for a 1-yr project total.

2.3. Institutional Commitments

Private agencies and foundations frequently request that an investigator's institution contribute in some way to the financing of a project. Often this commitment is in the form of salary support for the principle investigator or other members of the research team. Another way of demonstrating an institutional commitment is to provide a waiver of a portion of the indirect cost recovery allowance. Finally, an institution may show commitment by donating services such as mailing, telephone, or photocopying services. When considering institutional commitments, an investigator must know precisely what a granting agency is willing to fund and then budget accordingly.

3. CALCULATING COSTS AND JUSTIFYING THE BUDGET

In addition to a budget proposal, an investigator is also expected to submit a budget justification. This portion of the grant is just as important as the budget itself and therefore should command serious attention. The budget justification provides a brief explanation and rationale for each line item in the budget and how each was calculated. It usually follows the same outline as the budget itself.

3.1. Personnel

After a researcher has determined the type and number of personnel needed for a project, he or she must allocate the amount of time each employee will spend on specific tasks. A useful tool for envisioning how the project will proceed is a time line bar graph. By using a time line, the investigator can plan each step of the project and determine the commitment of personnel and other resources on each specific task, or set of tasks. The construction of such a time line will also be useful in the development of project-specific job descriptions.

Research personnel often work on a number of different studies concurrently, sometimes across departments. The percent of effort expended on each is charged to the individual projects in proportion to the time spent. For example, a data manager who works 5 d per week, might work on one study for 2 d and on another project for the remainder of the week. If the institution considers 5 d per week to be full time, then the data manager is considered to be a 1.0 FTE (full-time equivalent). The first project would budget the manager for 40% effort plus that share of the fringe benefits, and the second for 60% effort plus that share of the fringe benefits. Experienced research personnel are well worth higher personnel costs, especially if the investigator is new to research or will not have a great deal of time to devote to the project. Some grants are subject to salary limits. That information may be obtained from an institution's grants office.

To calculate the cost for each person for the initial year, multiply the institutional base salary by the percent effort on the project. Next, multiply the base salary allocated to the project by the appropriate fringe benefit rate. Add the two sums together to determine the Year 01 total for each person. Using appropriate increases in salary, and any changes to percent effort or a fringe benefit rate, calculate the remaining years of the proposed project in the same manner.

A budget justification should include the following information for each person requesting salary support starting with the principal investigator: name, degree(s), role on the project (e.g., principal investigator, project director), and percent effort for each of the years for which support is requested. A concise description of the experience that an individual will bring to the project should follow, along with the specific tasks that will be his or her responsibility. The investigator should carefully assess how much effort each employee will spend on the proposed project during each year of operation and should justify any changes in percent effort by year. How annual salary increases are determined, whether by negotiated contract or institutional guidelines, should be stated. Frequently, investigators anticipate hiring personnel only if a project is funded. These personnel should be listed as "to be named" and appropriate salary requests should be determined by the category of job description, or by institutional guidelines. Explain briefly the rate(s) used to calculate fringe benefits for each person. The budget justification should follow the outline of the research proposal and should permit a reviewer to locate easily the tasks assigned to each position being requested.

3.2. Consultant Costs

Consultant costs are usually calculated on either a per diem rate or some other appropriate unit. To determine the consultant costs, first determine the number of days each consultant is needed and then multiply this number by the daily rate that has been negotiated with the consultant. Add to that any travel or other agreed-on expenses. Calculate subsequent years in the same way, using any increase in daily rate, or change in number of days of consulting to be provided. Be certain to include an estimate of travel costs in subsequent years.

The budget justification should include the name and institution/organization for each consultant who will work on the research project. Furthermore, the researcher should provide a brief explanation concerning why this consultant was selected and the unique expertise he or she will bring to the project. Describe the services the consultant will provide and the agreed-on rate for the initial year. Explain any changes in rate, days, or services to be provided for subsequent years.

3.3. Equipment

Not all funding agencies will permit equipment purchases. Be sure to determine if equipment is allowed by the funding mechanism to which the application is being submitted. Cost estimates can usually be obtained using institutional guidelines. The budget justification should include an explanation describing the need for each piece of equipment being requested. Expensive items should be clearly identified as to their role in the research project.

3.4. Supplies

Institutional guidelines sometimes dictate how "supplies" and "other expenses" are differentiated. The instructions for grant preparation may also suggest how "supplies" and "other expenses" should be listed. To estimate total supply costs, first determine the purchase price for each item using institutional guidelines. Often there are preferred vendors for specific items. The investigator should check whether the institution has a central warehouse, or an onsite inventory. The budget justification should include explanations for each category of supplies and how each was calculated.

3.5. Travel

Requested travel funds should be separated into categories according to the purpose. Investigators frequently travel to present results, whereas research personnel travel to collect data or retrieve research material. Travel costs may include airfare, train fare, per diem, hotel ground transportation, mileage reimbursement for use of personal car, tolls, and parking fees.

Reasonable estimates should be used for future travel involving public transportation and overnight stays. Institutional guidelines will be useful in this category. More involved calculations will be needed for researchers who travel on a regular basis. Actual distances and number of trips per staff member need to be determined and then multiplied by the cost factor (e.g., per mile reimbursement for use of personal car). If a research project requires significant travel, a reviewer will want to see that careful thought has been given to the calculation of these expenditures. The budget justification should include who will be doing the traveling, the destination, the length of stay, the purpose of the travel, any related expenses that will be incurred, and how the costs were calculated. Some RFPs will require actual quotes for air fare and hotel accommodations, despite the fact that no funding is in place and no date can be assigned to the travel.

3.6. Patient Care Costs

There are many variables to be considered in calculating patient care costs. Some expenses will be shouldered by the research project, but other costs may be considered appropriate medical care and should be charged to third-party payers. When calculating patient care costs, researchers must consider when patient accrual will commence and how rapidly the appropriate number of study subjects will be enrolled. For inpatient care, determine the number of patient days, cost per day, and the cost for each test or treatment. For subsequent years, use the same formula, but use an appropriate inflationary factor. Also remember to include any changes in the number of days, costs per test or treatment, increases or decreases in the number of tests or treatments, and any changes in types of tests or treatments.

The budget justification should include the name of each hospital or clinic that will be providing patient care, the funds requested for each, and whether each has a current Department of Health and Human Services–negotiated research patient care rate agreement. If there is no such agreement, provide a detailed explanation concerning the proposed use of each facility and the number of patients expected and how each of the categories of costs itemized in the budget was calculated. For subsequent years, continue the detailed explanation making sure any changes are well documented. If study participants are reimbursed for expenses related to the research, explain in detail the rationale and how the reimbursement will be calculated.

3.7. Alterations and Renovations

This category of expense covers changes to a defined space or spaces required by the project. Changes may include repairs, partitioning, painting, and changes related to laboratory needs. Large research proposals may require extensive changes to facilities. One approach is to estimate a cost per square foot. If this is not possible, a detailed item-by-item cost should be provided. The budget justification should include the changes to be made, the essential nature of the changes, and a detailed explanation of how each of the costs was calculated. If a square foot cost can be determined, provide the basis for that calculation.

3.8. Other Expenses

The investigator's business office should be able to provide costs for institution-based expenses such as telephone lines, voice mail, computer network, and charges for other centralized functions, as well as estimates of fees for equipment maintenance contracts. Project-specific expenses such as postage, photocopying, and charges for retrieval of medical records need to be calculated based on the actual proposed usage. Unless careful attention is paid to detail, some "other expense" items can destroy an otherwise well-constructed research budget. Postage and photocopying charges are good examples of expense categories that are often poorly estimated. The budget justification should include an explanation of how the cost for each category of other expense was calculated.

3.9. Consortium/Contractual Costs

Each participating consortium/contractual organization must submit a separate detailed budget and budget justification for both the initial budget period and the entire proposed project period. Consortium arrangements may involve personnel costs, supplies, and other allowable costs, including Facilities & Administrative costs.

3.10. Facilities and Administrative Costs (Indirect Costs)

Facilities and administrative rates are negotiated between an institution and the federal government. The institution's grants and contracts office determines the appropriate rate to be used by an investigator. This rate is applied to most (but not all) categories of funds requested in a proposed budget. The investigator's business office may be a resource in this area. The budget justification should state the negotiated rate being applied to the particular project.

3.11. Institutional Commitments

Institutional commitments to the proposed project should be clearly identified. If salary support is being provided to any or all of the research personnel, this support should be detailed in the budget justification. Other support such as a decrease in the indirect rate or the provision by the institution of services such as postage or telephone use should be clearly identified.

4. RECRUITING AND MANAGING A RESEARCH STAFF

The size and scope of a proposed project will dictate the number and kinds of staff required. Given the wide variety of clinical research, it is difficult to identify all the types of tasks that need to be completed. There are, however, some areas common to all research. They include protection of human subjects, data collection, data management and analysis, and manuscript preparation.

Protection of human subjects in research has come under increasing scrutiny over the last decade. New federal regulations under the Health Insurance Portability and Accountability Act, also referred to as the Privacy Rule, have had a major impact on clinical research. Therefore, the first staffing decision that should be made is whether the investigator, having ultimate responsibility for compliance with all regulations, will handle this important area or whether it will be delegated. Major institutions that are heavily involved in research may have a General Clinical Research Center, a clinical trials office, or some type of core research support facility that assumes the burden of appropriate documentation and reporting associated with protection of human subjects. If not, then the investigator must decide if his or her staff will include a person with the requisite training and experience for this responsibility. A task naturally associated with that position would be the construction and administering of consent documents.

In general, research associates involved with gathering data and the subsequent managing of those data should be compulsive and well-organized. Data collected in a research project are only as good as the people who collect and manage those data. An institution's human resources department will have generic job descriptions for data collection personnel, to which project-specific language usually may be added. Researchers hiring new staff should pay particular attention to applicants' resumes to determine if they have the proper qualifications. In some instances, specific licensed health professionals will be needed. Researchers should not ignore training or experience from outside the medical field.

An investigator should give careful thought to the amount of time he or she will be able to devote to the actual research itself, and, in particular, to managing a staff. For many investigators their primary effort is devoted to grant writing and manuscript preparation. The actual research work is done by a team of qualified personnel. For other types of clinical research, the investigator is present and is "hands-on" a good deal of time, with heavy interaction among research participants and staff. Other kinds of research exist on the fringes of a busy clinician's schedule and benefit from an alternative team leader. In that situation, and depending on the size of the proposed project, consideration should be given to an experienced project director.

The inclusion of a project director, with well-defined responsibilities including supervision of staff, can easily be made during budget preparation. Unfortunately, this is much more difficult to address half way into a project. Regularly scheduled meetings between the investigator and staff, especially in the early months of the project, are critical. Each staff member should be asked to prepare a succinct report of his or her area of involvement for informal presentation.

Researchers should encourage questions; research personnel learn from each other. Be alert to any staff member who does not ask questions. No matter how knowledgeable or experienced the individual, it is unlikely that he or she would be familiar with all facets of a new project.

A good research staff is a team, all working together toward a common goal—that of successful completion of the research project. One team member is as important as the next in reaching the goal. It is important that each team member know his or her worth to the project.

5. RESOURCES

For the investigator preparing a first grant application, some of the most useful information can be obtained from an institution's grants office and from the human resources department. Not only are there institutional guidelines in many areas, but there are differing institutional requirements. Departmental business offices may also be helpful. The human resources department can provide generic job descriptions, salary and wage classifications, as well as job posting and subsequent hiring information.

Finally, the set of instructions from the particular agency or funding mechanism to which the grant or contract will be submitted is a source of critical information for any investigator.

6. SUMMARY

Development of a comprehensive, well-justified budget takes time. It is a critical part of the overall preparation of a grant application and requires careful thought be given to all areas of the research plan. The process itself can identify missing steps or omissions in the body of the proposal. Before embarking on a budget preparation, an investigator should visit his or her institution's grants office and human resources department to collect the specific types of information referenced in this chapter, and to get an estimate of the timetable for institutional sign off. If the application is funded, the time and effort spent on budget preparation will pay off by removing one obstacle to the successful completion of a research project: running out of money!

Appendix 1: First-Year Budget Page of NIH Grant Applications (PHS Form 398)

Principal Investigator/Program Director (Last, First, Middle):

DETAILED BUDGET FOR INITIAL BUDGET PERIOD DIRECT COSTS ONLY						FROM		THROUGH
PERSONNEL *(Applicant organization only)*		Months Devoted to Project				DOLLAR AMOUNT REQUESTED *(omit cents)*		
NAME	ROLE ON PROJECT	Cal. Mnths	Acad. Mnths	Sum. Mnths	INST.BASE SALARY	SALARY REQUESTED	FRINGE BENEFITS	TOTAL
	Principal Investigator							
	SUBTOTALS →							
CONSULTANT COSTS								
EQUIPMENT *(Itemize)*								
SUPPLIES *(Itemize by category)*								
TRAVEL								
PATIENT CARE COSTS	INPATIENT							
	OUTPATIENT							
ALTERATIONS AND RENOVATIONS *(Itemize by category)*								
OTHER EXPENSES *(Itemize by category)*								
CONSORTIUM/CONTRACTUAL COSTS							DIRECT COSTS	
SUBTOTAL DIRECT COSTS FOR INITIAL BUDGET PERIOD *(Item 7a, Face Page)*								**$**
CONSORTIUM/CONTRACTUAL COSTS							FACILITIES AND ADMINISTRATIVE COSTS	
TOTAL DIRECT COSTS FOR INITIAL BUDGET PERIOD								**$**

PHS 398 (Rev. 04/06) Page ___ **Form Page 4**

Appendix 2: Entire Budget Page of NIH Grant Applications (PHS Form 398)

Principal Investigator/Program Director (Last, First, Middle):

BUDGET FOR ENTIRE PROPOSED PROJECT PERIOD
DIRECT COSTS ONLY

BUDGET CATEGORY TOTALS		INITIAL BUDGET PERIOD *(from Form Page 4)*	ADDITIONAL YEARS OF SUPPORT REQUESTED			
			2nd	3rd	4th	5th
PERSONNEL: *Salary and fringe benefits. Applicant organization only.*						
CONSULTANT COSTS						
EQUIPMENT						
SUPPLIES						
TRAVEL						
PATIENT CARE COSTS	INPATIENT					
	OUTPATIENT					
ALTERATIONS AND RENOVATIONS						
OTHER EXPENSES						
CONSORTIUM/ CONTRACTUAL COSTS	DIRECT					
SUBTOTAL DIRECT COSTS *(Sum = Item 8a, Face Page)*						
CONSORTIUM/ CONTRACTUAL COSTS	F&A					
TOTAL DIRECT COSTS						

TOTAL DIRECT COSTS FOR ENTIRE PROPOSED PROJECT PERIOD $

JUSTIFICATION. Follow the budget justification instructions exactly. Use continuation pages as needed.

PHS 398 (Rev. 04/06) Page ___ **Form Page 5**

II

Clinical Research Design and Statistical Techniques

4 Nonrandomized Interventional Study Designs (Quasi-Experimental Designs)

David A. Axelrod, MD, MBA
and Rodney Hayward, MD

Contents

In contrast to observational study designs, interventional studies manipulate clinical care to evaluate treatment effects on outcomes. Although surgeons have often relied on observational studies to establish the efficacy and effectiveness of operative and perioperative interventions, observational studies (also referred to as case series) are limited to demonstrating the correlation between the outcome of interest and the procedure. Prospective controlled interventional trials will provide a higher level of evidence for a true cause-and-effect relationship.

Interventional studies may be categorized into two large classifications: true experimental designs and quasi-experimental designs. The randomized, blinded clinical trial (RCT) is the prototypical example of a true experimental design. In an RCT, patients are allocated to treatment arms in a prospective, random fashion in an attempt designed to ensure comparability between groups. The intervention and outcome are then administered and recorded, often with blinding of the interventionalist, the evaluator and the subject to reduce bias. This study design is discussed further in Chapter 5.

Unfortunately, surgical interventions often do not readily lend themselves to randomized blinded trials *(1)*. Consent for randomization is often difficult to obtain for surgical interventions because patients may have a preconceived notion of what treatment they wish to receive, blinding is often impossible (e.g., the patient and surgeon both know whether a cholecystectomy was performed laparoscopically or through open surgery),

From: *Clinical Research for Surgeons*
Edited by: D. F. Penson and J. T. Wei © Humana Press Inc., Totowa, NJ

and ethical concerns usually render sham surgery controls unacceptable *(2)*. Furthermore, the technical nature of surgery can make randomization difficult. Surgeons are more likely to be skilled in certain operations, results are likely to improve over time as a result of learning curve effects, and surgery often involves small incremental improvement rather than dramatic changes. As a result, quasi-experimental techniques, which do not require random assignment, are more often used in the assessment of surgical interventions than in the assessment of other medical treatments.

The most basic experimental research design is a comparison of outcome before and after a planned intervention without the use of a control group (also known as the pre/post design). Essentially, this is a systematic case series in which a new intervention or treatment is introduced during the period of study *(3)*. Unfortunately, interpretation of simple pre/post intervention studies is difficult. Changes in the outcome of interest may be due to the intervention; however, it may also reflect disease natural history (as the condition improves over time or clinical therapy improves with experience), patient selection (patients before and after the intervention may have differed in clinically important attributes), or placebo effects (because neither patient nor provider is blinded). In addition, there is a natural tendency for processes to regress to the mean, which may occur without intervention. Therefore, in this chapter we will examine alternative study designs, which, although not randomized, can often provide more rigorous evidence of a treatment effect than a simple pre/post design.

In this chapter, three principal interventional study designs will be considered:

- nonrandomly assigned control (or comparison) group
- time-series design with pre- and posttest comparisons
- preference allocation (patient, physician)

For each research design, we will consider appropriate research questions, basic design elements, allocation of subjects, outcome measurement, analytic techniques, and overall assessment of the strength of the design.

1. NONRANDOMLY ASSIGNED CONTROL (OR COMPARISON) GROUP STUDIES

In nonrandomly assigned control group studies, at least two separate groups are evaluated—one of which receives the intervention of interest and another that serves as a control or comparison group (Figure 1). Thus the nonrandom control group is similar in design to a RCT, except that patients are assigned to treatment groups in a nonrandom fashion. Quasi-experimental designs differ from that of an observational trial in that the patients are allocated to treatment groups by research protocol, whereas in an observational study the natural history of treatments is studied (i.e., there is no allocation to any intervention).

1.1. Identify Appropriate Questions

There are two main instances in which a nonrandomized control group trial is a good choice. The first is when an RCT would be ideal but practical considerations (e.g., costs, unacceptability to patients or providers) make a high-quality RCT infeasible. The second is when you are trying to establish the effectiveness of large-scale dissemination and implementation. Still, just as in observational studies, predictive variables need to be identified and measured to ensure comparability between the study groups. As is also true of observa-

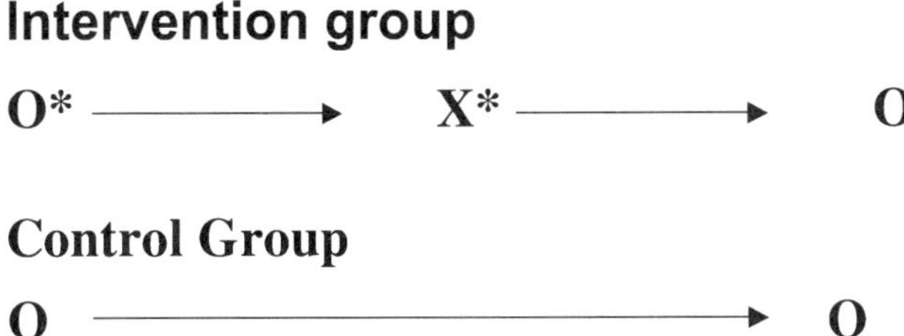

Figure 1: Nonrandomized control groups. *O represents observation; X* represents a study intervention.

tional studies, quasi-experimental studies are less desirable when studying outcomes that are multifactorial or are less well understood. The strength of these studies is in part determined by the investigator's ability to control for potential confounders using multivariable analysis, therefore, being able to identify and accurately measure these potential cofounders (such as patients' illness severity and comorbidities) is essential to minimizing the risk of bias.

1.2. Define Inclusion/Exclusion Criteria

Next, specific inclusion and exclusion criteria need to be established for the study population. Inclusion criteria must be identical for both the intervention and comparison groups. There is an inherent tension between using criteria that are broad enough to ensure that recruitment of an adequate sample and generalizability, but not so broad that meaningful comparison is not possible. For example, including all first-time uncomplicated hernias in men may be reasonable, whereas limiting the trial to asymptomatic hernias less than 2 cm may be unnecessarily restrictive. Exclusion criteria have two main purposes: (1) to exclude study subjects who present substantial risk to the scientific quality of the study (e.g., inability to follow up, plan to move out of state) and (2) to assure safe and ethical conduct of the study (e.g., inability to tolerate general anesthesia, contraindications to the one or both of the treatment arms, unable to give informed consent). Both clinical characteristics of potential study subjects and social/cognitive criteria should be considered when identifying exclusion criteria.

1.3. Estimate Sample Size

Finally, an estimate of the appropriate sample size needs to be determined. The size of the patient cohort is likely to be larger in a nonrandomized study to permit the application of multivariate regression techniques to adjust for differences in baseline characteristics. Although the mechanics of regression analysis are considered elsewhere in this text, it is important to remember that the number of independent variables that can be used in a regression model for a categorical outcome can be quite limited. About 10–20 outcomes (e.g., hernia recurrence) are required for each variable included in the regression analysis. Even if your study includes 1000 surgeries, if there are only 50 adverse outcomes you should include no more than 5 independent variables in the regression model *(4)*. For continuous outcomes variable (e.g., health-related quality of life or exercise tolerance) this restriction is considerably less (10–20 study subjects per independent variable). Clearly, the investigators need to ensure that there is adequate research support to recruit and follow a sufficient number of study subjects.

1.4. Allocate Subjects Between Groups

The selection of study sites and the allocation of subjects to treatment groups are among the most challenging issues in nonrandomly assigned control group studies. Subjects should be allocated to treatment groups in a manner that allows the groups to be generally comparable and to minimize the introduction of bias. For example, the investigator may choose to randomly select one of two comparable private hospitals as the intervention site and the other as the comparison site. Whether to randomize by hospital, ward or clinic, or physician will depend on feasibility, risk of contamination, and the nature of the intervention. The use of pseudo-randomization (e.g., every other patient) should be discouraged because it does not offer any great advantage over true randomization and is subject to manipulation by clinicians. The use of patient or physician preference to allocate patients to treatments can be used, but is less desirable. A discussion of preference trials is included later in this chapter.

1.5. Collect Baseline Data

In the nonrandomized controlled trial, it is crucial to collect a comprehensive dataset including all variables that can reasonably be expected to influence the outcome of the procedure. Baseline and follow-up (i.e., pre/post) measures should be collected at both the intervention site and the control sites. This data will allow the investigator to judge the comparability of the two groups and can also be used to statistically adjust for measured differences between the groups. In addition, pre/post data from the control sites can provide an external temporally synchronous control. Unfortunately, just as in observational studies, even with detailed data collection, it is still possible that unmeasured confounders will influence the study's results. This should always be reported as a limitation of this study design.

1.6. Measure the Outcome

The outcome of interest should be established before initiating the study and measured as accurately and reliably as possible. Because neither the patient nor the investigator is blinded to the nature of the procedure, the use of physician- and patient-reported outcomes can be quite problematic. The use of independent, blinded evaluators, imaging studies, or physiologic measurements (e.g., blood flow rate, degree, residual stenosis) may be less prone to bias than patient-reported outcomes. However, it is critically important that the outcome measure is clinically meaningful. For example, although measuring range of motion may be more objective and less prone to bias, it is also less clinically relevant than measures such as pain or ability to return to work. There may be some instances in which sham procedures or other placebo treatments are ethically acceptable and in such instance subjective outcome measured (such as pain, health-related quality of life) are likely to be much less subject to bias *(5)*.

1.7. Analysis of the Data

Although randomized controlled trials can be analyzed using relatively straight forward bivariate statistical analysis (e.g., *t*-tests, chi-squared statistics) when successful randomization is demonstrable, analysis of nonequivalent comparison groups generally requires multivariable modeling. First, the groups are analyzed to determine the degree of comparability using simple descriptive statistics. Bivariate statistics can be used, but it is critical to realize that both clinical and statistical significance of differences between intervention and control subjects should be considered. For example, if there are clinically substantive differences in an important preintervention patient attribute(s), then those variables should be adjusted for in the analyses even if the bivariate difference was not statistically significant.

Next, multivariate regression techniques are used to "control" or "adjust" for any observed differences in baseline characteristics. Treatment assignment is entered as an independent variable controlling for these potential confounders and the effect of treatment is determined from the regression coefficient. Just as in a true experiment, this variables signifies intention to treat (i.e., was the subject in the assigned intervention or the control group), not whether the subject received the treatment. Standard multivariate analysis assumes measurement of all potential confounding variables (although if you know the degree of measurement error, adjustments for low or moderate precision can be performed).

There are several threats to the interpretation of data from a nonrandomized clinical trial, of which unmeasured confounders is particularly prominent. For example, a hernia may recur more frequently in one hospital because the patients are more likely to be poor and must return to work earlier. Because patients are usually not "blinded" to the study intervention, there may also be differential degrees of placebo effects that may account for the clinical differences, especially of outcomes based on patient self-report. Finally, the investigator must consider issues that are relevant to any trial, including RCTs, such as the need for complete follow-up, the ascertainment of an unbiased outcome assessment, and concerns regarding the generalizability of the findings into a nonexperimental setting. The inherent uncertainty of achieving complete case-mix adjustment has left some experts to question whether we can rely on these statistical methods to account for differences in the characteristics of the comparison groups *(6)*.

1.8. Advantages of the Nonrandomized Controlled Trial

When a true experiment is not feasible, there are several potential advantages of including a control group (even nonrandomized) instead of relying solely on simple pre- and postintervention comparisons. The control group principally helps to account for threats to internal validity from temporal trends, regression to the mean, and the learning curve. A temporal trend bias is the potential that other advances or changes in clinical care, the nature of the disease, or patient population may account for observed changes. As long as these changes are reflected in both the control and experimental groups, they are likely to be identified using this design. Similarly, the impact of the learning curve has been widely established for new surgical procedures. Thus outcomes may improve over time, which must be accounted for in any analysis. Finally, the outcomes at the extreme are likely to moderate naturally over time, leading to a phenomenon of regression to the mean. Without controlling for this trend, observed effects may reflect chance rather than true clinical changes.

The use of a nonrandomized control group may also reduce the threats to external validity that limit the value of RCTs results. First, RCTs tend to be done at a few, highly selected sites, and are rarely done in community settings. Quasi-experimental designs can often involve more providers and settings, making the results more generalizable. Second, the lack of randomization often facilitates recruitment of a larger proportion of eligible patients, thus further increasing generalizability.

Intervention Group:

O*→ O→ O → X* → O→ O

Figure 2: Time-series analysis. *O represents observation; X* represents a study intervention.

1.9. Disadvantages of the Nonrandomized Clinical Trial

The principal disadvantage of this design is the potential for bias from confounding. The direction of this bias is unpredictable from study to study. For example, clinicians may differentially include the sicker patients in the intervention trial to provide the "best chance" for the patient, thus biasing the trial against the intervention. Alternatively, the healthiest patients may be included to ensure that the intervention has the optimal opportunity to work. Therefore, the investigator should try to preempt "hand-picking" study subjects who receive the intervention. Even when optimally conducted, this design can never ensure that unmeasured or imprecisely measured social, economic, cultural, or clinical variables do not account for the apparent treatment effect. Thus the results of these trials must be evaluated in a larger context, and internal and external validity may be best assessed through the replication of results in a variety of clinical settings.

2. TIME SERIES ANALYSIS

Time series analysis can provide a more robust method for addressing the problem of secular trends in clinical care. Essentially, the investigator measures the outcome of interest several times before initiating the experiment to establish a baseline value and trend in the data (Figure 2). After the intervention, the investigator will again measure the outcome several times to establish the impact of the intervention. This design differs from a standard cohort design because the investigator manipulates patient care to estimate the effect of the intervention and from a pre/post design because it can identify trends in the outcome rate that existed before the intervention.

2.1. Identify Appropriate Questions

Time-series experiments are useful in two clinical situations: first, when the intervention produces a rapid and sustained impact on the outcome of interest. For example, a time-series analysis has been used to determine the impact of laparoscopic techniques on the rate of bile duct injuries after cholecystectomy *(7)*. Interventions that produce a delayed or gradual change may be much more difficult to capture. Using a single pre/post comparison, but including sufficiently long follow-up, a multiple time-series design can improve the robustness of the statistical comparison. The validity of a time-series analysis can be further improved by conducting a similar analysis on a comparison (control) cohort (thereby combining a time series design with a nonrandom control group design).

2.2. Define Appropriate Inclusion/Exclusion Criteria

As in all studies, inclusion and exclusion criteria must be balanced to ensure adequate comparability of the study group and generalizability of the study results. In a time-series analysis, broad inclusion criteria should generally be used to ensure that there is little room for the investigators to differentially enroll patients into a study (e.g., all consecutive patients undergoing laparoscopic cholecystectomy). Thereby, as long as the under-

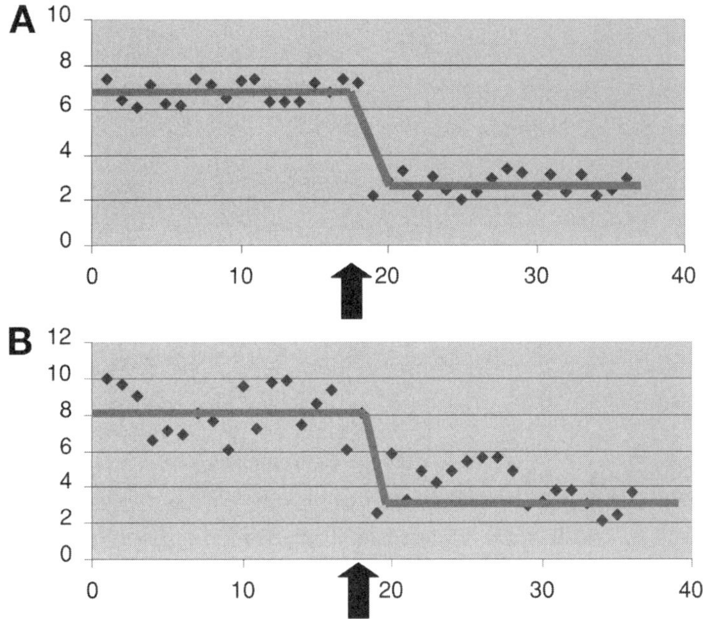

Figure 3: Impact of variation on stability in time-series trial. **(A)**. Time series with excellent precision in outcome assessment. **(B)** Time series with moderate precision in outcome assessment.

lying population does not change, patients included before and after the intervention should be similar. The exclusion criteria must include the patients' ability and willingness to remain in the study and have outcomes data collected for the length of the expected follow-up, because significant loss to follow-up is likely to bias the results of any longitudinal study. Clearly, it is also important to exclude patients from the study who would not have been candidates for procedures, including the cohort of potential study subjects preceding the introduction of the procedure. For example, there is a clear bias if outcomes from patients who are candidates for laparoscopic cholecystectomy are compared with outcomes for patients are candidates for open cholecystectomy, which may have broader eligibility criteria.

2.3. Estimate Sample Size

A vital feature of time series investigations is that there must be sufficient sample size to provide a stable estimate of the outcome incidence throughout the time series. In essence, the noise of random variation cannot be so great that it obscures the signal that you are trying to detect (the "true" outcome incidence rate). For example, in a study of risk adjusted mortality rates after cardiac surgery in the VA hospital system, investigators determined that 185 cases per 6-mo period were needed to produce a "statistical" stable (precise) estimate of surgical mortality *(8)*. Unfortunately, only one hospital during a single period achieved this case volume. Factors that influence the precision of statistical estimation are described elsewhere. To a degree, increasing the number of measurements (usually by increasing the follow-up time) and modern statistical methods for accounting for measurement error can help overcome moderate imprecision of the individual outcome rate estimates (Figure 3).

2.4. Collect Baseline Data

The number of baseline data collection points that need to be collected is in large part determined by the degree of temporal stability in the outcome rates before the intervention. As is often true in research, the exact number of data points is a balance between always wanting more data, but needing to consider incremental benefits and incremental costs of data collection.

2.5. Measure the Outcome

As in the case of the baseline data, data must be collected for a sufficient period to establish both a reliable postintervention baseline and to assess the durability of the response. It is particularly important to evaluate the potential that any observed effect merely regression to the mean.

2.6. Analysis of the Data

In a simple pre/post design, the outcome rate before and after the intervention is compared. This is also true for a true-series study, but a time-series analyses also compares the temporal changes within the pre- and postintervention periods. This is accomplished by fitting two multivariate regression models to the temporal trend in outcomes rates, one for the preintervention period and one for the postintervention period. If there is no effect of the intervention, the slopes of regression lines and their intercepts will be the same. A one-time effect will be reflected as an increase in the intercept of the regression line. Ongoing, longer term impacts will result in an acute change in the slope in the postintervention phase compared to the preintervention period (Figure 4).

2.7. Advantages of Time-Series Trial

Time-series experiments can allow the investigator to identify preexisting temporal trends in the outcome of interest and more effectively test causal influence. As shown in Figure 4, if the outcome of interest has been stable over several observation periods and then changes and persists at a new level at the time of the intervention, this provides strong inferential evidence of a treatment effect.

Time-series experiments have often been used to track quality of care and health care costs over time and assess the impact of practice changes (e.g., the impact of laparoscopic cholecystectomy on bile duct injury rates). They have also been applied to assess the impact of systemic changes on operative time, resource utilization, and throughput. They may also be particularly useful for single institution studies in which historical data are available to define precise base line values and temporal trends, thus improving on a simple pre/post intervention design without requiring a randomized control group. However, just like the pre/post design, a time-series analyses can usually benefit from adding comparison sites that are similar to the intervention sites except for the absence of the intervention. Also, pre/post and time-series studies should always include an evaluation of other changes that may have occurred at the study site at the same time as the intervention. Qualitative methods are often the preferred approach for collection this information on changes at the institution.

2.8. Disadvantages of the Time-Series Analysis

Time-series analyses are limited by the investigator's ability to completely control for potential confounders. Specifically, the population under study may change because of

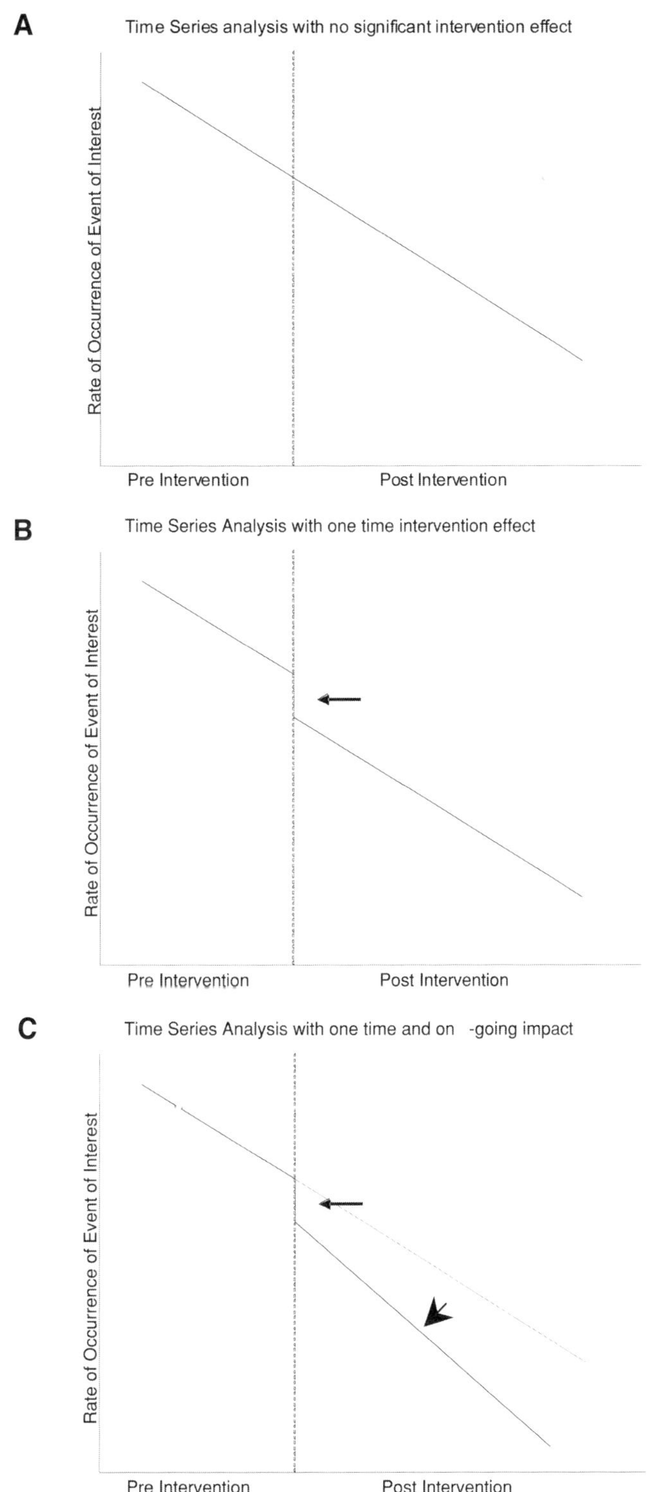

Figure 4: Time series analysis with no effect (**A**), one time effect (**B**; small arrow, reduction in intercept of plot line), and ongoing impact (**C;** small arrow, reduction in intercept of plot line; large arrow, reduction in slope of graph).

a contemporaneous phenomenon (e.g., change in neighborhood demographics or socioeconomic status), which may bias the results of the study. This threat can be minimized through prospective data collection with established entry criteria, the use of multiple pre- and postintervention measurements, and serial qualitative evaluations of the study site.

Although increasing the number of measurements is likely to reduce the threat of an unrecognized confounder, it will usually substantially increase the cost and complexity of a time series experiment. Investigators may seek to identify outcomes that can be tracked though administrative data or other existing systems to reduce the cost of the intervention. Thus health care environments that care for a more stable group of patients over time and have comprehensive medical information system data (e.g., staff model health maintenance organizations, VA hospitals) may be excellent venues for this type of analysis.

3. COMPREHENSIVE COHORT TRIALS/PATIENT PREFERENCE TRIALS

As a result of the invasive nature of surgical interventions, patients may be reluctant to agree to random assignment. Consequently, the representativeness of surgical RCTs may be substantially compromised, thus making extrapolation of results to the general population concerning. In the comprehensive cohort study (CCS) or patient preference trial (PPT) designs, patients who decline to participate in the randomized portion of a trial continue to be followed in their chosen therapeutic arm (Figure 5) *(9,10)*. At the conclusion of the trial, comparisons are made for four groups of patients: patients randomized into intervention A, patients who selected intervention A, patients randomized into intervention B, and patients who selected intervention B. The comprehensive cohort trial differs from a traditional cohort study because only patients who are considered appropriate for either treatment arm are enrolled and all patients undergo a uniform treatment as would occur in an RCT. In addition, for the nonrandomized subjects, you are specifically examining the impact of patient preferences under circumstances that minimize the impact of physician recommendations, access and economic barriers, and other nonclinical confounders that may influence treatment decisions in usual clinical practice.

For example, the Coronary Artery Surgery Study comparing coronary artery bypass surgery with medical therapy included patients who accepted randomization (780 of 2099 patients approached) and patients who refused randomization (1315 patients) *(11)*. Patients who underwent surgery within 90 d of their evaluation were considered surgical patients, and the remaining patients were assigned to the medical management arm. These patients were followed over time and analyzed according to an intention to treat methodology as described in the following section.

3.1. Identify Appropriate Questions

Comprehensive cohort trial designs have been used to augment a randomized controlled trial when clinicians and patients are likely to have strong preexisting treatment preferences and the outcome does not rely on patient reported outcomes. As originally described, the comprehensive cohort study should usually be used to evaluate techniques that are also available outside of the study, although it has been applied to settings in which treatment is limited to within the study environment *(12)*. If the intervention is limited only to randomized patients, no meaningful comparison can be made to the group followed in the CCS.

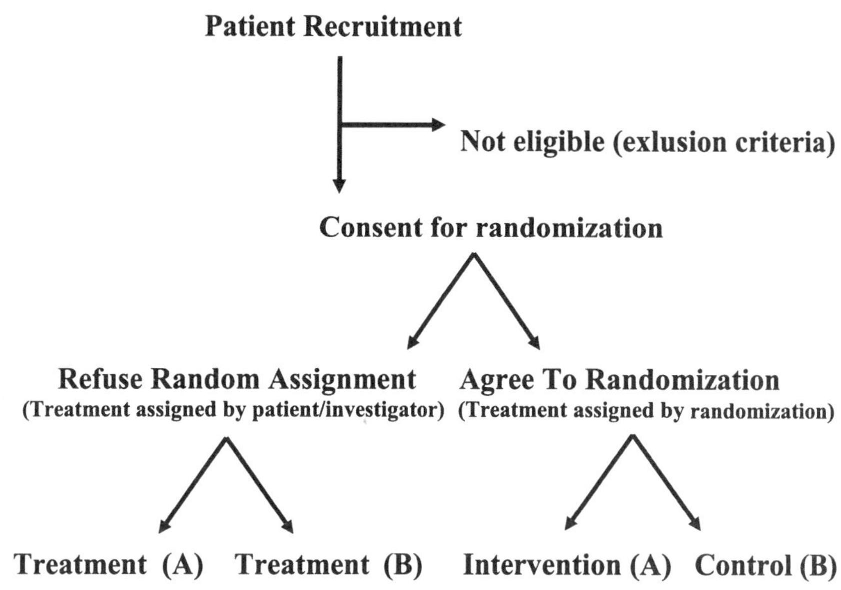

Analysis
1. **Intervention vs. Control**
2. **Treatment A vs. Treatment B**
3. **Treatment A+ Intervention vs. Treatment B+ Control**

Figure 5: Analysis scheme for comprehensive cohort trial.

3.2. Define Appropriate Inclusion/Exclusion Criteria

The inclusion and exclusion criteria for a CCS or PPT should be determined by the RCT component of the study. Patients who chose interventions that are not included in either study arm should be excluded from the analysis.

3.3. Estimate Sample Size

Sample size should be determined primarily by the underlying RCT and estimates of the number of patients who will agree to randomization. As described below (Section 3.7.), the evaluation of a CCS proceeds in stages and adequate recruitment into the RCT arm is necessary to determine the principal treatment effect.

3.4. Allocate Subjects Between Groups

In a CCS, all eligible subjects are initially approached and consent for randomization is requested. Patients who refuse randomization are then asked to consent to be included in the CCS follow-up study and are allocated to treatment groups based on their preferences. In a PPT, patients are initially asked whether they have a strong preference for a specific treatment. Those patients without a strong preference are then asked to consent to randomization. Patients who refuse randomization or have a preexisting strong preference are then assigned to their preferred treatment arm. Unfortunately, treatment assignment may be difficult for patients who seek care elsewhere or who delay initiating therapy. CCS trials should be analyzed using an intention to treat methodology, and criteria for treatment assignment (e.g., surgery within 90 d of evaluation) should be specified.

3.5. Collect Baseline Data

For the most part, considerations for baseline data collection are similar to those for other quasi-experimental designs (comprehensive collection of factors that may influence risk of the outcome). However, it may be particularly relevant to collect baseline information on the patients' perspectives on the treatment options, such as the strength of their preference, their expectations or optimism regarding outcomes, and the reasons for their treatment selection.

3.6. Measure the Outcome

A predefined, objective outcome measure should be used when possible and appropriate. The confounding between patient preferences and patient reported outcomes can make analysis of subjective outcomes (e.g., pain or health-related quality of life) problematic. Therefore, physiologic or clinical outcomes are preferred (e.g., death, stroke, strength testing). Nonetheless, a "subjective" measure may still be preferable to an "objective" measure that is not very clinically or socially compelling (e.g., range of motion) especially when the nature of the comparison interventions are similar (such as two different major surgical procedures).

3.7. Analysis of the Data

It is recommended that CCS and PPT trials be analyzed sequentially. The first analysis should compare outcomes for the patients in the randomized portion of the trial. Next, patients in the nonrandomized arm should be examined to determine if the treatment effect is consistent or inconsistent in the preference allocation cohort. Finally, all patients may be considered in a single multivariate regression analysis, including an indicator variable for randomization status as a covariate. Using this technique, the independent effects of treatment and patient preference and the interaction between patient preference and treatment choice can be determined *(13)*.

3.8. Advantages of CCS/PPT Design

CCS/PPT designs offer two principal advantages. First, the ability to choose treatment assignment may improve recruitment into a clinical trial and thus increase the sample size. Unfortunately, the availability of a CCS/PPT may limit patients' desire to enter the RCT portion of the trial and will, in the end, increase the duration and costs of the enrollment process necessary to obtain a sufficient number of randomized patients. Second, CCS/PPT trials may enhance the external validity of the study's main findings. Because patients who decline randomization may represent the majority of patients seen in clinical practice, a consistent finding in both the randomized and nonrandomized cohorts can provide some reassurance regarding generalizability. Furthermore, if the results between the randomized and nonrandomized cohorts are inconsistent, then one can describe the direction and magnitude of the bias introduced through the self-determination of treatment (or physician selection of treatment).

3.9. Disadvantages of the CCS/PPT Analysis

The addition of a CCS/PPT study to an RCT is likely to increase the cost and complexity of the trial. A CCS/PPT study will be larger and the follow-up may be more difficult if patients seek treatment outside of the study centers, but are still included in the cohort. Furthermore, the threat of residual unmeasured confounding is an inherent threat to the validity of the CCS/PPT even with state-of-the-art measures of baseline risk factors. Thus, although the CCS/PPT patients may be more representative of the general population, careful attention must be paid when adjusting for differences in baseline characteristics.

4. EMPIRICAL EVIDENCE COMPARING RANDOMIZED AND NONRANDOMIZED TRIALS

Although RCTs continue to be viewed as the gold standard for clinical research, a series of comprehensive evaluations has failed to demonstrate consistent differences in treatment effects found in high-quality randomized and nonrandomized investigations *(14–16)*. MacLehose and colleagues reviewed 14 articles involving 38 interventions in which the results from quasi-experimental studies were compared with those derived from RCTs. The authors concluded that there were no significant differences in the effect size or direction between high-quality quasi-experimental studies and RCT. However, in low-quality studies, the effect size appeared to be more extreme, but the direction varied in only one comparison. The low-quality studies were principally review articles that did not use appropriate meta-analytic techniques.

The difference in outcomes reported between randomized and nonrandomized trials may, sometimes, be just a reflection of the underlying patient characteristics. In a comprehensive review of RCT and non-randomized studies comparing coronary artery bypass grafting with angioplasty, Britton and colleagues determined that coronary artery bypass grafting was favored in the RCT and angioplasty in the nonrandomized cohort studies. However, after adjustment for patient characteristics, the differences were no longer statistically different *(15)*. Benson and Hartz reached a similar conclusion after examining 136 articles in 19 treatment areas. The effect estimates derived from RCT and nonrandomized trials were similar for 17 of the 19 conditions studied and there did not appear to be systematic bias in observational investigations *(15)*.

The impact of publication bias may differentially impact randomized and nonrandomized trials. Although adequately powered randomized trials that fail to demonstrate a significant difference between treatment arms are routinely published, it appears that nonrandomized studies demonstrating a similar conclusion are more often rejected and probably less often submitted for consideration by the authors as well. Thus, when RCT and nonrandomized trials comparing a single treatment are examined, the nonrandomized trials are more likely to demonstrate positive results. Perhaps, this publication inequality accounts for part of the perception that quasi-experimental and observational studies are intrinsically biased.

5. CONCLUSIONS

Quasi-experimental study designs offer surgical investigators a valuable tool to overcome many of the impediments to conducting a randomized clinical trial. The use of properly selected nonrandomized control groups can help to overcome threats to internal validity from temporal trends, surgical learning curve effects, regression to the mean, and the difficulty in obtaining equipoise among surgeons. Likewise, time-series analysis can be well suited for situations in which clinical practice outpaces research evaluation. Often combining these two methods (i.e., using both pre/post and contemporary comparisons) will be the optimal approach, making it possible to examine the impact of rapid clinical change in diverse patient populations and clinical settings.

The trade-off between RCT and a quasi-experimental study design is largely pragmatic. When feasible, an RCT is almost always preferred because it minimizes the risk that unmeasured confounding is biasing the studies conclusions. However, often a quasi-experimental study design can offer important insights into the care of surgical patients and can lead to more generalizable study results based on more representative patient populations. Surgical case-series reports continue to be very common; however, many of these case series could be readily redesigned to create rigorous and more scientifically-sound quasi-experiments. Quasi-experimental designs warrant careful consideration by surgical researchers and should be more widely used.

REFERENCES

1. McCulloch P, Taylor I, Sasako M, Lovett B, Griffin D. Randomized trials in surgery: problems and possible solutions. BMJ 2002;324:1448–1451.
2. Macklin R. The ethical problems with sham surgery in clinical research. New Engl J Med 1999;341:992–996.
3. Curry JI, Reeves B, Stringer MD. Randomized controlled trials in pediatric surgery: could we do better? J Pediatric Surg 2003;38(4):556–559.
4. Peduzzi P, Concato J, Feinstein AR, Holford TR. Importance of events per independent variable in proportional hazards regression analysis. II. Accuracy and precision of regression estimates. J Clin Epidemiol 1995:48:1503–1510.
5. Moseley JB, O'Malley K, Petersen NJ, et al. A controlled trial of arthroscopic surgery for osteoarthritis of the knee. N Engl J Med 2002;347:81–88.
6. Kunz R, Oxman AD. The unpredictability paradox: review of empirical comparisons of randomized and non-randomized clinical trials. BMJ 1998;317:1185–1190.
7. Flum DR, Dellinger EP, Cheadle A, Chan L, Koepsell T. Intraoperative cholangiography and risk of common bile duct injury during cholecystectomy. JAMA 2003;289:1639–1644.
8. Marschall G, Shroyer AL, Grover FL, Hammermeister KE. Time series monitors of outcomes: a new dimension for measuring quality of care. Med Care 1998;36:348–356.
9. Olschewski M, Scheurlen H. Comprehensive cohort study: an alternative to randomized consent design in a breast preservation trial. Methods Inf Med 1985;24:131–134.
10. Torgerson DJ, Sibbald B. Understanding controlled trials: what is a patient preference trial? BMJ 1998;316:360–361.
11. CASS Principal Investigators and their Associates. Coronary artery surgery study (CASS): a randomized trial of coronary artery bypass surgery. Comparability of entry characteristics and survival in randomized and non-randomized patients meeting randomization criteria. J Am Coll Cardiol 1984;3:114–128.
12. Olschewski M, Schumacher M, Davis KB. Analysis of randomized and non-randomized patients in clinical trials using the comprehensive cohort follow-up study design. Controlled Clin Trials 1992;13: 226–239.
13. Olschewski M, Scheurlen H. Comprehensive cohort study: an alternative to randomized consent design in a breast preservation trial. Methods Inf Med 1985;24:131–134.
14. MacLehose RR, Reeves BC, Harvey IM, et al. A systematic review of comparisons of effect sizes derived from randomized and non-randomized studies. Health Technol Assess 2000;4:34.
15. Britton A, McKee M, Black N, et al. Choosing between randomized and non-randomized studies: a systematic review. Health Technol Assessment 1998;2:13.
16. Benson K, Hartz AJ. A Comparison of observational studies and randomized controlled trials. N Engl J Med 2000;342:1878–1886.

5 Randomized Clinical Trials of Surgical Procedures

Michael P. Porter, MD, MS

Contents

One day when I was a junior medical student, a very important Boston surgeon visited the school and delivered a great treatise on a large number of patients who had undergone successful operations for vascular reconstruction. At the end of the lecture, a young student at the back of the room timidly asked, "Do you have any controls?" Well, the great surgeon drew himself up to his full height, hit the desk, and said, "Do you mean did I not operate on half of the patients?" The voice at the back of the room very hesitantly replied, "Yes, that's what I had in mind." Then the visitor's fist really came down as he thundered, "Of course not. That would have doomed half of them to their death." God, it was quiet then, and one could scarcely hear the small voice ask, "Which half?" —*Dr. Earl Peacock (1)*

The example above highlights some of the problems with performing randomized trials in the surgical disciplines. Surgeon bias for or against specific procedures, morbidity associated with surgery, and the acceptance of lesser forms of clinical evidence by the surgical community are all barriers to performing randomized trials. However, in fields such as oncology and cardiology, great strides have been made in patient care by using evidence from well-designed and well-executed randomized trials. In many cases, these trials have successfully randomized patients to potentially morbid and invasive therapies. Moreover, the trend toward evidence-based medicine is being embraced by not only the

From: *Clinical Research for Surgeons*
Edited by: D. F. Penson and J. T. Wei © Humana Press Inc., Totowa, NJ

medical community, but by patients and third-party payers. If the surgical disciplines are going to keep pace with the rest of medicine, a "cultural" shift away from the empiric and anecdotal evidence supplied by case series and case reports toward the more rigorous methodology of randomized interventions needs to occur.

The purpose of this chapter is to outline the randomized clinical trial, including its methodology, as it pertains to the surgical disciplines. Hypothetical examples and examples from published literature will be used to highlight the unique challenge randomized research poses for surgeons. This chapter will not cover advanced statistical topics or advanced trial designs. Entire books are published on randomized trials and their methodology; my goal is to cover the basic concepts. After reading this chapter, the reader should be able to plan a basic randomized trial and understand how to avoid the major potential pitfalls when designing and executing the study.

1. THE RANDOMIZED TRIAL: THE BASIC MODEL

A randomized trial is a comparative study between two or more interventions, where exposure to the intervention is determined by random allocation. The basic design of the randomized trial is illustrated in Figure 1. In general, Treatment A represents a new therapy (or surgical technique) and Treatment B represents the control group that may be the current standard or perhaps no therapy at all. There are many variations on the organization of a randomized trial, including more than two treatment groups, complex randomization strategies, and intentional crossover of patients between treatment groups. No matter what the variation, however, all contain the key step of random allocation.

Random allocation of subjects is the most important feature of the randomized trial. Randomization breaks the link between any unmeasured confounding variables and treatment status. This unique feature of the randomized trial is its biggest strength—by breaking this link, all differences in effect between treatment groups can be assumed to be a result of the differences in treatment. In other words, confounding should be absent. From a surgical standpoint, randomization can also be the biggest weakness of the randomized trial. Convincing patients to relinquish control of their care to a random process that determines whether or not they receive surgery, or which surgery they receive, is often the biggest challenge to completing randomized trials in surgery.

Randomized trials can usually be placed into one of two broad categories: *pragmatic trials* and *explanatory trials*. Pragmatic trials attempt to simulate clinical realities more accurately during patient recruitment, during formulation of the randomly allocated treatment groups, and during measurement of outcomes. By designing trials that more accurately parallel real-life clinical situations, practical information is gained that may be more generalizable and more easily accepted into clinical practice. Explanatory trials attempt to answer a more specific and narrow question. To maximize their ability to do this, eligibility criteria may seek a more homogeneous set of patients. Follow-up of patients and measurement of outcomes may be more intensive than in normal clinical practice. The information gained from such trials often effectively answers the narrow question of interest, but is less often immediately relevant to clinical practice. Another closely related concept that is often important in drug trials is effectiveness vs efficacy. *Effectiveness* is the ability of an intervention to accomplish its intended outcome in a population under real-life circumstances, whereas *efficacy* is the ability of an intervention to obtain its intended outcome under ideal situations *(2)*. Generally speaking, most randomized trials in surgery tend to be categorized as pragmatic, comparing surgical techniques and outcomes under usual clinical conditions.

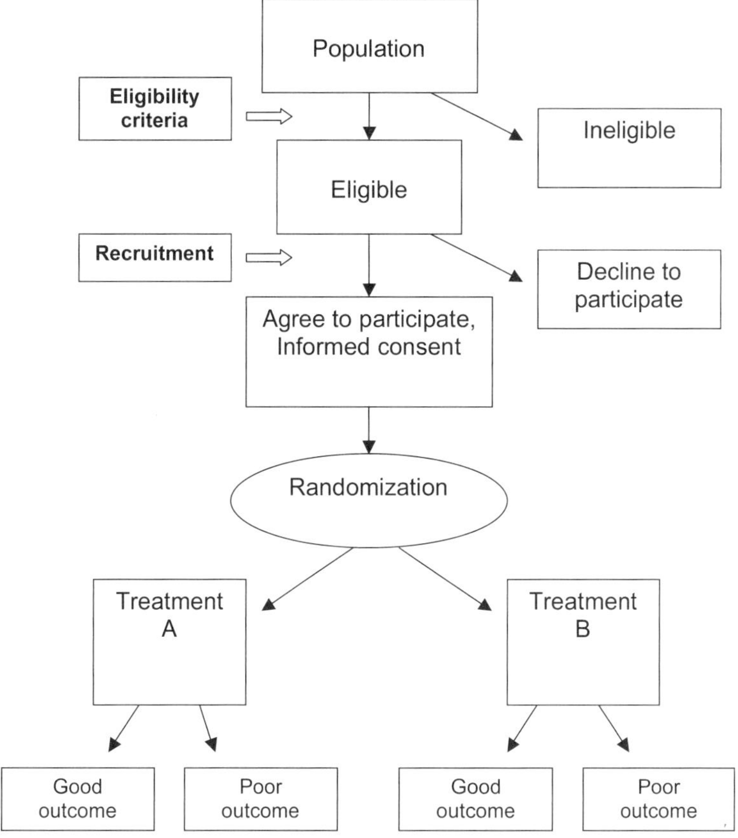

Figure 1: The basic design of the randomized trial.

2. BIAS AND VALIDITY

Random allocation offers great protection against confounding, but this does not mean that bias can not exist in a randomized trial. *Bias* in randomized trials is defined as systematic error within the study that results in a mistaken estimate of the effect of therapy on disease *(2)*. Bias can be introduced into any step of the process, including enrollment, randomization, and assessment of outcomes. The best protection against bias is the proper planning and execution of the trial, which ultimately results in a high degree of internal validity. In some instances, bias cannot be avoided; however, in such instances, attempts to measure the direction and magnitude of the bias should be undertaken. *Internal validity* is the ability of a trial to come to the correct conclusion regarding the question being studied. In other words, a study with high internal validity has been designed and executed in such a way that the observed differences between treatment groups, apart from sampling error, can only be due to differences in assigned treatments. The internal validity of a study can be compromised at any point in the design and execution of the trial, and low internal validity increases the opportunity for introduction of bias. Related to this concept is the external validity of a trial. *External validity* is the ability of a trial to produce results that are generalizable to the larger population of patients with the disease. External validity is insured mostly by proper subject selection, eligibility criteria, and outcomes selection. During the remainder of this chapter, the concepts of bias and validity will recur as different aspects of the clinical trial are discussed in greater detail.

3. FORMULATING THE RESEARCH QUESTION

For a randomized trial to be successful, careful thought must go into formulating a research question. Randomized trials should be designed to answer as narrow of a clinical question as possible, while still maintaining clinical relevance. With this in mind, there are certain research situations that tend to favor the randomized trial. Ideally, the disease process being studied is common enough to recruit the number of patients required to adequately power the study. The clinical question also needs to be important enough to justify the expense of the study, and legitimate uncertainty on the part of the medical community between the effectiveness of at least two therapies for the same disease process must exist. This latter concept is referred to as *clinical equipoise* and represents the ethical foundation on which another important concept is built—namely that patients should not be disadvantaged by their participation in a clinical trial. Finally, physicians and patients participating in the study need to be willing to relinquish control of choice of treatment, sometimes setting aside deeply held beliefs regarding the most appropriate form of treatment. As previously stated, this last hurdle is often the most difficult to overcome in surgical trials and in many ways highlights some of the major differences between randomized trials in the surgical specialties and the medical specialties. As we have all experienced, many patients find their way to a surgical clinic due to preexisting beliefs about the benefits of surgery, and many patients avoid surgical consultation from equally strong beliefs about the dangers of surgery. When competing medical and surgical therapies exist, it is the rare patient that presents with no preconceived beliefs about the best treatment. Anecdotally, these preconceived beliefs are most likely to be highlighted when two forms of alternative therapy are the most different (e.g., surgery vs medical treatment or surgery vs observation). However, if legitimate uncertainty exists regarding the best treatment, this challenge should not be allowed to stand in the way of a well-designed trial. With this in mind, the ideal problems to study with randomized trials are diseases where (1) the morbidities of the procedures under investigation are very similar (e.g., two surgical procedures that use similar incisions); (2) the disease has high morbidity, lacks effective therapy, and a new surgical technique has legitimate promise of offering improved outcomes; and (3) diseases with currently effective therapies that result in potential long-term morbidity in which patients may be willing to expose themselves to the short-term morbidity of a more invasive procedure for potential improvement in long-term side effects (e.g., coronary artery stents vs long-term medical management of coronary artery disease).

From a methodologic standpoint, there are several important issues that need to be decided during the formulation of the research question. First, which end points are the most clinically relevant? Second, can these end points be accurately and reliably measured? And if not, what potential surrogate end points are available, and at what expense to the internal validity and external validity of the study? The answers to these questions in large part will determine the practicality, expense, and feasibility of a randomized trial *(3)*. The answers to these questions will also lead the investigator to choose the most appropriate end-point for the trial, which in turn will determine the remainder of the trial design. Many types of clinical, economic, and patient-oriented end points can be used as end points for a clinical trial *(4)*. The selection of end points will in part be based on whether the trial is pragmatic or explanatory. Pragmatic trials will generally have patient relevant end points such as morbidity, mortality, and functional status. A single end point may not be adequate for many pragmatic trials, and often multiple end points are measured. Explanatory trials tend to have end points that are less directly clinically relevant, such as radiographic evidence, physiologic parameters, and serum markers. These rules for end points however, are not universal, and there is no clear-cut line that determines whether a trial is pragmatic or explanatory and which end points are required. The only iron-clad rule for end points is that they need to represent a measure of disease status that directly relates to the research question.

Example 1: *In 2003, Thompson et al* (5) *reported on the results of a randomized trial designed to determine if a drug called finasteride, when taken daily, reduced the risk of prostate cancer. They enrolled more than 18,000 men 55 yr of age and older with no evidence of prostate cancer, and then randomly allocated them to receive either finasteride or placebo. The men were monitored closely for 7 yr, and participants who developed clinical evidence of prostate cancer underwent prostate biopsy. At the end of the study period, all participants in the study underwent prostate biopsy (whether or not there was evidence of cancer). The authors reported that that daily finasteride reduced the risk of prostate cancer by 24%, but high-grade cancer was more prevalent in the men taking finasteride.*

This large study represents an explanatory trial relevant to the field of urology and illustrates many of the points described previously. Though the trial did not randomize patients to an invasive therapy, it was successful in recruiting patients into a trial in which an invasive procedure (prostate biopsy) was a study requirement. The question of whether or not a daily dose of finasteride reduces the risk of prostate cancer is also clinically relevant enough to justify the cost of such a large study. The main study question was "Does daily finasteride reduce the risk of prostate cancer compared with placebo?" and main end point was histologic evidence of prostate cancer. If it had been designed as a pragmatic trial, the authors would have chosen a different end point, such as morbidity or mortality from prostate cancer. However, it is the immediate clinical relevance of the study question and end points to the disease process that makes this an explanatory trial, not the end point per se. Prostate cancer grows slowly, and a large subset of men who are diagnosed with the disease die of other causes before symptoms develop. Also, the men in this study were followed much more closely than is practical in real life, including an end of study biopsy. In contrast, if a similar study had been performed to evaluate the effect of a drug on the development of a rapidly fatal malignancy with a high case mortality rate, histology may have been a reasonable pragmatic end point.

4. ELIGIBILITY CRITERIA

Establishing precise eligibility criteria is an important step in the process of designing a randomized trial. Defining the appropriate subset of the population to study in large part determines the external validity of a trial. Careful thought at this stage of planning will avoid the potentially painful later realization that the study population is too narrow or too broad to yield the desired results. In general, pragmatic trials tend to have broader inclusion criteria, with an attempt made at replicating the clinical patient population of interest. Explanatory trials are narrower in focus, and by choosing a highly selected subset of patients, statistical advantage can be gained by reducing variability of the outcomes within the randomly allocated groups. However, this is done at the expense of generalizability. A good rule of thumb for all types of trials is that the eligibility criteria should be sufficiently specific and succinct to allow independent observers to come to the same conclusion regarding a patient's eligibility—little room should be left for subjective interpretation.

Example 2: *A vascular surgeon wishes to study a new minimally invasive technique for treating occlusive peripheral vascular disease of the lower extremities. The surgeon plans to enroll patients younger than 65 yr of age with clinical and radiographic evidence of occlusive disease requiring intervention. The patients are randomly allocated to receive the new minimally invasive therapy or the standard open bypass procedure. The planned end point is angiographic resolution of the occlusion.*

This hypothetical trial poses a couple of potential problems in its current form. First, the inclusion criteria may not necessarily reflect the typical patient with this disease by excluding patients older than 65. If positive results were found, then applying these results to older patients with peripheral vascular disease would require extrapolation, or a "leap of faith." This potential for lack of generalizability lowers the external validity of the study. Second, the eligibility criteria of the patients should be more objectively defined. Specific measure of angiographic severity of the lesion, objective measure of patient symptoms measure by validated methods, and specific exclusion criteria should all be determined in detail in advance. Finally, the planned end point (radiographic resolution of the lesion) may not be the best determinant of therapeutic success. Additional end points such as patient symptoms, activity level, and wound healing may be more appropriate and more clinically relevant.

5. SAMPLE SIZE

After a study question, end points, and eligibility criteria have been determined, the process of patient recruitment begins. But how many patients need to be recruited for any given study? The answer to this question involves several factors. To understand the factors that influence sample size, it is necessary to be familiar with the table in Figure 2. This 2 × 2 table represents all of the possible outcomes of a randomized trial. The columns represent "the truth," while the rows represent the conclusions of the study. Ideally, the conclusion of the study is concordant with the truth. That is, if no difference exists, the study concludes that no difference exists. Likewise, if a difference does exist, the study would ideally conclude that a difference exists. However, two other possibilities exist. The truth may be that the treatments do not differ, but the study may conclude that they do (in other words, the null hypothesis is incorrectly rejected). This is referred to as a *Type I error*. The probability of a Type I error occurring is designated as α. This value is usually set arbitrarily at 0.05, and the sample size does not have a direct affect on the probability of making a Type I error because the significance level is set by the investigator at the beginning of the study. However, the significance level does affect the sample size needed, with smaller significance levels requiring larger sample sizes. The final possible outcome occurs when the study concludes that the treatments do not differ when in fact they do differ (in other words, failure to reject the null hypothesis when it should have been rejected). This is referred to as a *Type II error*, and the probability of it occurring is designated as β. Unlike Type I errors, Type II errors are directly related to sample size, and the sample size is directly related to the power of the study. The *power* of a study is the probability that a study will detect a difference between two (or more) treatments when in "truth" a difference does exist. This probability, or power, is defined

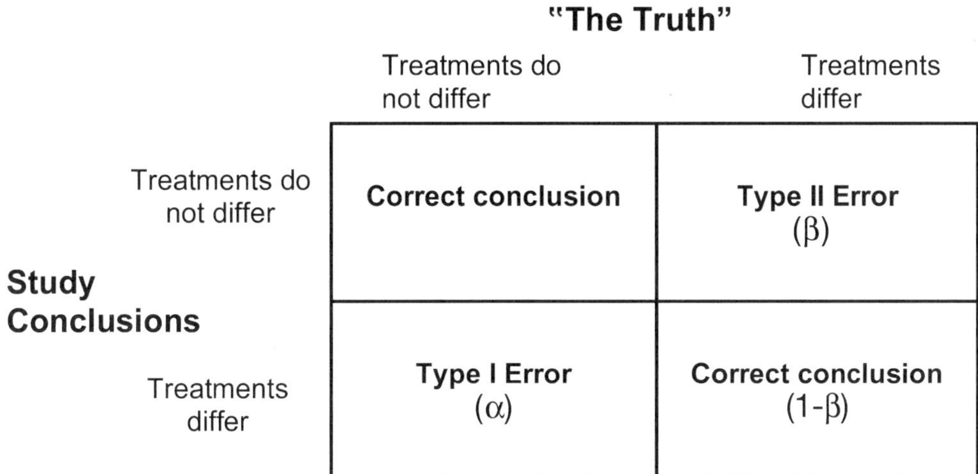

Figure 2: Possible outcomes of a randomized trial.

as 1-β. Notice that 1-β occurs in the table in Figure 2 in the cell that represents the study correctly determining that the treatments differ. In general, most well-designed studies strive to achieve a statistical power between 0.80 and 0.90, which in turn reduces the Type II error rate to 0.20 and 0.10, respectively.

The *sample size* of a study is determined by four factors. The first is the desired power. The more subjects are enrolled into a study, the higher the power of the study (i.e., the smaller the probability of a Type II error). However, the relationship is not linear. For example, the additional number of patients that need to be recruited to increase the power from 0.70 to 0.80 is far less than the additional number required to increase the power from 0.80 to 0.90. At some point, the incremental cost of recruiting more patients does not justify the small increase in power that is gained.

The second major determinant of sample size is the anticipated difference in outcomes between the randomized study groups. The magnitude of this difference is inversely related to the statistical power such that very small differences in outcomes between treatments are associated with very low statistical power, which often can only be overcome by increasing the number of subjects. It is therefore important to go to some lengths to accurately estimate the probable treatment effect when designing the trial, usually by conducting pilot nonrandomized observational studies. At the very least the estimated difference should not be less than what is considered to be "clinically relevant."

The third major determinant of sample size is the desired α, or significance level. As previously mentioned, this is usually set at 0.05, but there are occasions in which more stringent significance levels are set. The smaller the α, the larger the sample size needs to be for any given power.

The final determinant of sample size is variability in the outcomes data. This is often measured as the variance, standard deviation, or standard error from a pilot study. Data that tend to be more variable (e.g., a tumor marker with a large range of normal values) requires larger numbers to obtain significance compared with less variable data. It is for this reason that explanatory trials often attempt to recruit as homogenous of a patient population as possible, because clinically similar patients tend to have less variable outcomes.

In the planning stages of a randomized trial, these factors all need to be taken into account. A basic understanding of significance, power, and sample size will allow the surgeon-researcher to formulate more feasible study questions and end points. To determine the appropriate sample size required to complete the study, statistical consultation is mandatory. This is usually best accomplished early in the course of the study design.

6. CONTROLS AND LEVELS OF EVIDENCE

The topic of selecting appropriate controls is generally a more relevant issue in observational studies, but there are some concepts surrounding control selection that are important to understand. Generally, studies with the weakest level of evidence are studies without controls, often referred to as *case series*. Case series tend to be retrospective and they are a frequent finding in the surgical literature. Because there is no control group, it is difficult or impossible to conclusively infer that the outcome of interest was a direct result of the intervention being presented. This does not mean that case series are without merit. Case series are adequate to provide descriptive analysis of outcomes after a procedure; for example, quantifying morbidity and mortality after an extirpative cancer operation. A large series of patients undergoing the procedure may be the best method of defining these types of outcomes. The problem arises when such studies attempt to use these types of data as evidence of superiority to another competing approach to the same disease. Without a control group, this level of evidence is weak, because the potential for bias is almost overwhelming. However, when a new intervention has drastically different outcomes than established approaches, the case series may be the first level of evidence needed to plan more appropriate controlled studies.

Some case series attempt to circumvent the problem of lack of controls by using *historical controls*. For example, if a laparoscopic procedure has gained popularity, a comparison may be made between 20 recent consecutive laparoscopic procedures to 20 consecutive open procedures that were performed several years prior at the same institution. This method of control selection may eliminate some potential for bias; for example, the procedures may have been performed by the same surgeon on what are assumed to be relatively similar patient populations. However, the temporal separation between these two "case series" makes the validity of comparison suspect. Changes in hospital care and technology are impossible to account for, and the assumption that the patients in each series are similar may be false. In short, historical controls are only a short step above case series in level of evidence.

Another concept that is important to consider in studies without controls is the placebo effect. The *placebo effect* is the effect (usually beneficial) produced by a medical intervention that is not from the intervention itself, but from the expectations of the patient. The placebo effect may be substantial, and it is usually most apparent when disease outcomes are subjective. For obvious reasons, studies without controls are most vulnerable to the placebo effect, but studies in where treatments vary markedly or the subjects are privy to their treatment assignment (that is, the subjects are not "blinded" to their treatment), the placebo effect can also be problematic.

There are many valid and effective nonrandomized ways to choose controls and perform good research without randomization. Controls can be established prospectively, retrospectively, or recruited at the same time as case subjects. This multitude of *observational study* designs and the strengths and weaknesses of each make up a large part of the field of epidemiology and are beyond the scope of this chapter. However, there is a very important concept that needs to be considered whenever a study uses nonrandomized controls: *confounding*. Confounding occurs in observational studies when a third variable (confounding variable) is related to both the exposure of interest (treatment group) and the outcome of interest. Confounding introduces error and can cause the results of a study to be inaccurate. Confounding has the potential to occur whenever a third variable is related to both the exposure variable (treatment vs control) and the outcome of interest. Confounding was discussed in more detail in Chapter 1, but it is important to note that special efforts need to occur to control for the effects of confounding in all observational studies. Differences in the characteristics between the treatment and control group are accounted for by statistical adjustment during analysis of the results. Characteristics that are commonly adjusted for are age, comorbidities, disease stage, and sex, but any characteristic that is different between the groups being compared has the potential to introduce error through confounding. Perhaps most importantly, unmeasured differences between the comparison groups cannot be controlled during analysis. And because it is impossible to measure every characteristic between comparison groups, the potential for *residual confounding* exists in all observational studies. This is one reason why randomized trials are considered to be above observational studies on the hierarchy of evidence.

The randomized clinical trial selects its control group from the same pool of patients as the treatment group. This pool is established by the eligibility criteria and recruitment process. The patients are then randomly allocated to one of the possible treatment/control groups. As mentioned at the beginning of the chapter, this random allocation severs the link between potentially confounding characteristics and the treatment the patients receive. In other words, all characteristics of the patients, measured and unmeasured, should be equally distributed between the different treatment groups if the randomization process was successful. From a statistical standpoint, this prevents confounding and allows for analysis without adjustment. However, the superior level of evidence ascribed to a randomized clinical trial is contingent on a successful randomization.

7. RANDOMIZATION

On the surface, randomization seems like a straightforward and easy process to accomplish. However, this stage of the randomized trial represents a major potential threat to the internal validity of the study. If the randomization process can be predicted, the possibility of selection bias is introduced into the trial. *Selection bias* is the systematic difference in characteristics between patients chosen for a trial and those who are not (other than the characteristics defined in the eligibility criteria) or systematic differences between patient characteristics in different treatment groups because of investigator interference. One recent study showed that when researchers failed to adequately conceal randomization from the investigators, an average 41% percent increase in treatment effect occurred compared with trials where the randomization process was concealed appropriately *(6)*. Introduction of selection bias and the subsequent increase in treatment effect is a more dangerous form of bias than that which occurs in observational studies because of confounding. In observational studies, statistical correction of confounding is attempted, and results are interpreted in the context of nonrandomization. In a randomized trial that has been distorted by selection bias, statistical correction is often not possible. For these reasons, the randomization process must be truly unpredictable and concealed from the investigator or person who is enrolling the study subject. The exact mechanism of the randomization process is then determined by the specific logistics and practical constraints of the study.

Example 3: *In a study to determine which of two competing treatments for a specific type of extremity fracture was more effective, patients were allocated to receive either surgical intervention or immobilization. The treatment assignment was determined by the hospital number assigned to the patient on registration in the emergency department hospital numbers ending in an odd number received surgery and those ending in an even number received immobilization with a cast. A preliminary review of the data revealed that twice as many patients were being assigned to the surgical intervention. On review of the process, it was determined that a resident with a particularly strong desire to operate was convincing a friend at the front desk to change patient numbers whenever a particularly well-suited surgical candidate was assigned to immobilization.*

This example illustrates a randomization process with two major flaws. The process is predictable as well as unconcealed from the investigators. Though this example may seem striking and obvious, any time a randomization strategy can be deciphered, the door is open to selection bias. A better randomization system would allocate the patient to a treatment after the patient consented to participate, and the investigator would not be able to predict the assignment before randomization. A commonly employed strategy is to use a computer or random number table to generate a sequence for allocation. Individual treatment assignments are then enclosed in opaque, serially numbered envelopes. After a patient is enrolled, the next envelope in the sequence is opened, revealing the treatment. This method meets the criteria for unpredictability and concealment.

Example 4: *In a busy trauma center, a randomized trial was being performed to assess whether abdominal ultrasound or diagnostic peritoneal lavage was more accurate in diagnosing intra-abdominal injury after blunt trauma. After being admitted to the emergency department, clinical indicators were used to determine the need for rapid abdominal assessment. If clinical criteria were met, the surgical resident would open the next opaque, serially numbered envelope stored in a designated folder in the emergency department. The contents of the envelope indicated which assessment would be used. As is the case in busy trauma centers, two or more patients often presented simultaneously. This afforded the busy resident an opportunity to open more than one envelope at once and subsequently decide which patient received which assessment. At the end of the trial, it was noted that, more often than not, diagnostic peritoneal lavage was performed on thinner patients.*

In this example, the concealment of the randomization mechanism could be breached whenever more than one patient presented at the same time. As with the previous example, the process was corrupted by a third party motivated by self interest, likely not understanding the potential effects on the outcome of the trial. However, an investigator could just as easily open envelopes before clinic and then assign treatments based on personal bias. Incredulous as this seems, it should be remembered that randomized trials often address questions to which there are strongly held and competing points of view. These strongly held opinions can often influence behavior, whether it is flagrant subversion of the process or subtle, subconscious differences in recruitment. In the words of one investigator, "Randomized trials appear to annoy human nature—if properly conducted, indeed they should" *(7)*.

What, then, is the ideal method of randomization? Probably the most difficult to compromise system is distance randomization *(8)*. This approach is similar to the opaque envelope, except that the order of randomized allocation is stored remote from the enrollment site. The investigator (or recruiter) then calls the randomization site after informed consent has been obtained from a willing patient. After basic demographic data about the patient are obtained, the treatment allocation is disclosed to the investigator. There are many potential subtle variations, including use of the Internet, but the key component is lack of investigator access to the allocation sequence.

For randomization to result in an equal distribution of patient characteristics between treatment groups, sufficient numbers of patients need to be recruited to assure that chance variations in characteristics do not result in a biased analysis. The more subjects that are recruited, the less likely that major discrepancies will exist between the two treatment groups by chance alone. But even with sufficient numbers, there is no guarantee that discrepancies will not exist between the treatment groups. The following are modifications of the randomization process that can help to decrease the potential for imbalance between treatment groups, particularly for studies with smaller sample sizes.

Block randomization is a process that can be used to assure equal sized groups or equal distribution of a specific important trait between groups. Block randomization is a process that groups subjects into "blocks" or clusters of a fixed size, and then randomizes individuals within the block equally to one of the treatment groups. For example, if the block size was four and there were two treatment groups, two patients would be randomized to each group within every block of four subjects. The same process would occur in the next four consecutively recruited patients (the next block). By alternating treatment assignments within blocks, equal numbers in treatment groups are assured, which is particularly useful for smaller studies. A major potential drawback of block randomization is the potential ability to predict the next treatment assignment. If an investigator knows that the block size is four patients, it becomes possible to predict the next assignment by keeping track of previous assignments. This can be prevented by randomly altering the block size and using larger block sizes, typically four to eight per block (Figure 3). Practically speaking, only the study staff designing the randomization schema (typically the statistician) needs to know the block sizes. Revealing that information to other study staff and investigators only increases the potential for the allocation sequence to be revealed.

Stratified randomization is a process that is similar to block randomization, but is used only to assure that certain traits are equal among the treatment groups. Stratified randomization is a process where patients are first grouped by a specific trait (stratified), and the patients within each group (or stratum) are then randomized to the different treatment groups. This may be important if a particular trait portends a better or worse prognosis, such as morbid obesity in a surgical trial. Such a trait needs to be distributed equally among treatment groups in order to conduct an unbiased analysis. By stratifying first, the opportunity for a "bad" or "unequal" randomization is minimized. Probably the most common use of this is in multicenter trials (Figure 4). If a particular surgical procedure is performed with a better outcome at one center than another, unequal distribution of treatment groups between the centers could bias the results for or against the treatment. By stratifying by treatment center, the probability that treatment groups are equal within each center increases. If block randomization is also used, equal numbers in the treatment groups within each center can be guaranteed. This reduces or eliminates bias that may occur from outcomes that vary by center, not necessarily by treatment per se. More than one variable can be used for stratification, and in general all key variables that have are known to be strongly associated with the study outcome should be stratified.

Block 1= E, E, C, C, E, C

Block 2= E, C, C, E

Block 3= C, E, C, E, E, C, C, E

Block 4= E, E, C, E, C, C

Block 5= C, C, E, C, E, C, E, E

.

.

.

Figure 3: Block randomization. Each block consists of a random assignment sequence (E, experimental group and C, control group). Note that if the process were stopped after 5 blocks, there would be 16 patients in each group. Also note that the sequence is different in each block and the blocks are of varying length.

How do you stratify or guarantee equal distribution for traits that you cannot measure? The answer is that you cannot. The effectiveness of the randomization process can be assessed by comparing important variables that have been shown to be associated with outcome in the different treatment groups, such as age, sex, and major comorbidities. From a statistical standpoint, these variables should be selected at the outset of the study *(9)*. However, even if all these traits appear equal among the randomized groups, there is no guarantee that some unmeasured trait that may influence response to therapy is equally distributed between the treatment groups. This is simply a property of random chance, and it is one reason that even apparently well-executed randomized trials should be viewed with a degree of skepticism if the results are surprising or counterintuitive, realizing that sometimes medical science sometimes cannot advance until results are duplicated by independent studies.

8. BLINDING

Blinding refers to the process of concealing the results of the random allocation from each subject, or each investigator, after the randomization process has occurred. Traditionally, a *single-blinded* study refers to a study where the treatment assignment is concealed from the subject only. In a *double-blinded* study, the treatment assignment is concealed from both the subject and the investigator who ascertains the study end point. Blinding is one of the major hurdles that need to be addressed when planning a randomized clinical trial of a surgical intervention, and it represents the key difference between planning a trial of competing medical interventions and a trial of competing medical–surgical interventions. In medical trials comparing different pharmacologic therapies for

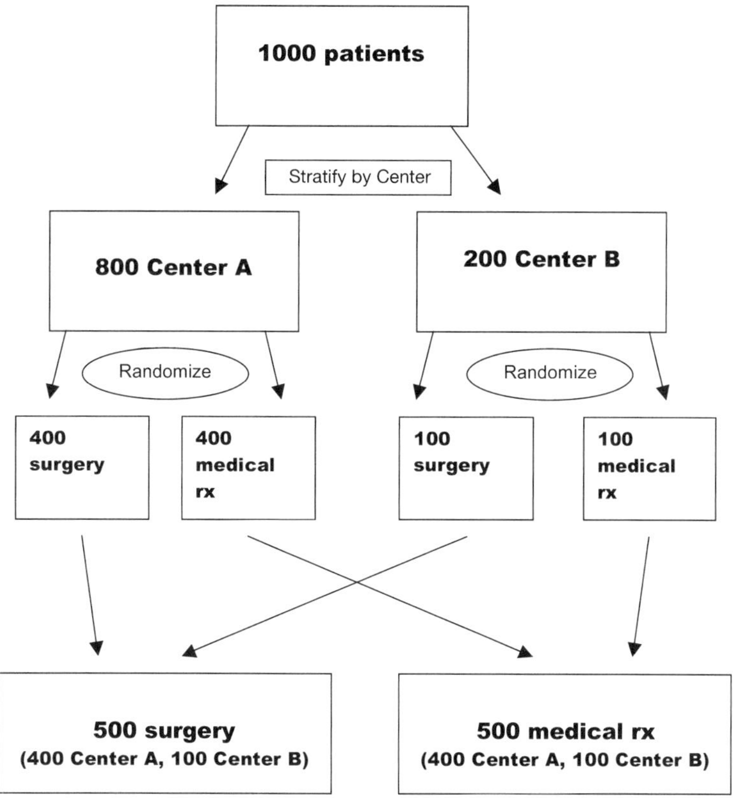

Figure 4: Stratified randomization in a hypothetical multicenter trial.

a disease, blinding can often be easily accomplished by dispensing similar appearing drugs (placebo tablets) with only a code on the label. Though sometimes the blinding process in medical trials can become quite intricate, it rarely poses the ethical and logistic challenges of designing a trial of surgical therapy that is properly blinded. Blinding is important in randomized trials as a means of reducing or preventing observation bias and the placebo effect. *Observation bias* is systematic variation between the true outcome and the outcome observed. It is more likely to occur when the investigator is unblinded to the treatment allocation. A researcher who knows the treatment the patient is receiving may consciously or unconsciously interpret results of treatment to favor the experimental treatment. As mentioned previously, the *placebo effect* can be problematic in controlled trails in which patients are not blinded. A patient who is assigned to a new form of experimental therapy may be more likely to exaggerate the effects of the treatment based on their enthusiasm and expectations, rather than on the actual effect of the intervention. It is for these reasons that double-blinded trials are preferred, assuming that they are logistically and ethically possible to complete.

Example 5: *In 1959, Cobb et al.* (10) *reported a randomized trial designed to determine if internal mammary artery ligation was effective in the treatment of angina. The researchers enrolled 17 patients with classic angina symptoms and recorded preoperative severity of symptoms as well as stress test electrocardiogram data. The patients then underwent surgery. After the internal mammary arteries were isolated, an envelope was* *opened that randomly directed the surgeon to either ligate the vessels or simply end the operation with the vessels intact. The patients were informed preoperatively that the ligation procedure was experimental, but they were unaware of the randomized nature of the study (i.e., the possibility of sham surgery). Postoperatively, data were collected, and it was found that the patients undergoing the sham procedure had greater symptomatic relief than those undergoing ligation, and one patient in the sham surgery group had reversal of stress test electrocardiogram abnormalities. Overall, patients in both groups showed only modest improvement.*

This example illustrates several important points. The first is the concept of sham surgery. Though the methods used may be considered unethical by today's standards, the use of sham procedures is an important tool in surgical randomized trials. This example also illustrates the importance of controls. If this study had been performed as a case series, with all 17 patients undergoing ligation of their internal mammary arteries, the authors may have incorrectly concluded that the procedure had merit in the treatment of angina. But because they randomized the patients, blinded the patients, and performed sham surgery on half, only three conclusions are possible: (1) either the placebo effect caused the improvement in both groups, (2) something other than ligation of the internal mammary artery occurred during the procedure that improved the angina, or (3) the observed benefit was part of the natural history of the disease.

Example 6: *In 2002, Moseley et al.* (11) *performed a randomized trial designed to evaluate the efficacy of arthroscopy in treating osteoarthritis of the knee. They recruited 180 patients with moderate to severe osteoarthritis of the knee and randomly allocated them to receive arthroscopic lavage, arthroscopic debridement, or placebo (sham) surgery. The placebo treatment consisted of arthroscopic port skin incisions, but the instruments were not inserted into the joint. Instead, the surgeon simulated arthroscopic lavage while the patient was sedated with short-acting intravenous anesthetic agents. The treatment allocation was assigned randomly at the start of the operation. All patients spent the night in the hospital and had the same postoperative management. All patients were aware of the randomized nature of the study and the possibility of receiving sham surgery, and all patients remained blinded. Outcomes assessment was performed by a blinded third party. Patients in all three groups had some modest improvement in symptoms, but there was no difference in symptom improvement between groups.*

This contemporary study is an excellent example of a randomized, double-blinded surgical trial. Until this trial, only case series had existed, and most of the case series showed that arthroscopy improved the symptoms of osteoarthritis. The investigators were able to maintain the double blind nature of the study by not revealing treatment status to the patients and by using a blinded third party (not the operating surgeon) to perform outcomes assessments on the study patients. Though not mentioned in the example, the investigators also verified the effectiveness of blinding by asking all study participants to guess which treatment they had received. This is generally a good practice, because patients are often able to infer which treatment they received based on medication side effects, pain severity, or other factors that are not necessarily anticipated or preventable. In this study, the patient's guesses were no better than random chance, verifying that blinding remained successful.

This study was able to exploit the low morbidity of the procedure and the similar surgical approaches to effectively blind participants. Oftentimes, it is not possible to perform a randomized trial in a blinded fashion. Such trials may compare radically different surgical techniques or approaches requiring different incisions, or they may compare surgical therapy to observation or medical intervention. This does not necessarily mean that the trial should not be performed. As previously mentioned, the main advantage of blinding is the prevention of observation bias (blinding the investigator who ascertains the outcomes) and the placebo effect (blinding the patient). However, both of these forms of bias can be minimized if objective outcomes are assessed. Unfortunately, purely objective outcomes my not necessarily suit many studies, and they may not be very pragmatic. However, standardizing the ascertainment of outcomes, attempting to quantify outcomes as objectively as possible, and using a neutral third party (as opposed to the investigator) to ascertain the outcomes minimizes the potential for observation bias and placebo effect. In some research situations, this is the best that can be done, and the results need to be interpreted in light of the study design.

9. PATIENT CROSSOVER AND INTENTION-TO-TREAT ANALYSIS

Despite carefully planning and execution, randomized trials are often plagued by methodologic problems such as patient dropout, noncompliance, missing data, and patient desire to receive the therapy that they have not been assigned to. Though it is beyond the scope of this chapter to discuss the statistical ramifications associated with all of these problems, the problem of patient crossover deserves special mention, because it is a common phenomenon in surgical randomized trials. *Patient crossover* occurs whenever a patient receives therapy in an arm of the trial that he or she was not randomized to. This can occur by patient or provider choice. For instance, a patient may change his or her mind about undergoing surgical therapy after randomization and therefore receive medical therapy. Alternatively, the investigator may determine that the patient's condition has changed and that he or she may do better with the therapy he or she was not assigned to. In either case, a dilemma exists when it comes to analyzing the data.

The best way to handle crossovers in the final analysis is often debated, and no definite correct answer exists. Ideally, the number of patients who crossover is small, and no matter how they are analyzed, it does not affect the final conclusion of the study. Sometimes, however, the numbers of patients who cross over is large, and the problem needs to be addressed.

Example 7: *From 1972 through 1974, the Veterans Administration Coronary Artery Bypass Surgery Cooperative Study Group enrolled patients into a randomized study comparing coronary artery bypass surgery with medical therapy in patients with angina and radiographic evidence of coronary artery disease* (12). *Patients were randomized to receive either surgical or medical therapy. After 14 yr of follow-up, 55% of patients assigned to receive medical therapy had crossed over to receive surgical treatment, whereas 6% of the patients assigned to receive surgical treatment decided to not undergo the procedure and were thus managed medically.*

This study illustrates the problem of patient crossover and the subsequent dilemma in the data analysis. In general, there are two major, opposing approaches that could be used to handle this situation: analyze the outcomes based on the original randomized assignments, regardless of treatment actually received, or analyze the data based on the treatment actually received, not by the random allocation. The first method described is known as *intention-to-treat analysis*.

In the intention-to-treat approach, outcomes are analyzed based on the initial treatment that the patient was randomized to, regardless of actual treatment received. This may seem counterintuitive, but one must remember that the strength of the randomized trial rests in the randomization process. When a patient crosses over to another treatment arm, it does not occur by chance, and if he or she was analyzed by the treatment actually received, the randomization is broken, thus allowing the introduction of selection bias. How might this work in the above example? When the data were analyzed using the intention-to-treat approach, there was no difference in survival between the surgical and medical groups. When analyzed by treatment actually received, the surgery arm had a survival advantage *(13)*. Though at the time the proper way to interpret this was hotly debated, a couple of factors other than the treatment itself likely explain at least some of the difference. First, the average patient crossed over after 5 yr of medical management, thus automatically adding an initial period of survival prior to surgery (unlike those randomized to surgery whose clock started at the time of surgery). Second, patients or providers that decided to switch by definition needed to still be surgical candidates at the time of crossover, perhaps a healthier group on the whole than those who remained in the medical management group who were likely to survive longer regardless of which arm they were in (selection bias). Regardless of what nonrandom event caused patients to switch to surgical therapy, there is a possibility that this event somehow was related to survival and confounded the conclusions of any analysis that did not use an intention to treat approach.

Unfortunately, when crossover rates this high occur, the validity of the study comes into question regardless of the final analysis used. Even more unfortunate is the fact that trials that compare very different therapies are more likely to suffer from large numbers of crossovers (e.g., trials comparing surgical with medical intervention). Multiple statistical methods, more complex than the two simple approaches explained here, have been developed to deal with the problem with crossovers. It is generally accepted, however, that the intention to treat analysis is the most valid analysis in randomized trials. This is based on preserving randomization and preserving the validity of the statistical analysis. Whenever a randomized trial is encountered that does not analyze the outcomes on an intention to treat basis, the reader must wonder what nonrandom factors caused patients to crossover. The reader should then wonder how those nonrandom factors may have biased the study results.

10. CONCLUSION

As the health care environment evolves, greater levels of medical evidence are expected from policy makers, caregivers, and patients alike. This has resulted in an increase in the publication of randomized trials. Surgeons have fallen behind the rest of the medical community, partly because of the challenges required to complete a valid randomized trial of surgical therapy, and partly because of our acceptance and reliance on lesser forms of evidence. This can only be changed by a commitment to evidence-based medical practice, and by persistence and application of novel approaches to overcoming difficult methodologic hurdles.

REFERENCES

1. Tufte ER. Data analysis for politics and policy. Englewood Cliffs NJ: Prentice-Hall, 1974.
2. Last JM, Abramson JH, International Epidemiological Association. A dictionary of epidemiology. 3rd ed. New York: Oxford University Press, 1995:xvii.

3. Greenberg RS. Medical epidemiology. 3rd ed. New York: Lange Medical Books/McGraw-Hill: 2001:xiii, 215.
4. Roland M, Torgerson D. Understanding controlled trials: what outcomes should be measured? BMJ 1998:317(7165):1075.
5. Thompson IM, Goodman PJ, Tangem CM, et al. The influence of finasteride on the development of prostate cancer. N Engl J Med 2003:349(3):215–224.
6. Schulz KF, et al. Empirical evidence of bias. Dimensions of methodological quality associated with estimates of treatment effects in controlled trials. JAMA 1995;273(5):408–412.
7. Schulz KF. Subverting randomization in controlled trials. JAMA 1995;274(18):1456–1458.
8. Torgerson DJ, Roberts C. Understanding controlled trials. Randomisation methods: concealment. BMJ 1999;319(7206):375–376.
9. Roberts C, Torgerson DJ. Understanding controlled trials: baseline imbalance in randomised controlled trials. BMJ 1999;319(7203):185.
10. Cobb LA, Dillard DH, Merendino KA, Bruce RA. An evaluation of internal-mammary-artery ligation by a double blind technique. N Engl J Med 1959;260(22):1115–1118.
11. Moseley JB, O'Malley K, Petersen NJ, et al. A controlled trial of arthroscopic surgery for osteoarthritis of the knee. N Engl J Med 2002;347(2):81–88.
12. Eleven-year survival in the Veterans Administration randomized trial of coronary bypass surgery for stable angina. The Veterans Administration Coronary Artery Bypass Surgery Cooperative Study Group. N Engl J Med 1984;311(21):1333–1339.
13. Peduzzi P, Detre K, Wittes J, Holforo T. Intent-to-treat analysis and the problem of crossovers. An example from the Veterans Administration coronary bypass surgery study. J Thorac Cardiovasc Surg 1991;101(3):481–487.

6 Use of Observational Databases (Registries) in Research

Deborah P. Lubeck, PhD

CONTENTS

The outcomes and costs associated with medical care are critical issues for society. Interventions, treatments, and health care providers are required to be both effective and cost-effective. More and more the cumulative effects of disease, treatment, and outcome are becoming the standard for evaluation of effectiveness and cost-effectiveness. Although randomized clinical trials are the "gold standard" for comparing alternative treatments, results may not be generalizable to usual clinical care nor reflect treatment effectiveness in community practice. The discrepancy between clinical trials and studies of actual effectiveness has been pointed out a number of times over more than 30 years *(1–4)*. Key design elements of clinical trials, such as strict selection criteria, double blinding of patients and providers, and treatment protocols, are specified to isolate differences resulting from treatment. As a result, other sources of variability, including practice heterogeneity, patient heterogeneity, comorbid illness, and imperfect adherence to treatment regimens, limit the generalizability of results to usual clinical care.

There are several alternatives to the conventional randomized clinical trial that may yield results more generalizable to clinical practice, but that still provide rigorous measurement of outcomes. These include pragmatic clinical studies that randomize patients to usual care, retrospective cohort studies, and prospective multicenter cohort studies *(4–6)*. These studies measure outcome over time and can capture the impact of long-term

From: *Clinical Research for Surgeons*
Edited by: D. F. Penson and J. T. Wei © Humana Press Inc., Totowa, NJ

illness and evaluate the role concomitant disease or treatment play in long-term effectiveness and are a sensitive indicator of treatment effect and have been shown to have similar results to clinical trials *(7–9)*. This chapter will discuss the design of observational studies.

1. WHAT ARE LONGITUDINAL OBSERVATIONAL STUDIES?

Observational, prospective cohort studies, also called registries, evaluate the actual experience of persons after the identification of a specific event, such as a disease diagnosis, clinical milestone, or initiation of medical or surgical treatment. Sequential measurement of clinical and patient-reported outcomes, obtained at regular intervals, is an essential component of these studies. Longitudinal, observational studies are useful in evaluating a breadth of data in a timely fashion; especially patient reported outcomes, resource utilization, costs, and clinical outcomes in community settings because there is no assignment of patients to specific treatment protocols.

Although the term *registry* is used widely to describe longitudinal studies, it is most often used to describe prospective cohort studies and not registries as defined in epidemiologic studies. The true registry, exemplified by the Surveillance Epidemiology and End Results (SEER) registry for cancer, is population based and records incident events. SEER provides basic information on multiple cancers from various geographically diverse areas of the United States *(10)*.

There are several successful observational databases in chronic disease that have yielded significant research findings. ARAMIS (Arthritis, Rheumatism and Aging Medical Information System) is now more than 25 years old and includes patients with rheumatologic conditions and community populations followed through patient self-report. ARAMIS investigators have published hundreds of peer-reviewed articles in the areas of treatment strategies, health status assessment, costs of care, and radiologic outcomes *(11)*.

Observational databases also have been used extensively in clinical research. This chapter will focus on use of observational studies in prostate cancer to evaluate the longitudinal outcomes associated with surgical and radiation therapy. There are three prostate cancer databases, the Prostate Cancer Outcomes Study, the Cancer of the Prostate Strategic Urologic Research Endeavor (CaPSURE) database *(12)*, and the Department of Defense Center for Prostate Disease Research. The CaPSURE database was established in 1995 and includes both clinical variables and patient-reported outcomes. Patients are recruited from community sites and three academic medical centers throughout the United States *(12)*. Evidence from CaPSURE suggests that the results of the diagnostic biopsy contribute significantly to accurate risk assessment among patients with newly diagnosed prostate cancer and that the incidence of positive surgical margins after prostatectomy is associated with adverse outcomes *(13,14)*.

2. WHAT ARE THE OBJECTIVES OF OBSERVATIONAL DATABASES?

There are multiple objectives for observational databases. The first goal is to accumulate and document a large, heterogeneous patient experience over time. These studies allow access to large samples of patients treated by a broad base of community practitioners. Clinical data, outcomes, survival, resource utilization, workforce participation, health-related quality of life, and patient satisfaction with care and treatment may all be collected over time.

Another goal of observational studies is to use this experience to identify and prioritize the key issues for medical effectiveness research, including aiding in development of clinical trials to address gaps in clinical knowledge (e.g., practical clinical trials). A special focus is to improve the quality of clinical research studies, especially for rare and time-delayed treatment or disease-related events.

The collection of information on patient quality of life and resource use is also a common objective of observational studies, often mandated by decision makers, such as for formulary approval *(15)*. The focus of these studies is to identify changes in clinical practice over time and evaluate the impact of these changes on patient outcomes, including quality of life, costs, and survival. This includes treatment comparisons not evaluated in clinical trials, such as over-the-counter alternatives.

Last, observational databases may be designed to obtain information on practice patterns over time by type of provider and geographic variation. Randomized clinical trials are done under well-defined protocols with formal evaluation of treatment compliance. In usual clinical care, new technologies are introduced into practice and treatment compliance may be poor, thus producing different outcomes, or treatments may be combined. The only way to determine how treatments are introduced into clinical care and how they influence practice is to focus specifically on longitudinal changes in practice patterns and possible comparison against recommended treatment guidelines, if they exist.

3. WHEN ARE OBSERVATIONAL STUDIES USEFUL?

Registries provide a parallel body of evidence to randomized clinical trials. There are conditions and environments in which disease registries are especially useful. The first is for disease management. Evaluation of a specific treatment may be a focus of a disease registry, but complete evaluation of multiple health interventions and their outcomes, such as in the case of treatment for localized prostate cancer, is an ideal setting for a registry. For example, over the past decade, diagnosis and treatment of localized prostate cancer has changed significantly. Earlier diagnosis, made possible by the advent of prostate-specific antigen screening, has facilitated the increased use of aggressive local treatment, including radical prostatectomy, external-beam radiation, and interstitial radiotherapy (brachytherapy) *(16 18)*. At the same time, all these available treatments can negatively affect patient quality of life, including impotence and incontinence. In prostate cancer, the long-term survival after diagnosis has increased, whereas the potential for reduced quality of life associated with treatment is a significant factor in treatment decision making *(19)*.

Within this same context, registries are ideal for determining how treatment practice has changed over time. Cooperberg and colleagues *(19)* have noted multiple changes in treatment of prostate cancer among low-risk patients. Choice of initial observation, or watchful waiting, has decreased by more than 50%. Use of external-beam radiation has also declined from 13% to 7%, whereas brachytherapy has increased rapidly (from 4% to 22%).

In a similar vein, the ARAMIS databases have highlighted trends in treatment and diagnosis in rheumatoid arthritis. Published ARAMIS studies have repudiated the nonsteroidal anti-inflammatory drug strategy for persons with rheumatoid arthritis and suggested a strategy of early and persistent treatment with disease-modifying and disease-remitting agents. Data have also accumulated to provide an evidence-based approach to optimal drug sequencing in treatment of rheumatoid arthritis—data that are not obtainable from clinical trials *(20–22)*. Radiologic outcomes have improved as a result of several comparative studies of alternative techniques and comparison of early findings with later joint destruction and later onset of disability and need for total joint replacement *(23–25)*. Also, postmarketing studies have confirmed that gastropathy associated with nonsteroidal anti-inflammatory medications is a significant problem associated with death and hospitalization; that risk varies with age, dose, prior side effects, and type of anti-inflammatory medication *(26–28)*.

There are a number of other instances in which randomized clinical trials do not provide the needed study design for economic and patient outcomes research, and these are also instances in which disease registries provide alternative sources of information. First, clinical end points used in clinical trials may not be suitable for use as the measure of benefit in economic evaluation; of importance to payers and government decision makers. Second, when there are multiple technologic alternatives, including surgical and medical interventions, registries may be the only research format in which sufficient numbers of patients can be compared across practice settings.

Finally, there are sometimes important events, such as uncommon or rare risk factors (children born with cleft palate) or moderate, but long-term, treatment effects that contribute to increased morbidity, resource utilization, and reduced quality of life. These are difficult to capture in a clinical trial because of an unachievable sample size or short follow-up. Similarly, persons excluded from clinical trials (e.g., those on concomitant medications, with comorbidities, of a specific age) are often the most intense users of health care resources and have more quality-of-life impairment. Registries are a practical approach for capturing their clinical outcomes *(29, 30)*.

4. DATA TO BE COLLECTED IN A REGISTRY COME FROM MULTIPLE SOURCES

Simplicity and brevity are critical to ensure prospective and complete data collection in longitudinal observational studies. Data collection is optimized when there are serial measurements of important variables and when variables are updated over time to reflect changes in clinical practice. A sparse dataset with many possible variables is of less value than a complete dataset of core elements. Data collection may be initiated with a core set of modules and variables as illustrated in Table 1 and then expanded over time as new technology or changes in practice dictate.

An important rule for observational studies is that all data collection should be built on specific modules that use standardized data collection forms. Forms are completed at the time of a routine or emergency care visit or other patient contacts. It is important to note that required protocol visits are not a part of observational studies, but that standardized and high-quality data collection is essential. It is optimal to have data collection forms reviewed annually by an advisory panel of providers to reflect treatment advances or changes in practice, so that the data collection tool does not become obsolete. Data modules for a clinical study are likely to include: laboratory; treatment and treatment outcomes (reported by physician at each event); hospitalization admission and discharge dates, procedure and diagnosis codes, status at discharge; medical history reported by physician; death: cause and location of death, date of death (which may be obtained from administrative information); summary of referrals and consulting physicians; and patient-reported outcomes as obtained by baseline and serial questionnaires. The frequency of serial questionnaires is dependent on the study focus. For example, if one is interested in studying the early recovery of urinary incontinence after radical prostatectomy, it would be important to have several patient contacts to collect data in the first months after

Table 1
Data Modules by Source for Observational Databases

Patient Reported	*Clinic/Physician*	*Hospital*	*Administrative*
Background information:	**Medical history**	**Per admission**	**Per event**
Questionnaire date	Date of visit	Admission date	Date of status change (e.g., death)
Date of birth	Date of diagnosis	Location (ER, ICU, etc.)	Date of death
Other demographics	Severity of diagnosis	Diagnosis codes	Cause of death
Height/weight (BMI calculated)	Blood pressure	Procedure codes	Location of death (e.g., hospital, home)
Insurance	Contributing causes	Specialized care	
Employment	Weight	Relevant labs	
Serial questionnaires	Respirations	Medications	
Lifestyle:	Symptom codes	Discharge status	
Exercise	Comorbidities		
Relevant dietary	Treatment plan		
Blood pressure	**Office visits:**		
Smoking	Date of visit		
Alcohol	Reasons for visit		
Quality of life:	Diagnosis codes		
Physical function	Symptom codes		
Emotional function.	Treatment		
Disease specific	Medications		
Work/productivity loss	Procedures		
Resource use:	**Laboratory results:**		
Doctor visits	Hematocrit		
Hospital visits	Hemoglobin		
ER visits	White count		
Disability days	Platelet count		
Unpaid/paid help	Urine		
Diagnostic tests	Etc.		
Side effects			
Medications			

surgery, when there are often huge changes in continence; however, if one is interested in studying the prevalence of debilitating urinary incontinence requiring additional surgery, then the frequency of questionnaires should be less frequent early on and routine (e.g., semiannual) for first 2 yr.

Observational databases are an ideal format for capturing patient-reported outcomes, including quality of life, satisfaction with care, overall resource use, and disease and treatment symptoms. The specific quality of life instruments should include both generic and disease-specific questions and questions germane to treatment and treatment side effects. For example, in prostate cancer, disease-specific instruments focus on incontinence and impotence associated with treatment, whereas in arthritis, disease-specific instruments emphasize pain and functional activities such as walking, climbing, and reach.

5. IMPLEMENTATION

Appropriate implementation of disease registries is critical to their success to ensure adequate and appropriate data collection, data management, and data analysis. An interdisciplinary advisory or steering committee of clinical experts, a number of site participants, and other technical representatives are crucial for developing appropriate data collection forms to review annually for change to forms (this is critical to keep abreast of clinical practice). Study physicians and nurses must be committed to continuing recruitment of patients and to data entry of clinical data. Equally important is the attainment of Institutional Review Board approval and compliance with Health Insurance Portability and Accountability Act regulations. Failure to do so, even in noninterventional studies, may result in premature study closure and investigator censure.

Study data may be collected electronically or via scanned paper forms to facilitate quality assurance and data reporting. Using secure intranet technology, clinical forms can be completed via a web interface. This has many advantages: less costly data collection, simultaneous quality assurance checks, routine computation of calculated variables, and timely graphic summaries available to the physician to be used in patient care or for benchmarking.

Participant questionnaires can be mailed and scanned into the same secure intranet so that comparative data on laboratory values, quality of life, and other treatment milestones are readily tracked. New scanning technology allows for large, easy-to-read questions and response areas, and the ability to write in numbers or text responses that can be scanned, thus reducing error and time to data availability for analysis. Patients may also access similar questionnaires via the web if accessible. However, Internet access is variable across the United States and across patient demographic characteristics.

Patient-reported information in observational studies should not be short-changed, even though it may require additional effort and costs. Successful longitudinal databases have provided rigorous studies of patient outcome based on descriptors and interventions provided by patient questionnaires, and do not rely solely on the medical history. Key steps necessary in planning an observational study are summarized in Table 2.

6. METHODS OF ANALYSIS AND DATA QUALITY ASSURANCE

Although there is not a mandated visit protocol in registries, there are standardized data collection forms and rigorous quality assurance protocols. Data should be reviewed for quality with regard to out-of-range responses, missing data, scorability of questionnaires, and data entry errors. Cleaned, standard, analytic files are prepared by the data management group regularly (quarterly or semiannual) must be prepared so that individuals conducting research are using similar definitions of episodes of care and treatment intervals. These standard analytic files are the basis of research analyses and are maintained for further evaluation.

The intention of many large, observational studies is to go beyond descriptive data to draw causal inference about treatment impact and efficacy. However, observational data also introduce their own biases that must be acknowledged when conducting analyses. These data are usually based on individuals who select to join in the database, and they have experiences that occur before the start of data collection that may not be completely available. They may also have characteristics that unexpectedly influence outcomes. For example, persons who participate in longitudinal databases may be better educated, more

Table 2
Necessary Steps in the Planning of an Observational Study

Develop study question into specific aims
Consider scope of study, cost of data collection and funding
Identify appropriate study population
Determine inclusion/exclusion criteria
Determine whether accrual over study duration be feasible
Consider statistical power for hypothesis testing
Determine whether sampling be necessary (e.g., oversample minority group that may otherwise occur in small numbers and result in poor statistical power)
Determine whether stratification be necessary
Determine appropriate independent and dependent variable
Include demographics, baseline (pre-event) data, operative data, postoperative data, end points (outcomes such as death, morbidity, quality of life)
Use validated measures where such exists (it is worth the effort to take time to properly develop measures if they do not exist)
Determine data sources
Develop study forms, questionnaires, and database
Avoid collecting data that are unlikely to be used in the analysis; each data point collected costs money
Ensure that each data point has clear bounds
Anticipate programming time because is expensive
Determine whether data be patient self-reported, web-entered, via telephone interview, or collected in clinic
Identify staff to identify subjects, obtain consents, and collect data
Consider if chart reviews will play a part in data collection
Consider use of skilled clinical research team with project management, data analyses program, and quality assurance expertise (highly desirable)
Obtain Institutional Review Board approval
Collect data
Data cleaning (out of range responses, missing data, scorability of questionnaires, and data entry errors)
Prepare standard analytic files
Develop a statistical analysis plan
Frame tables, plots of anticipated data
Involve statistician
Plan for reporting
Abstract deadlines
Develop writing committee

likely to be retired, and more likely to be female. All of these characteristics may affect treatment and disease outcomes. When patients drop out of longitudinal studies, their outcomes may not be readily available.

There are a number of techniques used to analyze data with right or left censoring or other biases, incomplete, or episodic data entry or follow-up. These include mixed models, random effects model, proportional hazards regression, generalized estimating equations, and a number of nonparametric approaches *(31, 32)*. Larger sample sizes allow for inclusion of critical covariates in the analyses—an approach not often used in clinical trials in which evaluable patients or intent-to-treat patients are compared. There are specific approaches for addressing selection bias in analyses, including propensity scores and instrumental variables *(33, 34)*. Through these approaches, patients may be matched on clinical or demographic characteristics for a more stand cohort comparison. Because the type of errors present in observational studies vary widely across databases, it is critical to have the participation of a data analyst familiar with issues of selection bias, censoring, and missing data in discussions of analytic design and methods.

7. WHAT ARE THE LIMITATIONS OF REGISTRIES?

Medical researchers have always been taught that clinical trials have high internal validity compared with registries because differences between treatment groups are unlikely to be biased *(35)*. This is a shortcoming of observational studies. Clinicians' and patients' decisions regarding treatment are not random, creating opportunity for bias. For example, certain treatments may be reserved for individuals with worse prognoses or those individuals who have failed prior treatments. Other external factors, unreported in the registry, may affect clinical and quality-of-life outcomes for registry participants. On the other hand, clinical trials usually have relatively low external validity and generalizability, whereas observational studies are often quite easy to generalize.

Well-designed scientifically rigorous registries require large samples sizes and significant costs are involved when collecting standardized data collection in a registry. The length of time it takes to have sufficient numbers of patients to evaluate critical outcomes may be longer than in a focused clinical trial. All of these factors need to be considered when deciding to start a disease registry.

8. SUMMARY

There are many types of studies that can be completed with registry data. Questions to be addressed are dependent on heterogeneity of patients and clinical sites, length of follow-up, and completeness of data collection *(36–41)*. Examples that may be used in surgical and medical registries include:

- Economic studies (costs of illness, cost-effectiveness of treatment, cost utility, measurement of incremental or side effect costs)
- Impact of treatment on patient outcome (clinical and quality of life)
- Impact of comorbidity on treatment and clinical outcomes of disease
- New clinical markers
- Treatment efficacy based on type of provider, setting of care, and volume of patients treated.
- Patient and caregiver satisfaction with treatment.

- Impact of concomitant therapy on clinical and patient reported outcomes, including use of complementary and alternative medicines.
- Changes in standard of care over time and adherence to clinical guidelines.

Thus the goals of observational studies are multifaceted, but all focus on improving health and health care delivery.

REFERENCES

1. Schwartz D, Lellouch J. Explanatory and pragmatic attitudes in clinical trials. J Chronic Dis 1967; 20:637–648.
2. Sackett D, Gent M. Controversy in counting and attributing events in clinical trials. N Engl J Med 1979;301:1410–1412.
3. Chassin M, Brook RH, Park RE, et al. Variations in the use of medical and surgical services by the Medicare population. N Engl J Med 1986;314:285–290.
4. Tunis SR, Stryer DB, Clancy CM. Practical clinical trials: increasing the value of clinical research for decision making in clinical and health policy. JAMA 2003;290(12):1624–1632.
5. Oster G, Borok GM, Menzin J, et al. A randomized trial to assess effectiveness and cost in clinical practice: rationale and design of the Cholesterol Reduction Intervention Study (CRIS). Controlled Clin Trials 1995;16:3–16.
6. Drummond MF. Experimental versus observational data in the economic evaluation of pharmaceuticals. Med Decision Making 1998;18(2 Suppl.):S12–S18.
7. Fries JF, Williams CA, Morfeld D, Singh G, Sibley J. Reduction in long-term disability in patients with rheumatoid arthritis by disease modifying antirheumatic drug-based treatment strategies. Arthritis Rheum 1996;39:616–622.
8. Concato J, Shah N, Horwitz RI. Randomized, controlled trials, observational studies, and the hierarchy of research designs. N Engl J Med 2000;342:1887–1892.
9. Benson K, Hartz AJ. A comparison of observational studies and randomized, controlled trials. N Engl J Med 2000;342:1878–1886.
10. Kelsey JL, Whittemore AS, Evans AS, Thompson WD. Methods in observational epidemiology. 2nd ed. New York: Oxford University Press, 1996.
11. Ramey D, Fries J, Singh G. The health assessment questionnaire 1995—status and review. In Spilker B, ed. Quality of life and pharmacoeconomics in clinical trials. 2nd ed. Philadelphia: Lippincott-Raven Publishers, 1996:227–237.
12. Lubeck DP, Litwin MS, Henning JM, et al. The CaPSURE database: a methodology for clinical practice and research in prostate cancer. Urology 1996;48:773–777.
13. Grossfeld GD, Chang JJ, Broering JM, et al. Impact of positive surgical margins on prostate cancer recurrence and the use of secondary cancer treatment: data from the CaPSURE database. J Urol 2000;163:1171–1177.
14. Grossfeld GD, Latini DM, Lubeck DP, et al. Predicting disease recurrence in intermediate and high-risk patients undergoing radical prostatectomy using percent positive biopsies: results from CaPSURE. Urology 2002;59:560–565.
15. Sullivan SD, Lyles A, Luce B, Grigar J. AMCP guidance for submission of clinical and economic evaluation data to support formulary listing in US health plans and pharmacy benefits management organizations. J Managed Care Pharm 2001;7:272–282.
16. Jani AB, Vaida F, Hanks G, et al. Changing face and different countenances of prostate cancer: racial and geographic differences in prostate-specific antigen (PSA), stage, and grade trends in the PSA era. Int J Cancer 2001;96:363–371.
17. Stanford JL, Stephenson RA, Coyle LM, et al. Prostate cancer trends 1973–1995, SEER program. Bethesda, MD: National Cancer Institute, 1999. (NIH publication number 99-4543).
18. Wei JT, Dunn RL, Sandler HM, et al. Comprehensive comparison of health-related quality of life after contemporary therapies for localized prostate cancer. J Clin Oncol 2002;20:557–566.
19. Cooperberg MR, Broering JM, Litwin MS, et al. The contemporary management of prostate cancer in the Unites States: lessons from CaPSURE, a national disease registry. J Urol. In press.
20. Fries JF, Williams CA, Morfeld D, Singh G, Sibley J. Reduction in long-term disability in patients with rheumatoid arthritis by disease modifying antirheumatic drug-based treatment strategies. Arthritis Rheum 1996;39:616–622.

21. Fries J, Spitz P, Kraines R, Holman H. Measurement of patient outcome in arthritis. Arthritis Rheum 1980;23:137–145.
22. Fries JF, Williams CA, Singh G, Ramey DR. Response to therapy in rheumatoid arthritis is influenced by immediately prior therapy. J Rheumatol 1997;24:838–844.
23. Pincus T, Larsen A, Brooks RH, et al. Comparison of 3 quantitative measures of hand radiographs in patients with rheumatoid arthritis: Steinbrocker stage, Kaye modified sharp score and Larsen score. J Rheumatol 1997;24:2106–2112.
24. Pincus T, Fuchs HA, Callahan LF, Nance EP, Kaye JJ. Early radiographic joint space narrowing and erosion and later malalignment in rheumatoid arthritis: a longitudinal analysis. J Rheumatol 1998;25:636–640.
25. Wolfe F, Sharp JT. Radiographic outcome of recent-onset rheumatoid arthritis. Arth Rheum 1998;41:1571–1582.
26. Singh G, Ramey DR. NSAID induced gastrointestinal complications: the ARAMIS perspective—1997. J Rheumatol 1998;25(Suppl 51):8–16.
27. Singh G, Ramey DR, Morfeld D, et al. Gastrointestinal tract complications of nonsteroidal anti-inflammatory drug treatment in rheumatoid arthritis. Arch Intern Med 1996;156:1530–1536.
28. Fries JF. Current treatment paradigms in rheumatoid arthritis. Rheumatol 2000;39(Suppl 1):30–35.
29. Fries J. Toward an understanding of NSAID-related adverse events: the contribution of longitudinal data. Scand J Rheumatol 1996;25(Suppl. 102):3–8.
30. Etzioni R, Riley GF, Ramsey SD, Brown M. Measuring costs: administrative claims data, clinical trials, and beyond. Med Care 2002;40(6 Suppl.):III-63–72.
31. Byar DP. Problems with using observational databases to compare treatments. Stat Med 1991;10:663–666.
32. Katz BP. Biostatistics to improve the power of large databases. Ann Intern Med 1997;127(Suppl. 8):769.
33. Rubin DB. Estimating causal effects from large data sets using propensity scores. Ann Intern Med 1997;127:757–763.
34. McClellan M, McNeil B, Newhouse J. dome more intensive treatment of acute myocardial infarction in the elderly reduce mortality? Analysis using instrumental variables. JAMA 1994;272:859–866.
35. Pryor DB, Califf RM, Harrell FE, et al. Clinical data bases: accomplishments and unrealized potential. Med Care 1985;23:L623–L647.
36. Cohen CJ, Iwane MK, Palensky JB, et al. A national HIV community cohort: design, baseline, and follow-up of the AmFAR observational database. J Clin Epidemiol 1998;51(9):779–793.
37. Pilote L, Tager IB. Outcomes research in the development and evaluation of practice guidelines. BMC Health Services Res 2002;2:7–18.
38. Peniket AJ, Ruiz de Elvira MC, Taghipour G, et al. An EMBT registry matched study of allogeneic stem cell transplants for lymphoma: allogeneic transplantation is associated with a lower relapse rate but a high procedure-related mortality rate than autologous transplantation. Bone Marrow Transplant 2003;31:667–678.
39. Miller RG, Anderson FA, Bradley WG, et al. The ALS patient care database: goals, design and early results. Neurology 2000;54(1):53–57.
40. Madsen JK, Bech J, Jorgensen E, et al. Yield of 5,536 diagnostic coronary arteriographies: results from a data registry. Cardiology 2002;98(4):191–194.
41. Spencer FA, Santopinto JJ, Gore JM, et al. Impact of aspirin on presentation and hospital outcomes in patients with acute coronary syndromes (The Global Registry of Acute Coronary Events [GRACE]). Am J Cardiol 2002;90:1056–1061.

7 Risk Adjustment

William G. Henderson, MPH, PhD
and Shukri F. Khuri, MD

CONTENTS

Risk adjustment is the process by which outcome measures of health care interventions are corrected for differences in patient characteristics so that comparisons between groups can be made more fairly. The groups to be compared could be types of interventions, health care workers providing treatment, or institutions providing the interventions, to name a few. Outcomes of a health care intervention can be affected by several factors: (1) baseline characteristics or risk factors of the patients receiving the health care intervention; (2) quality of the processes and structures of care; or (3) random chance *(1)*. Risk adjustment attempts to account for the characteristics of the patients having the health care interventions, whereas the statistical analyses that are performed account for the random variation. Statistical analyses of patient risk-adjusted outcomes can then become indicators of variations in the quality of the processes and structures of care.

Risk adjustment is considerably more important in observational studies than in experimental studies. In experimental studies, such as the randomized controlled clinical trial, the clinician randomizes patients to treatments so that there are no selection biases in the formation of the comparative groups. The comparative groups should be fairly equal in all measured and unmeasured baseline patient characteristics. However, in observational studies, assignment of patients to treatment groups or health care providers or institutions is not done randomly. There is the potential for large selection biases and differences in the measured and unmeasured patient baseline characteristics between comparative groups, so that statistical adjustment must be made before any comparisons. It is for these reasons that the randomized controlled clinical trial is considered to be the "gold standard" of clinical research, providing Level I evidence to guide clinical decision making. Although risk adjustment can improve comparisons in observational studies, these studies are considered to provide Level II evidence and cannot rise to the Level I evidence provided by the randomized controlled clinical trial.

From: *Clinical Research for Surgeons*
Edited by: D. F. Penson and J. T. Wei © Humana Press Inc., Totowa, NJ

Risk adjustment in observational studies is important, because if differences in outcomes of care are found between different treatments, health care providers, or health care institutions, these differences might solely be due to differences in the patient characteristics of the comparative groups. As an example, in 1986 the Health Care Financing Administration made a public release of supposedly risk-adjusted hospital level mortality data for Medicare patients. According to their analyses, 142 hospitals had significantly higher death rates than predicted, whereas 127 hospitals had significantly lower rates. At the facility with the worst risk-adjusted outcomes, 87.6% of the Medicare patients died compared with a predicted 22.5%. This facility happened to be a hospice taking care of terminally ill patients, and the Health Care Financing Administration risk adjustment methodology did not adequately account for patient risk.

Comparison of outcomes of health care episodes is by no means a new phenomenon, although there has been increasing attention to this over the past 20 years. Leading practitioners of this approach include Florence Nightingale in the mid-nineteenth century and Ernest Codman in the early twentieth century. Florence Nightingale was troubled by the observation that hospitalized patients died at higher rates than those treated elsewhere, even in comparisons of patients with the same diseases. She also observed that the mortality rates between hospitals varied greatly, and she related these variations to differences in sanitation, crowding of patients, and distances from sewage disposal. She also recognized that the comparisons in the mortality rates should be adjusted for differences in patient characteristics; that some hospitals might be penalized in their mortality statistics by the transfer of the sickest patients to them; and that statistics should also concentrate on recovery and the speed of recovery, as well as mortality *(2)*. Ernest Codman was a surgeon at the Massachusetts General Hospital in the early twentieth century. He became devoted to the ideas of following patients after their surgery to determine the outcomes, of comparing the outcomes from one surgeon to another and from one institution to another, and of determining the reasons for bad outcomes. When he ran into resistance of his ideas at the Massachusetts General Hospital, he started his own, small end-results hospital to champion his ideas *(3)*.

In this chapter, we will first discuss some of the fundamental considerations in developing a risk-adjustment outcomes system for a health care intervention. We will then describe the development and functioning of a risk-adjustment outcomes system in a large, national healthcare system (the Department of Veterans Affairs National Surgical Quality Improvement Program).

1. DEFINING AND MEASURING THE QUALITY OF CARE IN SURGERY

Health care quality may be assessed in terms of structure, process, and outcomes of the health care episode *(4)*. Structures of care refer to the environment in which the health care occurs, such as physical plant, equipment, number and training of health care personnel, and culture of the organization. Structure variables are generally not measured at the patient level, but at the organizational level. Processes of care refer to the individual procedures that are done for each patient, such as use of preoperative antibiotics, surgical procedure and anesthetic chosen, and other elements of preoperative, intraoperative, and postoperative care. Processes of care are usually measured at the patient level. Outcomes of care are measures of the patient's health status after the episode of care, such as postoperative mortality, morbidity, functional status, or health-related quality of life.

These are also measured at the patient level. There is considerable debate about whether processes of care or outcomes of care are the better measures of quality of care *(5)*. In situations in which process measures are supported by Level I evidence and outcomes require long-term follow-up (e.g., annual retinal examinations in patients with type II diabetes to prevent blindness), process measures instead of outcome measures are reasonable indicators of quality of care. For surgical interventions, processes supported by Level I evidence are relatively scarce, but important outcomes can be observed with relatively short-term follow-up. In these instances, outcome measures might be more reasonable measures of quality of care than process measures. The topic of quality of care is discussed more thoroughly in Chapter 15.

2. IMPORTANT ASPECTS TO CONSIDER IN DEVELOPING A SYSTEM FOR COMPARING RISK-ADJUSTED OUTCOMES

2.1. Defining the Patient Population

As in all randomized controlled trials and observational studies, it is important to start by clearly defining the target patient population. For surgical studies, the patient population should be minimally defined in terms of gender, racial/ethnic groups, age, and type of operation. In establishing a risk-adjustment program for surgery, one of the major decisions is whether to include all operations vs only those for which the typical patient is at relatively high risk for an adverse postoperative event. If only major operations are to be included, how should "major" be defined? If only certain high-volume "indicator" operations are to be included, how should these be selected?

The definition of the patient population also has large implications on the needed resources to conduct the program. If data are to be collected on all major and minor operations, will there be enough manpower available to capture the data? Alternatively, if the resources available will permit the collection of data only on a sample of patients, it is important to clearly specify the sampling scheme to make sure that a random, representative, and unbiased sample of patients will be entered into the study. Furthermore, if it is known that a certain type of operation is done very frequently (e.g., inguinal hernia repair or transurethral resection of the prostate), it might be useful to limit the number of these operations entering the database, otherwise the results of the study might be dominated by a few very common operations.

2.2. Selection of Outcomes

There are many outcomes that could be considered in a surgical study. These include: postoperative mortality, postoperative morbidity, long-term survival, functional status, health-related quality of life, patient satisfaction, postoperative length of stay, and costs, to name a few. All of these outcomes theoretically could be risk-adjusted.

Postoperative mortality is an important outcome and relatively easy to ascertain reliably. Central databases, such as the National Death Index or the Beneficiary Identification and Records Locator Subsystem in the Veterans Administration (VA) *(6–8)*, can be used to augment local sources in finding deaths. Some studies use in-hospital mortality, but as length of stay declines, this is not the most objective measure. We recommend using 30-d postoperative mortality inside or outside the hospital from any cause as a standard measure to compare surgical programs. The main drawback to postoperative mortality is that, for some surgical subspecialties and operations, the event rate is so small that the risk-adjusted outcome measures lack precision, and other measures of outcome must be considered as a replacement for postoperative mortality (e.g., routine ophthalmologic procedures, transurethral resection of the prostate, orthopedic procedures).

Postoperative morbidity has the advantage that the events are frequent enough to enable the attainment of reliable estimates of risk-adjusted outcomes. However, the ascertainment of morbidity is more problematic than the ascertainment of mortality because the occurrence of these events is not always clear. A uniform definition of each morbid event that can be accepted by clinicians is important to establish before the study data are collected, and it is also important to develop a tracking system to ascertain these events in patients after discharge (e.g., by using morbidity and mortality (M & M) conferences, chart review, and contacting the patient or family on or near the 30th postoperative day).

In calculating risk-adjusted morbidity outcomes, one must decide whether to combine the morbidity outcomes in some fashion (e.g., using as the outcome variable patients with no postoperative complications vs those with one or more postoperative complication) or to model each complication individually. The advantage of the latter approach is that specific processes are often related to specific outcomes (e.g., use of β-blockers to prevent intraoperative or postoperative myocardial infarction), but the disadvantage is that this can produce an unwieldy number of risk-adjusted outcomes for which to compare providers or institutions.

Long-term survival is probably not a very meaningful quality measure in many surgical settings, because the patients' primary disease process and burden of illness are probably more influential on this measure than the specific operation that the patient is undergoing. Exceptions to this might be surgery for cancer or cardiovascular disease, in which long-term survival is generally the objective.

Functional status, health-related quality of life, and patient satisfaction are also very important measures of outcome to the patient. Several instruments exist to measure these dimensions which have been tested for reliability and validity. The SF-36 or SF-12 *(9,10)* has been used in many studies as a generic measure of health-related quality of life. Their applicability to assess postoperative surgical outcomes, however, may be limited by their generic nature. Disease-specific instruments have also been developed for many diseases, and these instruments sometimes are more sensitive to change in health status from disease-specific interventions than the more generic tools *(11–13)*. The Patient Satisfaction Questionnaire *(14)* is a well accepted, reliable, and validated measure of patient satisfaction with health care. An important drawback to these outcomes is that large resources are required to capture the data completely and reliably.

Postoperative length of stay and costs of care are becoming increasingly more important indicators of quality of care. Improved quality of care can lead to reduced postoperative complications that can, in turn, lead to reduced postoperative length of stay and health care costs. Length of stay and costs of care can be estimated using administrative databases. In recent years, health economists have spent considerable effort at defining and standardizing proper methodology for cost studies *(15,16)*.

2.3. Selection of Risk Factors

The first decision that should be made with respect to collection of risk factors is whether to collect generic risk factors, disease-specific risk factors, or a combination of the two. If the target patient population is characterized by many different diseases and surgical operations, then generic risk factors should be collected. If the patient population is more homogeneous in regard to the disease and operation, then disease-specific or a combination of generic and disease-specific variables should be collected. Before a decision is made about which specific risk factors to collect, a thorough literature review should be performed to identify which variables are most frequently reported as related to the outcomes of interest.

The risk factors should be chosen on the basis of clinical relevance, reliability of data collection, noninvasiveness to the patient (if at all possible), and availability and ease of data collection. The risk factors should include only patient characteristics that are collected preoperatively. Variables that occur intraoperatively or postoperatively should not be included in the risk-adjustment models, because they could be influenced by the quality of the care delivered. Preoperative laboratory tests should be collected as close to the time of the operation as possible. Age, gender, race/ethnicity, and body mass index should always be among the preoperative risk factors collected.

2.4. Risk-Adjustment Methods

The most common statistical techniques used for risk-adjustment are logistic regression for dichotomous outcome variables *(17)* and general linear regression for continuous outcome variables *(18)*. In these models, the dependent variable is the outcome of interest (e.g., dead or alive at the 30th postoperative day; increase in the SF-36 score for physical functioning) and the independent variables are the risk factors. The statistical software that performs these analyses calculates an intercept and a β coefficient and *p* value for each risk factor (and odds ratio, in the case of logistic regression), which measures the relationship of the risk factor and the outcome of interest independent of the other risk factors in the model. If one is interested in the most parsimonious model, a stepwise regression procedure can be used to select only those risk factors that are independently and statistically significantly associated with the outcome. The advantages of these techniques are that they are commonly used and understood, excellent statistical software is available, and the results are readily interpretable. The software often has features to allow for risk factors to be arbitrarily "forced" into the models by the person operating the program, but we would discourage this type of analysis because it is subjective and difficult to replicate.

The intercept and β coefficients of the logistic regression models can be used to calculate the probability of postoperative mortality or morbidity for the individual patients. These probabilities can then be summed to obtain an expected mortality or morbidity for a given patient population categorized by treatment, provider, or institution. An observed-to-expected (O/E) mortality or morbidity ratio can then be calculated in which the observed mortality or morbidity is the number of patients observed with the event, and the expected number is derived as described previously. If the O/E ratio is statistically significantly greater than 1, this means that the targeted patient population has experienced more adverse events than would be expected based on the preoperative severity of illness of the patients in that population. If the O/E ratio is statistically significantly less than 1, this means that the targeted patient population has experienced fewer adverse events than would be expected based on the preoperative severity of illness of the patients in that population.

The c-index is used as a measure of how well the logistic regression model is predicting outcome. The c-index is the proportion of all possible pairs of patients with and without an event, for which the individual with an event has a higher probability of the event than the individual without the event. If the model is not predictive at all, the c-index will be close to 0.50. As the model improves in predictability, the c-index will get closer to 1.00. Alternatively, if the risk adjustment involves multiple linear regression, an r^2 value can be used as a measure of predictability, which is the proportion of the variance of the continuous outcome variable explained by the risk factors. More predictive models will have a higher r^2 value but most clinical models seldom exceed 0.3–0.4 range given the significant influence of unexplained and unmeasured factors. Although the c-index and the r^2 both speak to the ability of the statistical model to predict an outcome, they are not equivalent terms and therefore cannot be compared.

It is sometimes useful to develop point scores for the significant risk variables as a function of the β coefficients produced by the regression analysis and then sum the scores to develop a more clinically usable risk index to predict mortality or morbidity *(19)*. The clustering effect of multiple observations within each health care provider or institution also should be accounted for in the regression analysis *(20)*.

2.5. Data Collection

One of the most important aspects of a surgical quality improvement program based on risk-adjusted outcomes is reliable data that can be believed by the clinicians and health care administrators. Administrative data are often used for risk-adjustment purposes because they are readily available and inexpensive, but they are poor substitutes for reliable and valid clinical data collected by trained nurses using a standardized protocol and definitions. Administrative data are hampered by a number of factors: (1) limitations of the ICD-9-CM (International Classification of Diseases-9-Clinical Modification) coding for surgical operations; (2) limitations of the discharge abstracts for distinguishing between preoperative diagnoses and postoperative complications; (3) inconsistency of coding and lack of standardized definitions across sites; and (4) lack of clinical variables to allow for good risk adjustment *(21)*. A recent study in the Department of Veterans Affairs showed that the sensitivity and positive predictive value of administrative data in comparison to clinical data collected in the VA's National Surgical Quality Improvement Program (NSQIP) were poor *(22)*. ICD-9-CM coding was only available for 13 (45%) of the top 29 preoperative risk variables. In only three (23%) was sensitivity and in only four (31%) was positive predictive value greater than 0.500. There were ICD-9-CM codes for all 21 postoperative complications collected in the NSQIP, but in only 7% was sensitivity and only 4% was positive predictive value greater than 0.500.

To ensure reliable clinical data in a multisite surgical quality improvement program, we recommend the following steps, which have been incorporated into the NSQIP: (1) development of an operations manual detailing selection of patients and clear definitions of all preoperative risk factors, intraoperative variables, and postoperative outcomes; (2) a qualified surgical nurse or clinical reviewer to collect the data; (3) in-depth, face-to-face training of the nurse reviewers with a plan to renew the training periodically; (4) central data monitoring and validation; and (5) annual interrater reliability site visits. The cost of reliable data collection and analysis and oversight of the NSQIP has been estimated at about $12 per case for total surgical volume, or $38 per case assessed in the program *(23)*. Although this adds to the total cost of surgical care, it is a small price to pay for good-quality data to support proper patient management.

3. THE NATIONAL SURGICAL QUALITY IMPROVEMENT PROGRAM (NSQIP)

Several national and regional consortiums collect risk-adjusted surgical outcomes for quality improvement and research purposes. These have been mainly in cardiovascular surgery *(24–28)*. NSQIP, started in the US Department of Veterans Affairs in 1991, is the largest program to address noncardiac surgery. NSQIP was started as a response to a Congressional mandate in 1986 that called on the VA to compare its surgical outcomes with those in the private sector. A retrospective study was first attempted, comparing the mortality outcomes of selected operations in the VA system using administrative databases compared to outcomes from the private sector reported in the literature *(29)*. This study suffered from small sample sizes for some of the operations, nonuniformity of definitions and follow-up for outcomes, and lack of risk-adjustment. The authors of this investigation concluded that only a well-designed system of prospective data collection would provide data to satisfy the Congressional mandate.

The National VA Surgical Risk Study was started in 1991 at the 44 largest VA medical centers that performed both cardiac and noncardiac surgery. In Phase I of this study (October 1, 1991, to December 31, 1993), the protocol and operations manual were developed; a surgical nurse reviewer was recruited at each of the 44 VA medical centers (VAMCs) and trained in the data collection process; software was developed and exported to each of the sites to input the data; software was developed at the data coordinating center to edit and manage the data and produce feedback reports; data reliability site visits were conducted to all 44 sites by two traveling nurse coordinators; data were collected on 87,078 major operations performed at the 44 centers; statistical modeling was performed for mortality and morbidity for all operations combined and for eight major subspecialty areas (general, vascular, orthopedics, neurosurgery, ear-nose-throat, urology, thoracic, and plastic surgery); and risk-adjusted outcomes were fed back to the chiefs of surgery at the 44 centers. The protocol paper was published in 1995 *(30)*, and the mortality and morbidity results articles were published in 1997 *(31,32)*.

Two validation studies were conducted, one by site visits *(33)* and one by chart reviews *(34)*, to determine whether differences in risk-adjusted outcomes were true indicators of differences in quality of surgical care. In the site visit study, evaluators were sent to 10 VAMCs with low O/E ratios for mortality or morbidity and 10 VAMCs with high O/E ratios for mortality or morbidity. The evaluators included a chief of surgery, an operating room nurse, and a health services researcher. The evaluators and the institutions they visited were masked as to whether the site was a low or high outlier. They evaluated a number of dimensions of quality of care, including technology and equipment, technical competence, interface with other services, relationship with affiliated institutions, monitoring of quality of care, coordination of work, leadership, and overall quality of care. Mean scores for all of these dimensions were better for the low outlier hospitals compared with the high outlier hospitals, and the differences were statistically significant ($p < 0.05$) for technology and equipment and overall quality of care. The differences were of borderline statistical significance ($p < 0.10$) for technical competence, relationship with affiliated institutions, and monitoring quality of care. The evaluators were also asked to guess whether the institution was a high or low outlier hospital. The evaluators correctly guessed for 17 of the 20 centers (85%, $p = 0.002$).

The chart review validation study was less definitive. In this study, 739 charts from low and high outlier VAMCs for patients undergoing general, vascular, or orthopedic surgery were reviewed and graded for quality of care. Ratings of overall quality of care did not differ significantly between patients from the high or low outlier hospitals. However, at the patient level of analysis, those patients who died or developed postoperative complications and had a low predicted risk of mortality or morbidity were rated lower on quality of care compared to those who died or developed a postoperative complication and had a high predicted risk of mortality or morbidity. It is not known why the site visit study tended to validate the risk-adjusted outcomes more than the chart review study, but one reason might be that site visits may be more discernible of quality of care issues than chart reviews.

In Phase II of the NSQIP (January 1, 1994, to August 31, 1995), the program was implemented in all 132 VA Medical Centers that perform major surgery. Regional training meetings were held for the chiefs of surgery and local surgical nurse reviewers. The computerized data collection system was implemented at each hospital, and data were collected on an addition 107,241 major operations in the eight subspecialty areas. Statistical models were developed using the Phase II data and compared with the Phase I models. The risk-adjustment models have remained remarkably stable over time, with the same major risk variables appearing significant in the all operations model and most subspecialty models and in these models over time *(23)*. Risk-adjusted outcomes were fed back to the chiefs of surgery at each of the 132 VA Medical Centers.

In Phase III of the NSQIP (October 1, 1995, to the present), the reporting system has been put on a federal fiscal year basis (October 1 to September 30 of each year). The NSQIP executive committee reviews the risk-adjusted outcomes from the VA centers each January and makes recommendations regarding levels of concern for high outlier centers or commendations for centers who are low outliers or who have improved their risk-adjusted outcomes. Annual reports are sent to the chiefs of surgery and the hospital and Veterans Integrated Service Networks' directors to enable them to compare their risk-adjusted outcomes with others in the system. Those centers who are low outliers or who have improved their risk-adjusted outcomes are asked to provide feedback about best surgical practices that might have accounted for these results, and these practices are shared with all chiefs of surgery. The NSQIP executive committee also offers to organize site visits to high outlier hospitals to help them with quality improvement efforts, if requested. Also if requested by the participating hospital, lists of patients who have low probabilities of adverse events but who experience an adverse event are sent to the hospital for local analysis. These quality improvement efforts have been coincident with a decline in the unadjusted 30-d postoperative mortality rate in the VA system from 3.16% in Phase I to 2.14% in fiscal year 2002 (a 32% decline), and a decline in the unadjusted 30-d postoperative morbidity rate from 17.44% in Phase I to 9.98% in fiscal year 2002 (a 43% decline). The NSQIP was recently cited by an Institute of Medicine Report as "one of the most highly regarded VHA initiatives employing performance measures" *(35)*.

3.1. Data Collection

The NSQIP patient population consists of all major operations performed under general, spinal, or epidural anesthesia. Minor operations and some operations with known low morbidity and mortality rates are excluded. Some very common operations (e.g., inguinal hernias and transurethral resections of the prostate) are limited to five in each 8-d cycle. In most of the VAMCs, all major operations are included. There are a few

VAMCs that have too high a volume of major operations to be collected by one nurse reviewer. In those hospitals, the first 36 consecutive cases are sampled in each 8-d cycle, with the cycles starting on different days of the week for each cycle. Seventy preoperative risk factors are collected for each case, including demographics, lifestyle variables, functional status, American Society of Anesthesiologists (ASA) class, emergency operation, preoperative laboratory values, and comorbidities. Eleven intraoperative variables are collected, including the Current Procedural Terminology (CPT) codes of the principal operation and concomitant procedures, times of operation, postgraduate year (PGY)-level of surgeon, wound classification, and blood loss and transfused. Twenty-four postoperative outcomes are collected, including vital status, postoperative complications, return to operating room, and length of stay. Selective postoperative laboratories are also captured. With nearly 100,000 operations added each year, the NSQIP database reached a total of more than 1 million cases by September 30, 2003.

3.2. Risk-Adjustment Models

Risk-adjustment models are developed each year for all operations combined, and the eight major subspecialties. Each CPT code is assigned a complexity score from 1 = least complex to 5 = most complex, to take into account variations in the complexity of the operation, above and beyond the risk factors that are brought into the operation by the typical patient having that operation. These scores were established by panels of subspecialists convened in the early years of the NSQIP. These complexity scores are entered into the risk models for mortality and morbidity for the all-operations model and each of the subspecialty models. In recent years, with the advent of new surgical procedures (e.g., laparoscopic procedures) and changes to the CPT codes, 10–20% of the CPT codes lack complexity scores. Preliminary investigations have revealed that there is a high correlation (0.70–0.80) between the NSQIP complexity scores and work Resource Based Relative Value Units, so consideration is being made to replace the complexity scores with work Resource Based Relative Value Units. Also, in the all-operations model, subspecialty of the operation is included among the predictor variables to adjust for differences between medical centers in surgical subspecialties represented.

The NSQIP data are very complete, except for the preoperative laboratory values, where there is considerable variability in completeness rates between type of laboratory test, participating sites, subspecialties, and individual operations. A regression technique *(36, 37)* is used to impute missing laboratory data.

The c-indexes for models for mortality and morbidity (patients with no complications vs one or more complications) are given in Table 1 for various groups of operations. In general, the c-indexes are excellent for predicting mortality for all operations combined and for the individual subspecialties; moderate to good for predicting individual postoperative morbidities; moderate for predicting overall morbidity for all operations combined and for the individual subspecialties; and weakest for predicting mortality and overall morbidity for individual operations. The probable reason for the models being less predictive for overall morbidity compared with mortality is that the overall morbidity variable is a combination of 21 different heterogeneous postoperative complications; the probable reason for the models being weakest for predicting mortality and morbidity for individual operations is that the risk factors are generic and not disease-specific.

Important risk variables have remained very stable over time. Table 2 lists the top predictors of 30-d mortality and overall morbidity from 1991 to 1997 *(23)*. Some of the

Table 1
C-Indexes for National Surgical Quality Improvement Program Models Predicting 30-d Postoperative Mortality and Morbidity From Preoperative Patient Characteristics

Group of Operations	*Mortality*	*Overall Morbidity*	*Specific Morbidities*
All operations	0.889	0.777	Respiratory failure, 0.846 Pneumonia, 0.805
General surgery	0.892	0.783	Wound dehiscence, 0.731
Orthopedics	0.913	0.763	
Urology	0.861	0.729	
Vascular	0.794	0.689	
Neurosurgery	0.896	0.762	
Otolaryngology	0.906	0.793	
Thoracic	0.766	0.717	
Plastic	0.912	0.752	
Proctectomy	0.755	0.684	
Pulmonary resection	0.729	0.623	
Nephrectomy	0.741	0.64	
Below-knee amputation	0.81	—	
Above-knee amputation	0.79	—	
Gastrectomy	0.735	0.722	
Esophagectomy	0.69	0.62	
Hip replacement	—	0.654	
Knee replacement	—	0.633	
Pancreaticoduodenectomy	0.692	—	

variables have been important for predicting both mortality and morbidity (serum albumin, ASA class, emergency operation, age, blood urea nitrogen >40, operation complexity score, weight loss >10% in past 6 mo, functional status, and white blood cell count >11,000). Some variables are predictive of mortality but not morbidity (disseminated cancer, do-not-resuscitate status, serum glutamic-oxalacetic transaminase (SGOT) >40), whereas some variables are predictive of morbidity but not mortality (history of chronic obstructive pulmonary disease, hematocrit <38, and ventilator dependency). Operation complexity score and patient's functional status tend to be more important for predicting morbidity than mortality.

Figure 1 shows the distribution of hospital O/E ratios for mortality for the all-operations model for the 44 hospitals in Phase I of the NSQIP. Six hospitals had O/E ratios statistically significantly greater than 1, meaning that they were experiencing more operative deaths than would be expected on the basis of the severity of illness of their patients. Seven hospitals had O/E ratios statistically significantly less than 1, meaning that they were experiencing fewer operative deaths than would be expected on the basis of the severity of illness of their patients *(31)*.

Table 2
Top Predictors of 30-d Mortality and Overall Morbidity in National Surgical Quality Improvement Program, 1991–1997 (Average Rank Entering the Models)

Risk Factor	*Rank for Mortality*	*Rank for Morbidity*
Serum albumin	1	1.3
ASA class	2	2
Disseminated cancer	3.3	—
Emergency operation	4.3	4
Age	5	8.3
Blood urea nitrogen >40	7	20.3
Do-not-resuscitate order	7.3	—
Operation complexity score	11	2.8
SGOT >40	11.3	—
Weight loss >10% in last 6 mo	11.5	13.3
Functional status	12.3	5
White blood cell count >11,000	14	10
History of chronic obstructive pulmonary disorder	—	7.5
Hematocrit <38%	—	9.5
Ventilator dependent	—	16.5

ASA, American Society of Anesthesiologists; SGOT, serum glutamic oxalacetic transaminase.

Figure 2 shows the changes in the rankings of the hospitals before and after risk adjustment. If risk adjustment did not make any difference, the figure would look like a ladder or railroad track with steps or ties that are completely horizontal. Ninety-three percent of the hospitals changed rank after risk adjustment, 50% by >5 ranks and 25% by >10 ranks *(31)*. An analysis of the mortality O/E ratios of the 123 VA medical centers performing major surgery in 1997 revealed that outlier status of the hospitals would have been ascribed incorrectly 64% of the time if it had been based on unadjusted mortality rather than risk-adjusted mortality *(23)*.

3.3. Uses of the NSQIP Database

Since 1991, the NSQIP has created a rich database of more than 1 million operations that can be used to explore important topics in clinical and health services research in surgery. To access the database, the investigator must have a VA appointment or be a participant in the NSQIP Private Sector Initiative (*see* Future Directions). The eligible investigator writes a brief proposal describing the background and rationale for the study, research objectives, methods, and data needed (types of operations, time period, and data elements). The proposal is reviewed by the investigator's local Institutional Review Board and the NSQIP executive committee. If approved, the Denver Data Analysis Center either performs the analysis for the investigator or sends the investigator deidentified data for local analysis. The NSQIP executive committee must review and approve all abstracts and manuscripts emanating from the studies before submission.

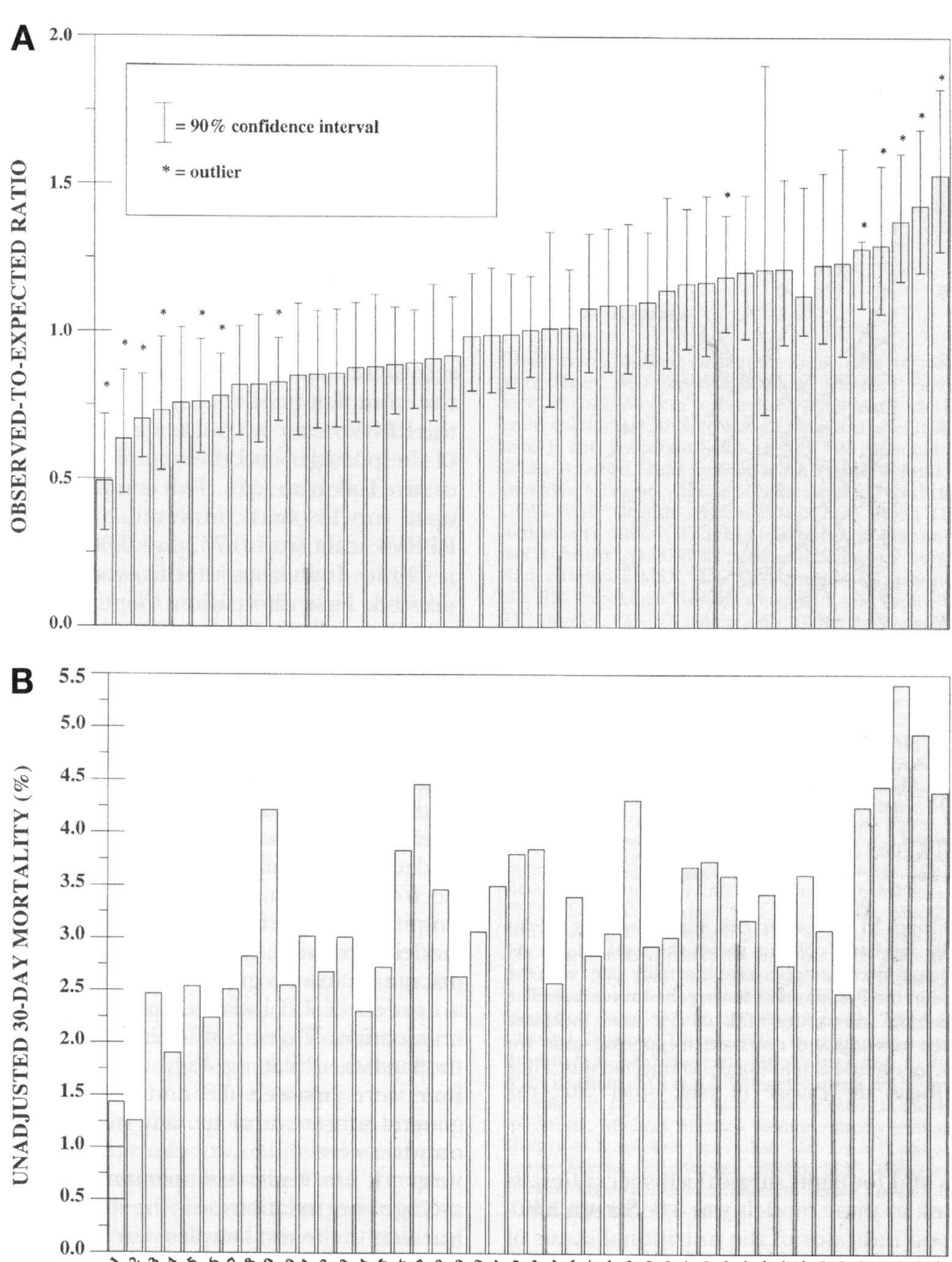

Figure 1: (**A**) Observed-to-expected mortality ratios of 44 VA medical centers in phase I of NSQIP. (**B**) Unadjusted mortality rates of 44 VA medical centers in phase I of NSQIP in same order as hospitals in (A).

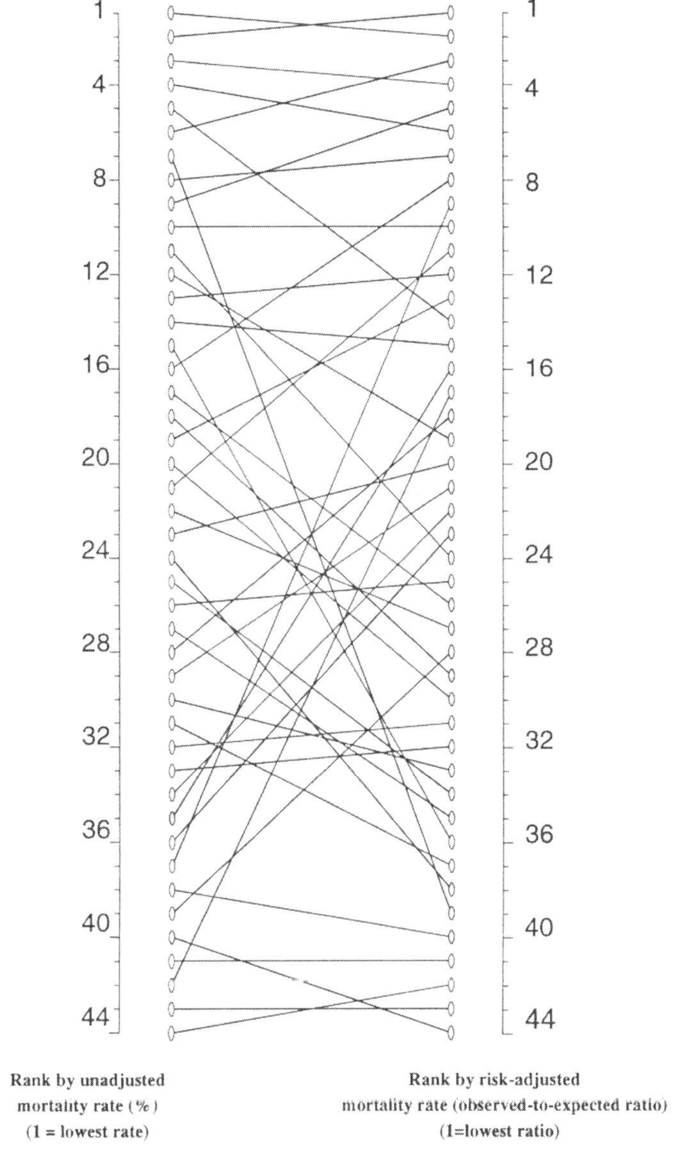

Figure 2: Rankings of 44 VA medical centers in phase I of NSQIP based on unadjusted mortality (left side) and risk-adjusted mortality (right side).

As of October, 2003, 47 manuscripts have been published from the database by 29 different first authors from 14 different academic/health center institutions. Types of research include modeling of postoperative mortality and morbidity *(31,32)*, relating surgical processes and structures of care to risk-adjusted outcomes *(38–41)*, risk factors for and outcomes of specific operations *(42–56)*, effect on outcomes of specific risk factors *(57,58)*, predictors of outcomes other than 30-d mortality and overall morbidity *(59–63)*, surgical outcomes in special populations *(64–68)*, and risk-adjustment methodologic issues *(22,69)*. Table 3 presents some selected findings from research using the NSQIP database.

Table 3
Some Selected Findings from Research Using the
National Surgical Quality Improvement Program (NSQIP) Database

1. Surgical services with a high degree of feedback and programming had lower morbidity observed-to-expected (O/E) ratios, but not mortality O/E ratios *(39)*
2. Growing use of laparoscopic cholecystectomy in a globally funded system such as the Veterans Administration (VA) did not result in increase in total volume of cholecystectomies, in contrast to the private sector where increase in laparoscopic cholecystectomy has led to an increase in the total volume of cholecystectomies *(43)*
3. Preoperative serum albumin has an inverse linear relationship to operative mortality and morbidity and is the best predictor of these outcomes, and yet it is only done in 60% of surgical cases in the VA *(57)*
4. There is no relationship between surgical volume and risk-adjusted outcomes in eight major types of operations in the VA system *(40)*
5. Sensitivity and positive predictive value for surgical risk factors and outcomes using administrative data compared to NSQIP data are poor *(22)*
6. NSQIP risk-adjusted mortality and morbidity are indicators of differences in surgical quality of care *(33)*
7. NSQIP methodology can be implemented and generates reasonable predictive models of postoperative mortality and morbidity in non-VA hospitals *(70)*
8. Modest increase in preoperative serum creatinine is a significant predictor of risk-adjusted morbidity and mortality after general surgery *(58)*

3.4. Future Directions

There has been considerable interest among surgeons outside of the VA to participate in the NSQIP. In 1999, three academic medical centers (Emory University, University of Michigan at Ann Arbor, and University of Kentucky at Lexington) joined an NSQIP Private Sector Initiative as "alpha sites" to test the NSQIP methodology outside of the VA. One of the challenges was to develop a data capture system that would enable the participation of independent sites with diverse information technology systems. This was accomplished by the development of an Internet-based data collection system by an outside contractor (QCMetrix, Inc, Tewksbury, MA). Preliminary analyses suggest that the statistical models developed in the VA are applicable in the non-VA setting *(70)*.

In 2001, a grant supported by the Agency for Healthcare Research and Quality was funded to apply the NSQIP methodology to an additional 11 non-VA "beta sites" to improve patient safety in surgery *(71)*. This grant was a collaborative effort between the Department of Veterans Affairs and the American College of Surgeons. As part of this grant, the relationships between selected processes and structures of the surgical services, including mechanisms to promote patient safety, and risk-adjusted outcomes will be studied. The American College of Surgeons has developed plans to offer the NSQIP to additional private sector surgical services at a cost sufficient to support the infrastructure of the program.

In spite of its successes, the NSQIP has been limited by its initial design. Quality improvement efforts depend on retrospective reports fed back to the participating centers. The next step in the development of the NSQIP will be to provide the surgical team with prospective risk information and suggested evidence-based guidelines to help guide the management of the medium- to high-risk patient about to be operated.

Further work also needs to be done on outcomes other than short-term mortality and morbidity. Functional status, health-related quality of life, patient satisfaction with care, and cost-effectiveness of the interventions are other important outcomes that need to be addressed.

3.5. Limitations of the NSQIP

The primary limitations of the NSQIP include the following. (1) The program is labor-intensive, requiring dedication on the part of the participating site, surgeons, and nurse reviewers to collect complete and reliable data on which to base decisions about quality improvement efforts in surgery. However, after the concepts of the program are fully accepted and implemented, the data can provide critical empowerment to improve processes and structures of care for the improvement of care of the surgical patient. (2) In the smaller surgical programs and in some subspecialty areas or individual operations, there are not enough major surgical cases and adverse events to generate reliable risk-adjusted outcomes for mortality and major morbidity. In these cases, the NSQIP methodology needs to be applied to other, important patient outcomes, such as functional status and health-related quality of life. (3) The NSQIP risk factors are primarily generic, so that risk-adjustment models for specific operations have limited usefulness. Further work is needed to add some selected disease-specific variables for important high-frequency operations in the subspecialty areas. (4) The NSQIP does not collect process and structure variables. As methods are refined, there are plans to collect important process and structure variables as they are identified.

REFERENCES

1. Iezzoni LI, ed. Risk adjustment for measuring health care outcomes. Ann Arbor, MI: Health Administration Press, 1994.
2. Nightingale F. Notes on Hospitals. 3rd ed. London: Longman, Green, Longman, Roberts, and Green, 1863.
3. Codman EA. A study in hospital efficiency as demonstrated by the case report of the first five years of a private hospital. Boston: Thomas Todd Company, 1917.
4. Donabedian A. Explorations in quality assessment and monitoring. The definition of quality and approaches to its assessment. Vol. 1. Ann Arbor, MI: Health Administration Press, 1980.
5. Brook RH, McGlynn EA, Cleary PD. Quality of health care. Part 2: measuring quality of care. N Engl J Med 1996;335:966–970.
6. Fisher SG, Weber L, Goldberg J, Davis F. Mortality ascertainment in the veteran population: alternatives to the National Death Index. Am J Epidemiol 1995;141:242–250.
7. Page WF, Braun MM, Caporaso NE. Ascertainment of mortality in the US veteran population: World War II veteran twins. Mil Med 1995;160:351–355.
8. Cowper DC, Kubal JD, Maynard C, Hynes DM. A primer and comparative review of major US mortality databases. Ann Epidemiol 2001;11:2–7.
9. Ware JE Jr, Sherbourne CD. The MOS 36-item short-form health survey (SF-36). I. Conceptual framework and item selection. Med Care 1992;30:473–483.
10. Ware JE, Kosinski M, Keller SD. SF-12: how to Score the SF-12 Physical and Mental Health Summary Scales. 2nd ed. Boston: The Health Institute, New England Medical Center, 1995.
11. Barry MJ, O'Leary MP. Advances in benign prostatic hyperplasia. The development and clinical utility of symptom score. Urol Clin N Am 1995;22:299–307.

12. Bellamy N. Pain assessment in osteoarthritis: experience with the WOMAC osteoarthritis index. Semin Arthritis Rheum 1989;18:14–17.
13. Spertus JA, Winder JA, Dewhurst TA, et al. Development and evaluation of the Seattle Angina Questionnaire: a new functional status measurement for coronary artery disease. J Am Coll Cardiol 1995;25:333–341.
14. Ware JE Jr, Snyder MK, Wrigh, WR, Davies AR. Defining and measuring patient satisfaction with medical care. Eval Program Plan 1983;6:247–263.
15. Weinstein MC, Siegel JE, Gold MR, Kamlet MS, Russell LB for the Panel on Cost-Effectiveness in Health and Medicine. Recommendations of the panel on cost-effectiveness in health and medicine. JAMA 1996;276:1253–1258.
16. Siegel JE, Weinstein MC, Russell LB, Gold MR, for the Panel on Cost-Effectiveness in Health and Medicine. Recommendations for reporting cost-effectiveness analyses. JAMA 1996;276:1339–1341.
17. Hosmer DW, Lemeshow S. Applied logistic regression. New York: John Wiley and Sons, 1989.
18. Neter J, Wasserman W. Applied linear statistical models. Homewood, IL: Richard D. Irwin, Inc., 1974.
19. Le Gall JR, Lemeshow S, Saulnier F. A new simplified acute physiology score (SAPS II) based on a European/North American multicenter study. JAMA 1993;270:2957–2963.
20. Panageas KS, Schrag D, Riedel E, Bach PB, Begg CB. The effect of clustering of outcomes on the association of procedure volume and surgical outcomes. Ann Intern Med 2003;139:658–665.
21. Iezzoni LI. Data sources and implications: administrative databases. In: Iezzoni LI, ed. Risk adjustment for measuring health care outcomes. Ann Arbor, MI: Health Administration Press, 1994.
22. Best WR, Khuri SF, Phelan M, et al. Identifying patient preoperative risk factors and postoperative adverse events in administrative databases: results from the Department of Veterans Affairs National Surgical Quality Improvement Program. J Am Coll Surg 2002;194:257–266.
23. Khuri SF, Daley J, Henderson W, et al, and the participants in the National VA Surgical Quality Improvement Program. The Department of Veterans Affairs' NSQIP. The first national, validated, outcome-based, risk-adjusted, and peer-controlled program for the measurement and enhancement of the quality of surgical care. Ann Surg 1998;228:491–507.
24. Hammermeister KE, Johnson R, Marshall G, et al. Continuous assessment and improvement in quality of care. A model from the Department of Veterans Affairs Cardiac Surgery. Ann Surg 1994;219:281–290.
25. Hannan EL, Kilburn H Jr, Racz M, et al. Improving the outcomes of coronary artery bypass surgery in New York State. JAMA 1994;271:761–766.
26. O'Connor GT, Plume SK, Olmstead EM, et al. A regional intervention to improve the hospital mortality associated with coronary artery bypass graft surgery. JAMA 1996;275:8441–8446.
27. Jones RH, Hannan EL, Hammermeister KE, et al. Identification of preoperative variables needed for risk adjustment of short-term mortality after coronary artery bypass graft surgery. The Working Group Panel on the Cooperative CABG Database Project. J Am Coll Cardiol 1996;28:1478–1487.
28. Ferguson TB Jr, Dzuiban SW Jr, Edwards FH, et al. The STS national database: current changes and challenges for the new millennium. Committee to Establish a National Database in Cardiothoracic Surgery. The Society for Thoracic Surgeons. Ann Thorac Surg 2000;69:680–691.
29. Stremple JF, Bross DS, Davis CL, McDonald GO. Comparison of postoperative mortality in VA and private hospitals. Ann Surg 1993;217:277–285.
30. Khuri SF, Daley J, Henderson W, et al, and participants in the National Veterans Administration Surgical Risk Study. The National Veterans Administration Surgical Risk Study: risk adjustment for the comparative assessment of the quality of surgical care. J Am Coll Surg 1995;180:519–531.
31. Khuri SF, Daley J, Henderson W, et al, for the participants in the National VA Surgical Risk Study. Risk adjustment of the postoperative mortality rate for the comparative assessment of the quality of surgical care: results of the National Veterans Affairs Surgical Risk Study. J Am Coll Surg 1997;185:315–327.
32. Daley J, Khuri SF, Henderson W, et al, for participants in the National VA Surgical Risk Study. Risk adjustment of the postoperative morbidity rate for the comparative assessment of the quality of surgical care: results of the National Veterans Affairs Surgical Risk Study. J Am Coll Surg 1997;185:328–340.
33. Daley J, Forbes MG, Young GJ, et al, for the participants in the National VA Surgical Risk Study. Validating risk-adjusted surgical outcomes: site visit assessment of process and structure. J Am Coll Surg 1997;185:341–351.
34. Gibbs J, Clark K, Khuri S, et al. Validating risk-adjusted surgical outcomes: chart review of process of care. Int J Qual Health Care 2001;13:187–196.
35. Corrigan JM, Eden J, Smith BM, eds. Leadership by example. Coordinating government roles in improving healthcare quality. Washington, DC: Institute of Medicine of the National Academies, The National Academies Press, 2002.

36. Buck SF. A method of estimation of missing values in multivariate data suitable for use with an electronic computer. J R Stat Soc B 1960;22:302–307.
37. Roberts JS, Capalbo GM. A SAS macro for estimating missing values in multivariate data. SAS Users Group International Twelfth Annual Conference Proceedings. Dallas, TX: February 8–11, 1987.
38. Young GJ, Charns MP, Daley J, et al. Best practices for managing surgical services: the role of coordination. Health Care Manage Rev 1997;22:72–81.
39. Young GJ, Charns MP, Desai K, et al. Patterns of coordination and clinical outcomes: a study of surgical services. HSR: Health Services Res 1998;33:1211–1236.
40. Khuri SF, Daley J, Henderson W, et al, and the participants in the VA National Surgical Quality Improvement Program. Relation of surgical volume to outcome in eight common operations. Results from the VA National Surgical Quality Improvement Program. Ann Surg 1999;230:414–432.
41. Khuri SF, Najjar SF, Daley J, et al, and the participants in the VA National Surgical Quality Improvement Program. Comparison of surgical outcomes between teaching and nonteaching hospitals in the Department of Veterans Affairs. Ann Surg 2001;234:370–383.
42. Longo WE, Virgo KS, Johnson FE, et al. Outcome after proctectomy for rectal cancer in Department of Veterans Affairs hospitals. A report from the National Surgical Quality Improvement Program. Ann Surg 1998;228:64–70.
43. Chen AY, Daley J, Pappas TN, Henderson WG, Khuri SF. Growing use of laparoscopic cholecystectomy in the National Veterans Affairs Surgical Risk Study. Effects on volume, patient selection, and selected outcomes. Ann Surg 1998;227:12–24.
44. Harpole DH Jr, DeCamp MM Jr, Daley J, et al, and participants in the National Veterans Affairs Surgical Quality Improvement Program. Prognostic models of thirty-day mortality and morbidity after major pulmonary resection. J Thorac Cardiovasc Surg 1999;117:969–979.
45. Corman JM, Penson DF, Hur K, et al. Comparison of complications after radical and partial nephrectomy: results from the National Veterans Administration Surgical Quality Improvement Program. BJU Int 2000;86:782–789.
46. Longo WE, Virgo KS, Johnson FE, et al. Risk factors for morbidity and mortality after colectomy for colon cancer. Dis Colon Rectum 2000;43:83–91.
47. Weaver F, Hynes D, Goldberg JM, et al. Hysterectomy in Veterans Affairs medical centers. Obstet Gynecol 2001;97:880–884.
48. Feinglass J, Pearce WH, Martin GJ, et al. Postoperative and amputation-free survival outcomes after femorodistal bypass grafting surgery: findings from the Department of Veterans Affairs National Surgical Quality Improvement Program. J Vasc Surg 2001;34:283–290.
49. Feinglass J, Pearce WH, Martin GJ, et al. Postoperative and late survival outcomes after major amputation: findings from the Department of Veterans Affairs National Surgical Quality Improvement Program. Surgery 2001;130:21–29.
50. Grossman EM, Longo WE, Virgo KS, et al. Morbidity and mortality of gastrectomy for cancer in Department of Veterans Affairs Medical Centers. Surgery 2002;131:484–490.
51. Bailey SH, Bull DA, Harpole DH, et al. Outcomes after esophagectomy: a ten-year prospective cohort. Ann Thorac Surg 2003;75:217–222.
52. O'Hare AM, Feinglass J, Sidawy AN, et al. Impact of renal insufficiency on short-term morbidity and mortality after lower extremity revascularization: data from the Department of Veterans Affairs' National Surgical Quality Improvement Program. J Am Soc Nephrol 2003;14:1287–1295.
53. Margenthaler JA, Longo WE, Virgo KS, et al. Risk factors for adverse outcomes after the surgical treatment of appendicitis in adults. Ann Surg 2003;238:59–66.
54. Weaver F, Hynes D, Hopkinson W, et al. Preoperative risks and outcomes of hip and knee arthroplasty in the Veterans Health Administration. J Arthroplasty 2003;18:693–708.
55. Rentz J, Bull D, Harpole D, et al. Transthoracic versus transhiatal esophagectomy: a prospective study of 945 patients. J Thorac Cardiovasc Surg 2003;125:1114–1120.
56. Billingsley KG, Hur K, Henderson WG, et al. Outcome after pancreaticoduodenectomy for periampullary cancer: an analysis from the Veterans Affairs National Surgical Quality Improvement Program. J Gastrointest Surg 2003;7:484–491.
57. Gibbs J, Cull W, Henderson W, et al. Preoperative serum albumin level as a predictor of operative mortality and morbidity. Arch Surg 1999;134:36–42.
58. O'Brien MM, Gonzales R, Shroyer AL, et al. Modest serum creatinine elevation affects adverse outcome after general surgery. Kidney Int 2002;62:585–592.
59. Collins TC, Daley J, Henderson WG, Khuri SF. Risk factors for prolonged length of stay after major elective surgery. Ann Surg 1999;230:251–259.

60. Arozullah AM, Daley J, Henderson WG, Khuri SF, for the National Veterans Administration Surgical Quality Improvement Program. Multifactorial risk index for predicting postoperative respiratory failure in men after major noncardiac surgery. Ann Surg 2000;232:242–253.
61. Arozullah AM, Khuri SF, Henderson WG, Daley J, for the participants in the National Veterans Affairs Surgical Quality Improvement Program. Development and validation of a multifactorial risk index for predicting postoperative pneumonia after major noncardiac surgery. Ann Intern Med 2001;135:847–857.
62. Webster C, Neumayer L, Smout R, et al, and participants in the National Veterans Affairs Surgical Quality Improvement Program. Prognostic models of abdominal wound dehiscence after laparotomy. J Surg Res 2003;109:130–137.
63. Arozullah AM, Henderson WG, Khuri SF, Daley J. Postoperative mortality and pulmonary complications rankings: how well do they correlate at the hospital level? Med Care 2003;41:979–991.
64. Arozullah AM, Ferreira MR, Bennett RL, et al. Racial variation in the use of laparoscopic cholecystectomy in the Department of Veterans Affairs Medical System. J Am Coll Surg 1999;188:604–622.
65. Collins TC, Johnson M, Daley J, et al. Preoperative risk factors for 30-day mortality after elective surgery for vascular disease in Department of Veterans Affairs hospitals: is race important? J Vasc Surg 2001;34:634–640.
66. Collins TC, Johnson M, Henderson W, Khuri SF, Daley J. Lower extremity nontraumatic amputation among veterans with peripheral arterial disease. Is race an independent factor? Med Care 2002;40: I-106–I-116.
67. Horner RD, Oddone EZ, Stechuchak KM, et al. Racial variations in postoperative outcomes of carotid endarterectomy. Evidence from the Veterans Affairs National Surgical Quality Improvement Program. Med Care 2002;I-35–I-43.
68. Axelrod DA, Upchurch GR, DeMonner S, et al. Perioperative cardiovascular risk stratification of patients with diabetes who undergo elective major vascular surgery. J Vasc Surg 2002;35:894–901.
69. Hur K, Hedeker D, Henderson W, Khuri S, Daley J. Modeling clustered count data with excess zeros in health care outcomes research. Health Services Outcomes Res Methodol 2002;3:5–20.
70. Fink AS, Campbell DA Jr, Mentzer RM Jr, et al. The National Surgical Quality Improvement Program in non-Veterans Administration hospitals. Initial demonstration of feasibility. Ann Surg 2002;236: 344–354.
71. Henderson WG, Khuri SF, Daley J, et al. The National Surgical Quality Improvement Program: demonstration project in non-VA hospitals. Top Health Inform Manage 2002;23:22–33.

8 Basic Statistical Methods

David Etzioni, MD, MPH, Nadia Howlader, MS, and Ruth Etzioni, PhD

CONTENTS

Statistics is the language of data and numbers. It is a framework to understand the variables, measurements, and outcomes of research. Without this framework, it would be impossible to interpret the findings of medical research.

As a field, statistics is mystifying to clinicians because medical education, for the most part, has failed to provide a solid biostatistical foundation. Although it is quantitative and mathematical, it is not an exact science. A good statistical analysis requires an understanding of the medical problem and knowledge of a broad range of statistical techniques. To determine the most appropriate statistical approach, it is important to clearly specify the outcome of interest and to understand the nature of this outcome, in terms of the type of data it represents.

This chapter provides a review of statistical techniques commonly used in medical research. We begin by describing a basic nomenclature for classifying data. Drawing on this taxonomy, we examine several of the most commonly used statistical tests for comparing groups. First, we consider statistics for comparing groups with regard to a single variable. Next, we briefly describe the use of multivariate statistics in the context of continuous outcomes (linear regression) and binary outcomes (logistic regression). We also cover some basic concepts in the analysis of failure time (survival) data that are also discussed more thoroughly in Chapter 9.

Our goal is to provide guidance to the clinical researcher in selecting, conducting, and interpreting statistical tests. There is not always a single correct test to use—in practice, several tests may be appropriate. We hope to help the reader understand what questions

From: *Clinical Research for Surgeons*
Edited by: D. F. Penson and J. T. Wei © Humana Press Inc., Totowa, NJ

he or she should be asking to determine an appropriate test—and what the limitations of that test may be.

1. UNDERSTANDING DATA

1.1. Qualitative and Quantitative Data

Qualitative data are data classified into discrete groups. There are two types of qualitative data: ordered and nominal. *Ordered (or ordinal) data* represent a spectra of classifications, such as degree of improvement (improved, same, worse) or agreement (strongly agree, agree, unsure, disagree, strongly disagree). *Nominal data* represent data in groups that have no clear order relative to each other. Examples of nominal data include blood type (A, B, O), race (white, black, Asian, Hispanic), or geographic area (Northeast, Midwest, South, West). Dichotomous data are subgroups of nominal data in which there are only two choices; for example, gender (male, female) or vital status (alive, dead).

Quantitative data are reported in distinct units of measurement. There are two main categories: continuous and discrete. *Continuous data* represent real numbers in which there *are* intermediate values. Examples of continuous data include age, temperature, weight, and height. *Discrete data* are similar to continuous data except there are *no* intermediate values. Examples of discrete data include number of previous admissions, number of prior operations, number of family members with a specific disease, and number of comorbidities.

1.2. Dependent and Independent Variables

The dependent variable in an analysis is generally the outcome of interest. It is also often referred to as the response variable. Independent variables are the factors that are varied with the goal of determining the effect of that variation on the response variable. For example, in a comparison of radical prostatectomy versus external beam radiation therapy among localized prostate cancer patients, the dependent or response variable might be time to prostate-specific antigen recurrence, whereas the independent variables would include treatment and possibly also other factors such as disease grade, patient age, and race. The nature of the dependent variable (qualitative, quantitative, discrete, or continuous) will often determine which statistical analysis is appropriate for the problem.

1.3. Dependent and Independent Observations

Most statistical analyses assume that observations are independent. This condition is satisfied, for example, when the observations come from distinct individuals that do not have any relation to or influence on each other. However, it is not satisfied when the data consist of multiple measurements per individual, or the measurements are clustered, as might be the case when patients from several facilities are included in a single dataset.

1.4. Cross-Sectional and Longitudinal Data

Cross-sectional data consist of measurements that are taken at either a single point, or at distinct points, with the observations at different times being independent of each other. For example, CA-125 levels at diagnosis among a group of ovarian cancer patients would constitute a cross-sectional dataset. Longitudinal data, on the other hand, are measurements taken at different points in time that are not independent. Serial CA-125 patients taken on ovarian cancer patients would be an example of longitudinal data. The serial observations on a single woman would not be independent because they are from the same individual. These observations would tend to be more similar to each other than observations from another patient.

1.5. Features of the Data

We generally use two classes of descriptors (descriptors of the center also known as central tendency, and descriptors of the spread often measured as the variance, standard deviation or standard error) to summarize the distribution of the data. Another term for the center of the distribution is its location (on the number line). The spread of the distribution quantifies the uncertainty in the data. A large spread indicates a highly variable, highly uncertain dataset. Another way of thinking of location versus spread is as signal versus noise. A large spread is synonymous with a high level of noise in the data.

1.5.1. Descriptors of the Center

1.5.1.1. Mean

The mean refers to the average value of a set of numbers. It is important, especially when testing a hypothesis that the means between two groups are different (or not different). The mean is highly susceptible to the influence of data elements with very high or low values (outliers).

1.5.1.2. Median

The median also describes the center, but uses a slightly different method. To derive the median, a dataset is ordered from least to greatest; the middle is the median. The median is also referred to as the 50th percentile of the data because half of the values lie below it and half above it.

1.5.1.3. Mode

The mode is the most common value in the dataset. If a histogram of the data is drawn, the mode is represented by the peak of the histogram.

1.5.2. Descriptors of the Spread

1.5.2.1. Variance

The variance is the standard descriptor of the "spread" of values in a dataset. It as calculated as:

$$Variance = \Sigma (x_i - x_{mean})^2/(n - 1)$$

$$x_i = individual\ observation$$

where x_{mean} = average value of the sample; n = sample size

Therefore, the variance is essentially the average of the squared deviations of the observations from the mean of the dataset. The denominator of the variance is $n - 1$ rather than n for technical reasons that amount to conferring favorable theoretical properties on the variance.

1.5.2.2. Standard Deviation

The standard deviation is the square root of the variance.

1.5.3. Other Descriptors of the Distribution

1.5.3.1. Order Statistics and the Five-Number Summary

The order statistics are the percentiles of the distribution. The qth order statistic is the observation below which q percent of the data falls. The five-number summary consists of the minimum, maximum, median, and the 25th and 75th percentiles of the data. It is frequently used to summarize the distribution of the data.

1.5.3.2. Character of the Distribution (Modality and Symmetry)

Each set of data has a specific distribution in terms of the frequency with which specific values occur. Beyond the measures of center and spread described previously, we also characterize distributions in terms of their modality (number of peaks) and symmetry. A unimodal distribution has one peak. Symmetric distributions have equal probability on both sides of their center. To determine the character of the distribution, several graphical techniques are commonly used. These include histograms, stem-and-leaf plots, and box-and-whisker plots.

The most well-known statistical distribution, the normal or Gaussian distribution, is both unimodal and symmetric, with tails that descend exponentially to zero. However, many datasets are not normally distributed. For example, survival times (which are always non-negative) are generally not normally distributed. Similarly, health economic data are not normally distributed; for example, monthly medical care costs among cancer patients show a distinctly non-normal distribution with an extreme right skew and a mass of observations at zero, representing months in which no services were used *(1)*.

Before conducting any statistical analysis, the distribution of the dependent outcome variables should be determined. Different statistical techniques will be appropriate depending on this distribution. However, the normal distribution plays a central role in statistical analysis because the average of a group of numbers tends to be normally distributed. This result, termed the *central limit theorem*, is one of the most important results in all of statistics. It enables us to use a simple test, based on the normal distribution, to compare the means of different samples, regardless of the distributions of those samples. In the next section, we describe tests for comparing sample means. First, we outline the testing framework and its rationale.

2. STATISTICAL HYPOTHESIS TESTS

One of the most common exercises in medical research is to compare two (or more) groups of individuals (treated vs untreated; older vs younger) to determine whether an outcome or response of interest differs between the groups. As an example, consider the duration of response to therapy in a clinical trial of two anticancer drugs. Suppose that each group consists of 20 individuals. In the first group, which receives the standard treatment, the average response duration is 3 mo; in the second, which receives a novel treatment, it is 6 mo. Can we conclude from this result that the new drug is better than the standard? The answer is that it depends—on the uncertainty, or variance, in each of the groups. Statistical hypothesis testing provides a framework for quantifying this uncertainty.

2.1. Why Is It Important to Quantify Uncertainty in Conducting Statistical Hypothesis Tests?

It is important because even an apparently compelling result could have arisen by chance because of random variation. In the previous above, it is possible that the drugs are equally effective in inducing tumor response, but because of random variation, or "bad luck," the subjects in the first group happened to experience a shorter response on average than those in the second group. How likely is it that equally effective drugs could

Table 1

Sample Size Per Group	*p Value*	*Power*
5	0.37	16%
10	0.20	27%
20	0.07	48%
50	0.003	85%
100	0.000	99%
1000	0.000	100%
1000*	0.02*	61%*

*Power and *p* value computed to detect a 0.5-mo difference in response duration.

have produced the observed result by chance? The answer to this question is termed the *p value*.

2.1.1. What Is the *p* Value?

The *p* value is the central result of any statistical hypothesis test. The *p* value represents the likelihood that the observed results could have arisen by chance. Equivalently, the *p* value represents the probability that a conclusion of a difference between the groups under study is erroneous. A high *p* value indicates that we have relatively low confidence in making this conclusion; conversely, a low *p* value indicates a high degree of confidence in the result. By general consensus, a *p* value of less than or equal to 0.05 is considered the threshold to consider results as being "statistically significant." This equates to a 1 in 20 likelihood that the results would have occurred by chance. Why 0.05? Why not 0.10 or 0.01? The decision was originally arbitrary, but has become accepted as a research standard. Strict adherence to this threshold may result in under- or overemphasizing findings. The actual difference (in terms of research results) between a *p* value of 0.049 and 0.051 may be very small. Moreover, as we will see in the next section, a statistically significant *p* value may not be synonymous with a clinically significant difference between groups.

2.1.2. How Does the *p* Value Relate to the Uncertainty in the Data?

The *p* value is a direct consequence of the uncertainty or variability in the data. A large amount of uncertainty will generally produce a high *p* value, whereas a small amount of uncertainty will produce a low *p* value. In comparing the means of two samples, however, it is the uncertainty in the sample *means* that directly influences the *p* value. The uncertainty in the sample means is not the same as the uncertainty in the samples. Suppose that the sample variance is s^2 and the sample size is n; then the variance of the sample mean is s^2/n. Therefore, the uncertainty in the sample mean depends on both the uncertainty in the sample and the sample size. The uncertainty in the sample mean decreases as the sample size increases. Therefore, when comparing two sample means, larger sample sizes will generally lead to smaller *p* values. Table 1 illustrates this phenomenon. The table shows the *p* values that arise when comparing two samples with sample means equal to 3 and 6, variances equal to 5, and different sample sizes.

2.2. *What Is Power?*

Table 1 illustrates the general principle that larger sample sizes tend to produce statistically significant tests more readily than smaller samples. We say that the *power* of the test to detect a specified difference between groups increases with the sample size. In designing studies, the sample size is often selected so as to achieve a specific level of power, such as 80–90%. Suppose the power of a study to detect a difference *d* between two sample means is 80%. This means that if there is truly a difference *d* between the sample means, there is an 80% chance that the study will yield a statistically significant result. However, the sample size required to achieve this level of power may, in some cases, be prohibitive. Therefore, study design generally requires a careful balance of cost and power considerations. The last column of Table 1 gives the power corresponding to the sample means 3 and 6, a variance of 5 in each sample, and the various sample sizes given in the table.

2.2.1. Is There Such a Thing as Too Much Power?

Consider the second to last row of Table 1. The power to detect a 3-mo difference in the average response duration is 100% for the given sample size. Thus, with this sample size, it is practically certain that the study will yield a significant result and a conclusion that the second drug is superior to the first. Indeed, a 3-mo difference is highly clinically significant, when the average response time under the standard treatment is only 3 mo—this difference represents a 100% increase in the average response duration. However, it turns out that even for a 0.5-mo difference, this sample size yields a power of 61% (last row of Table 1). Thus even if there is only a 2-wk difference in average survival, a very large study may well yield a conclusion that the new treatment is superior to the standard treatment. If a 2-wk improvement is not clinically significant, then this represents a setting in which statistical significance and clinical significance are not synonymous. This example represents the following general principle: In a sufficiently large study, even small differences between groups will be statistically significant, even though they may not be clinically significant. Therefore, statistically significant results in large studies should always be confirmed for clinical significance. Although it may be tempting to "overpower" a clinical trial, the additional costs to the study and risks to the subjects are important considerations that may be limiting. Indeed, it would be considered unethical by most Institutional Review Boards to plan a clinical trial, associated with significant risks to subjects, for a clinically insignificant difference.

3. COMMONLY USED STATISTICAL TESTS FOR COMPARING SAMPLE MEANS

In this section, we describe some commonly used tests for comparing sample means. Unless otherwise stated, these tests apply to settings where there are independent observations, such as might be obtained from a cross-sectional study with one observation per individual. Options for testing fall into two broad categories: parametric and nonparametric *(2)*.

3.1. *Parametric Tests*

Parametric tests make assumptions about the distribution of the sample data. The most frequent assumption is that the data are normally distributed. In contrast, *nonparametric tests* do not make assumptions about the distribution of the data. If the assumptions made are valid, then the parametric test will tend to be more powerful than the nonparametric

Table 2
Data on Polyp Size by Gender

Patient No.	*Polyp Size (mm)*	*Gender*	*Age (yr)*
1	6	Female	49
2	15	Male	54
3	6	Male	53
4	6	Male	51
5	7	Female	48
6	16	Male	58
7	11	Female	53
8	13	Male	52
9	5	Female	47
10	6	Female	44
11	7	Female	45
12	14	Male	57
13	8	Female	51
14	4	Male	45
15	8	Female	55
16	12	Male	49
17	10	Male	53
18	12	Female	56
19	4	Female	46
20	9	Male	50

test. However, if the assumptions are not valid, then parametric tests may lead to biased results.

The *t-test* is the most commonly used parametric test. It is designed to be applied to continuous data that have a normal distribution. If the outcome variable is non-normally distributed, there may be a transformation that yields a normal distribution, but it is important to bear in mind that results will eventually have to be back-transformed to the original scale for interpretation purposes. Different versions of the *t*-test are appropriate depending on whether the sample variances are the same in the groups being compared or whether they are different. As an example, consider conducting a *t*-test to compare the sizes of polyps removed by colonoscopy from men vs women. The data are presented in Table 2.

The males have polyps with a mean size of 10.5 mm and a standard deviation of 4.2 mm (variance = 17.4). The females have a mean of 7.4 mm and a standard deviation of 2.5 mm (variance = 6.3). Histograms of the observations within each group are presented in Figure 1. The histograms show that the data appear normally distributed. In this case, the variances appear to be different (17.4 vs 4.2). Running the test under an assumption of unequal variances yields a p value of 0.062. Under an assumption of equal variances, the resulting p value is 0.059. Although both results are close, in some studies, it may be the difference between reporting a significant versus a marginally or insignificant p value. The more conservative approach would be to use the assumption of unequal variance unless there is clear evidence of equal variance.

The *paired t-test* is employed when there is a pair of observations on each subject. This test computes a test statistic based on the difference in value between the two observations

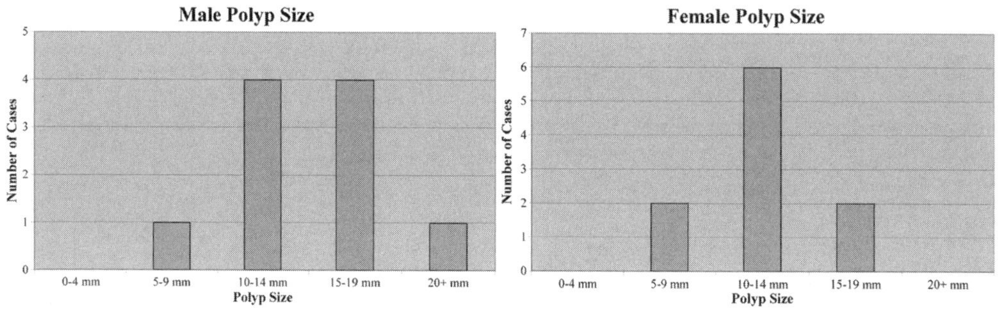

Figure 1: Histograms of polyp sizes by gender.

of each subject. Often, this test is used to compare outcomes before and after an intervention. Consider the example data presented in Table 3. We recently instituted an educational program stressing the importance of dietary fiber on cancer prevention and overall health. The research question is whether or not the fiber intake of our subjects increased during our study. The *t*-test in this case analyzes whether the values in the difference column are statistically different from zero. The *p* value is 0.038. What if we conducted this test as a standard (nonpaired) *t*-test? The resulting *p* value would be 0.37. As one can see, in this type of analysis the paired *t*-test can be a much more powerful test.

3.2. Nonparametric Tests

There are many different nonparametric approaches that can be used when comparing two groups of data. A comprehensive description of the entire range of nonparametric tests is beyond the scope of this chapter. We will briefly describe two of the most common approaches here and apply them to the polyp size data described earlier.

3.2.1. Mann–Whitney U Test

The Mann–Whitney U test, which may also be referred to as the Wilcoxon–Mann–Whitney test or the Wilcoxon rank-sum test, examines both parts of the dataset as a combined dataset. This can be considered the nonparametric equivalent to a *t*-test. In the Mann–Whitney test, each value in the combined dataset is ranked from 1 to 20. For example, rank number 1 would be assigned to the lowest value in the combined dataset (4 mm, for Patients 14 and 19); in practice, each of these cases would be assigned a rank of 1.5 (average of ranks 1 and 2) because they are tied observations. Rank number 20 would correspond to the highest value (16 mm, for Patient 6 in the closed group). The U test then analyzes whether lower ranked observations are more frequently found in one group than in the other. The results are shown in Table 4. The z-score value is –1.672 with a corresponding *p* value of 0.09; remember, this probability is extracted from the table of z-scores. Because the result of this computation shows a z-score with such a high probability ($p > 0.05$), we must accept that the test cannot show that ranks for the male group differ from the ranks for the female group. From a research perspective, we conclude that the data do not show evidence that the polyp sizes do not differ by gender.

3.2.2. Kruskal–Wallis Test (H) Test

The H test is similar to the U test. In this variation, the datasets are once again combined and ranks are assigned. The ranks for each group are then totaled and weighted according

Table 3
Fiber Intake Before and After an
Educational Program Stressing the Importance of Dietary Fiber

Patient	*Fiber Intake*		*Difference in Fiber Intake*
	Before Education	*After Education*	
1	27	31	4
2	15	20	5
3	23	23	0
4	24	22	–2
5	25	23	–2
6	14	10	–4
7	18	20	2
8	12	13	1
9	31	37	6
10	26	33	7
11	30	35	5
12	16	13	–3
13	22	20	–2
14	25	26	1
15	22	30	8
16	15	18	3
17	23	20	–3
18	13	12	–1
19	25	33	8
20	11	19	8
Average	20.85	22.90	2.05
Standard deviation	6.11	8.05	4.11

Table 4
Mann–Whitney U Test Applied to the Polyp Size Data

Gender	*n*	*Rank Sum*	*Expected*
Female	10	83	105
Male	10	127	105
Combined	20	210	210

z = –1.672
p value = 0.09

to the number of observations in each group. The H test value will be higher if there are groups with a disproportionately large share of higher ranked observations. The advantage of the H test is that it can be used to compare results across more than two groups (the U test examines differences between two groups).

Both the Mann–Whitney U test (also called the Wilcoxon test) and the Kruskal–Wallis test are designed to be applied to independent groups. As with the paired *t*-test, there also exists a nonparametric test for the paired setting, the Wilcoxon signed-rank test.

Table 5
Wilcoxon Signed-Rank Test Applied to the Dietary Fiber Data

Sign	*n*	*Sum Ranks*	*Expected*
Positive	7	55	104.5
Negative	12	154	104.5
Zero	1	1	1
All	20	210	210

z = –1.852
p value= 0.06

3.2.3. Wilcoxon Signed-Rank Test

In the Wilcoxon signed-rank test *(3)*, the second observation within each pair is subtracted from the first observation in the pair. The absolute differences are then ranked and assigned a negative or positive sign according to the sign of the difference. The test compares the sum of the negative ranks with the sum of the positive ranks. The result of the Wilcoxon signed-rank test using the fiber data is in Table 5. In this example, the Wilcoxon two-sample paired signed-rank test here is used to test the null hypothesis that the population median of the paired difference of the two samples is 0. Because we have a $p > 0.05$, we conclude that the fiber intake the study subjects did not change during the study.

4. COMPARING PROPORTIONS

When considering a binary outcome (e.g., success/failure, response/nonresponse), comparison of proportions is often of interest. For example, suppose an outpatient clinic offers two types of colorectal cancer screening tests—namely, fecal-occult blood testing and flexible sigmoidoscopy (FS). An investigator might wish to compare the proportion of men and women selecting FS vs fecal-occult blood testing. A set of hypothetical data is presented in Table 6. Compared with women, men appear to have preferentially used sigmoidoscopy over fecal occult blood testing. Of 102 men, 58 used FS (56.8%), whereas of 105 women, only 45 selected FS (42.9%). Is this difference statistically significant?

There are at least two ways to test for comparability of proportions. The first approach is to observe that a proportion (e.g., the proportion of men using FS) is actually a sample mean. If we represent the observation for each man as a zero (fecal-occult blood testing) or a one (FS), the proportion selecting FS is simply the average of the sample data. Similarly, the proportion of women using FS is an average of the observations for women, in which each observation is either a zero or a one, depending on the screening test used. Recall that by the central limit theorem, sample averages are normally distributed so long as the sample size is sufficiently large. In this case, the sample size (>100 for both men and women) is considered sufficiently large to apply the central limit theorem. Therefore, in this case, we can use the standard test for comparing proportions across groups, which is based on the normal distribution. The *p* value for this test is less than 0.05, which

Table 6
Colorectal Cancer Screening Test Preferences by Gender

	Colorectal Cancer Screening Tests: Number of Times Selected by Gender		
	Fecal Occult Blood Test	*Sigmoidoscopy*	*Total*
Males	44	58	102
Females	60	45	105
Total	104	103	207

Table 7
Results of Chi-Square Test Applied to Colorectal Cancer Screening Data

	Colorectal Cancer Screening Tests: Number of Times Selected by Gender		
	Fecal Occult Blood Test	*Sigmoidoscopy*	*Total*
Males	44 (51.2)	58 (50.8)	102
Females	60 (52.8)	45 (52.2)	105
Total	104	103	207

indicates that there is a statistically significant difference in the frequency with which each test is used by men and women.

The second approach is the chi-square test. The chi-square test considers the entire two by two tables of the data, and asks the following question: Can we assume that gender and test selection are statistically independent? By statistically independent we mean that the probability of choosing FS is the same regardless of gender. What would we expect the data to look like under independence? The answer is that we would expect the same proportion of men and women to have had FS. Overall, 103/207 individuals (50%) had FS. Therefore, we would expect 51 men (50%*102 = 51) and 52 women to have had FS. These are the *expected* counts under the assumption of independence.

The chi-square test computes the difference between observed and expected counts to yield the following chi-square test statistic:

$$\chi^2 = \Sigma \text{ (observed value} - \text{expected value)}^2/\text{expected value}$$

Results from Table 6, with the computed expected values in parentheses are presented in Table 7. The chi-square test statistic, based on the formulas presented previously, is 4.06; this value can be converted to a p value using a standard table of the chi-squared distribution. In this case, the p value is < 0.05 and we conclude that males and females tended to use different screening tests.

For large samples, the two approaches are equivalent; however, in small samples, they can give different results. In situations in which the numbers of observations in different groups are highly uneven, or when the count in any one of the cells is less than 5, a variation of the chi-square test called Fisher's exact test is usually used.

5. REGRESSION ANALYSIS

All of the between-group comparisons we have discussed can be thought of as analyses to determine whether there is any association, or dependence, between an outcome and an independent variable. For instance, in the example that compared tumor response times across treatments, one could consider this as an analysis to determine whether tumor response time depended on treatment received. However, these comparisons are limited because they only describe the association between a single independent and a single dependent variable (bivariate analyses). Oftentimes, in clinical science, we wish to adjust for the effect of other covariates that are known to be influential to the dependent variable (multivariable analyses).

Regression analysis is a general approach for determining whether there is an association between a dependent variable and one or more independent variables. Regression analysis generalizes the between-group comparisons that we have already discussed in a number of ways. First, rather than restricting comparisons to two (or possibly more) discrete groups, we can now consider associations between a response variable and a continuous independent variable. Thus, for example, we could ask whether tumor response time increases with dose of a particular drug, where dose might take on a full range of values.

The regression framework allows one to include multiple independent variables (and even response variables, although discussion of this is beyond the scope of the present chapter). Thus, for example, we could ask whether tumor response time depends on treatment dose and other clinical factors such as comorbidity and age at diagnosis. Consideration of multiple independent variables is important because there may be associations between these variables that could affect the primary association of interest. In the drug dose-tumor response time example, suppose we find that dose is associated with tumor response time; patients receiving higher doses of the drug do tend to have longer response times. Now, suppose that comorbidity is also associated with drug dose; individuals with greater comorbidity tend to be able to tolerate lower doses of the drug. If greater comorbidity is independently associated with poorer tumor response, then the apparent benefits of higher drug doses may simply be an artifact, explained by the associations between comorbidity, drug dose, and the outcome. In statistical terms, the comorbidity variable would be termed a "confounding factor." A confounding factor is an independent variable that is associated with both the primary independent variable of interest and, independently, with the response variable. Inclusion of potential confounding factors in any regression analysis is important because they may explain some or even all of the apparent association between the primary independent variable of interest and the response variable.

Standard regression techniques are linear in that they model the association between the outcome variable and the independent variables as a linear relationship. What this means is that each unit increase in the independent variable is assumed to induce the same change in the dependent variable. Alternatives to linear regression include more flexible techniques such as smoothing functions and generalized additive models; however, these are beyond the scope of the present chapter.

5.1. Linear Regression

Linear regression is appropriate for outcome variables that are continuous and approximately normally distributed, and observations that are independent. For data

Table 8
A Sample Regression Analysis Dataset:
Procedure Duration by Gender and Body Mass Index (BMI)

Patient No.	*Duration of Operation (min)*	*BMI (kg/cm^2)*	*Gender*
1	152	28	Male
2	120	25	Female
3	150	31	Female
4	141	25	Male
5	175	38	Male
6	125	26	Male
7	155	32	Male
8	144	27	Female
9	130	24	Male
10	136	29	Female
11	140	31	Female
12	160	33	Female
13	170	35	Female
14	140	28	Female
15	145	29	Male
16	139	26	Male
17	141	30	Male
18	135	30	Female
19	165	34	Male
20	138	24	Male

that are not normally distributed, there are a number of options. The simplest approach is to use a transformation to convert the outcome data to be approximately normally distributed. Another alternative is to consider the class of generalized linear models *(4)*, which provides regression approaches for a large family of statistical distributions. For observations that are not independent, the standard linear regression approach has been extended to account for correlated observations. Generalized estimating equations *(5)* and generalized linear mixed models are classes of techniques for regression analysis of correlated data. These are available in a number of statistical software packages, such as SAS (version 8.2, SAS Institute Inc., Cary, NC), Splus (version 6.1.2, Insightful Corporation), and STATA (version 7, Stata Corporation, www.stata.com).

To illustrate the basic steps in a standard linear regression analysis, consider the following hypothetical example in which we are investigating the effect of patient body mass index (BMI) and gender on how long it takes to perform a rectal resection. In this example, the outcome (dependent variable) is duration of operation, and the predictor (independent) variables are BMI and gender. Our hypothetical dataset is presented in Table 8. To verify that the outcome variable (duration of operation) is normally distributed, we use a simple histogram. This is presented in Figure 2. In this case, it appears that the outcome data are reasonably normally distributed. To conduct a linear regression analysis, we can use any one of a number of computerized statistical software packages to run this type of a regression analysis. The output from this analysis would look something like the data presented in Table 9.

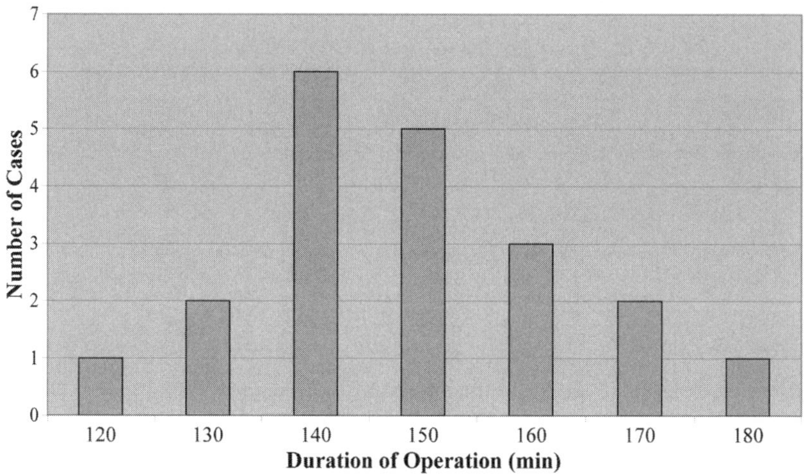

Figure 2: Histogram of procedure duration data.

Table 9
Results of Linear Regression Fit to Procedure Duration Data

Variable	*Parameter Estimate*	*Standard Error*	*t Value*	*p Value*
Body mass index	3.32	0.43	7.65	<.0001
Gender				
Female	—	—	—	—
Male	5.96	3.27	1.82	0.086
Model r^2	0.78			

In interpreting the model results, the following questions are of interest.

1. Is the model appropriate for the data? There are several components to this question, including: (1) Is the normality assumption satisfied? (2) Does the assumption of linearity hold? (3) Does the model provide a reasonable explanation for the data? To address items (1) and (2), one can use any of a number of techniques that are based on the *residuals* from the fitted model; the residuals are the observed values for the outcome variable minus the expected values predicted by the model.

   To address item (3), we look at the r^2 value. This number represents the amount of variation in the data that is explained by the predictor variables. In this example, the r^2 tells us that we can account for approx 78% ($r^2 = 0.78$) of the variation in duration of operation based on the values of our predictor variables (BMI and gender). When there is one predictor variable, the r^2 value is a direct function of the correlation between the predictor variable and the response variable.
2. What do the results tell us about the relationship between each independent variable and the outcome variable? Generally we are concerned with (1) whether the relationship is statistically significant (or, what is the p value associated with a specific variable) and (2) the magnitude and direction of the association (in particular, is it clinically significant?).

In this case, BMI has a p value less than 0.05, and a "parameter estimate" of 3.32. This p value is therefore interpreted as being statistically significant. The parameter estimate of 3.32 has a specific interpretation as well. It reflects how much longer we would expect an operation to take for each unit increase in BMI, namely 3.32 min. A negative value for this coefficient would suggest that operation times tend to decline as BMI increases.

Our model also evaluates the effect of male vs female gender using females as the reference group. In the analysis, male gender had a p value of 0.086, which is not usually reported as being statistically significant. Thus, in this model, we are unable to show a statistically significant relationship between gender and duration of operation. This may be due to low power for the male versus female comparison. However, the parameter estimate (approx 6 min longer for males) does not appear to be clinically significant. Therefore, the lack of a statistically significant result appears to be because (suggested by the data) if a difference exists in the duration of operations for males and females, it is likely to be small.

In practice, there may be many independent variables to be considered in the analysis. A standard approach is to first estimate the regression coefficients and p values for each independent variable separately. Those variables that are not associated with the independent variable in the individual analysis are generally not considered further unless they are potential confounding variables. To assess whether predictor variables might be potential confounders, a correlation analysis on the predictors is useful. A high positive (close to 1) or negative (close to –1) correlation between two predictors shows that they covary. Beyond knowledge of potential confounding effects, it is important to know how predictors covary, because if two predictors are very highly positively or negatively correlated, they may be almost synonymous in the model. In this case, it is wise to include only one of the synonymous predictors otherwise the model can become mathematically unstable. This is also important in the interests of parsimony. In other words, we often wish to report the statistical model with the fewest number of explanatory factors while still explaining most of the variance in the observed outcomes.

When there are many predictor variables, and relatively few observations, it may not be feasible to fit all the predictors in the model at one time. Insufficient observations relative to the number of variables can lead to overfitting as well as mathematical instability. In such a case, after preliminary analyses have yielded a set of predictors of interest, an automated procedure such as stepwise regression can be used to derive an optimal, parsimonious model.

5.2. Logistic Regression

In linear regression we model the impact of specific predictor variables on a *continuous* outcome variable (duration of operation). Logistic regression *(6)* is a technique for modeling the association between predictor variables on a binary outcome. A classic example of a binary outcome is the occurrence (or nonoccurrence) of an event of interest, in which a 1 denotes that the event occurred and a 0 denotes that it did not. In our hypothetical example below, the event of interest is wound infection after surgery.

To illustrate a logistic regression analysis we will use a hypothetical example of a study examining whether gender and age are related to the likelihood of having a postoperative wound infection. Our dataset contains 100 patients (53 women and 47 men); the actual dataset is not included here.

Let's first perform some basic analyses on our data. First, we'll look at what proportion of male vs. female patients had a wound infection:

Gender	*Wound Infection Rate*
Female	5/53 (9.4%)
Male	13/47 (27.7%)

It appears that men are considerably more likely to have a wound infection. Similarly, we now consider the average age of patients that had a wound infection vs those that did not.

Wound infection status	*Average age (yr)*
Wound infection	56.4
No wound infection	53.0

Based on this table, it appears that those patients with wound infections were older than those that did not. Thus both age *and* gender appear to be related to the likelihood of a wound infection.

We can now conceptualize a model in which both gender and age are potential risk factors for having a postoperative wound infection. In this analysis, the outcome variable was whether or not a wound infection occurred (yes or no); predictor variables are gender and patient age. Results are presented in Table 10. The logistic regression model tells us whether or not there is a statistically significant relationship between the likelihood of wound infection and the two predictors, gender and age. In this case, age does not appear to be an important predictor; the p value is 0.22. However, the p value for gender is statistically significant ($p = 0.017$). As with our previous example, the results are reported only for male gender; this is because we are comparing males to females. In this comparison, females are the reference group.

As with a linear regression, logistic regression analysis produces coefficients, or parameter estimates, which can then be used to determine the magnitude and direction of the association between the outcome and the predictor variables. However, for interpretation purposes, it is necessary to transform the coefficient estimates. For a given predictor with a parameter estimate b, exp(b) is interpretable as the increase in the odds of the event of interest for each unit increase in the predictor. Thus the model estimates above imply that for every year that age increase, the likelihood of a wound infection increase by 2.4%. Moreover, the likelihood of a wound infection for males is 4 times that for females. In this case, the female group is being used as the "reference" group, hence it has an odds ratio of 1.0 by definition but this is not typically reported in results tables by convention.

Logistic regression is subject to many of the same considerations as linear regression including model appropriateness, confounding, and the ratio of sample size to the number of predictors. To assess model appropriateness, a simple approach consists of discretizing the predictor variables that are included in the final model so as to create a relatively small number of categories for the observations in the data. Then, within each category, use the logistic regression model to predict the probability of the event of interest. The category-specific predictive probabilities can then be compared with the actual number of events within each category as a check of model adequacy.

In the simple case of a single binary predictor, and in large samples, logistic regression can be shown to be equivalent to the chi-square test and the standard test for comparing proportions described previously.

Table 10
Results of Logistic Regression of Wound Infection Status on Gender and Age

Variable	*Parameter Estimate*	*Odds Ratio (OR)*	*p Value*
Age	0.024	1.024	0.22
Gender			
Female	—	—	—
Male	1.40	4.05	0.017

6. CONCLUSIONS

This chapter has provided a catalog of basic statistical methods and concepts the surgeon scientist can use to decide which method might be most useful for the study he or she is performing.

Although we have summarized a broad array of methods, we recognize that there are many methods that we have not covered that go beyond the basic approaches presented here. We have tried to alert the reader to some of the most widely used approaches that may be appropriate when standard techniques are not applicable. These include mixed-effects modeling or generalized estimating equations for regression analysis with clustered or longitudinal data *(5)*, generalized linear models, when the dependent variable is not normally distributed *(4)*, and generalized additive models when the assumption of linearity is not satisfied. In survival analysis (*see* Chapter 9), we note that there are techniques for analyzing survival data when the proportional hazards assumption is not satisfied, and in the presence of competing risks that may not be independent of the risk of the event of interest. There is also a whole subfield of survival analysis dedicated to the analysis of multiple- or recurrent-event survival data, which occurs when the event of interest can be experienced more than once. Examples of recurrent survival data include the frequency of infections after bone marrow transplantation and the number of prostate-specific antigen tests conducted after a diagnosis of prostate cancer. In addition, we have not been able to cover the theory and application of diagnostic testing, much of which is discussed in Chapter 10. In conclusion, every statistical method is predicated on a set of assumptions, and it is critically important to ask the right questions to determine whether the assumptions underlying one's method of choice are satisfied.

REFERENCES

1. Etzioni R, Ramsey SD, Berry K, Brown M. The impact of including future medical care costs when estimating the costs attributable to a disease: a colorectal cancer case study. Health Econ 2001;10: 245–256.
2. Rosner B. Fundamentals of biostatistics. 5th ed. Pacific Grove, CA: Duxbury Press, 1999.
3. Conover WJ. Practical nonparametric statistics. 3rd ed. New York, NY: John Wiley, 1998.
4. McCullagh P, Nelder JA. Generalized linear models. 2nd edition. London, UK: Chapman and Hall, 1999.
5. Diggle PJ, Heagerty P, Zeger SL, Liang KY. Analysis of longitudinal data. Oxford, UK: Oxford Press, 2002.
6. Hosmer DW, Lemeshow S. Applied logistic regression. 2nd ed. New York, NY: John Wiley & Sons, 2000.

9 Survival Analyses

Rodney L. Dunn, MS *and John T. Wei,* MD, MS

Contents

INTRODUCTION

Survival analyses are used in many areas of surgical clinical research. As the name implies, they were initially developed in the analyses of survival after intervention to compare how long the average patient survived after having different interventions. The group of techniques commonly known as "survival analyses" was developed explicitly to address the unique features of this type of data. But what makes survival data unique? Why can't other commonly used analysis methods be used instead, such as the *t*-test, linear regression, or logistic regression? The answer is that survival data often include partial information on a subject. For example, a subject may enter a study that uses a survival outcome and still be alive 5 years later when the study closes. We know that this subject survived for at least 5 years after the surgery, but we don't know whether this subject eventually died the next day or 20 years later. Should this patient be discarded from the analysis dataset just because we don't know the exact date of death? If so, we would be throwing away valid, useful information obtained on that subject over the 5-year period. Survival analyses allow this partial information to be included in addition to data from subjects who have full information available (i.e., reached the study end point). These partial data are referred to as *censored* data and it is this aspect of the data that makes survival analyses unique.

Despite the commonly used terminology of "survival" analyses, there are many other applications of these techniques. One example of this is measuring the time until a biomarker reaches a certain level. A more specific example of this is from the prostate cancer literature, where time from surgery until a detectable prostate-specific antigen (PSA) level is a common end point. An example of another type of outcome is the time from when surgery is justified until the time the surgery is actually performed. Sometimes, the defined outcomes for survival analyses are combinations of two or more other outcomes. Recurrence-free survival is measured as the time from surgery or some other baseline until either recurrence or death, whichever happens first.

From: *Clinical Research for Surgeons*
Edited by: D. F. Penson and J. T. Wei © Humana Press Inc., Totowa, NJ

2. WHEN IS IT APPROPRIATE TO USE TIME-TO-EVENT ANALYSES?

An outcome is appropriate for survival analyses when it measures the time from some defined baseline event (e.g., surgery, birth, study entry) until the time that the outcome of interest occurs (e.g., death, recurrence, biomarker level). Because the outcome of interest does not necessarily need to be death, another common name for these types of analyses is "time-to-event" (TTE) analyses. Because this term is appropriate for all outcomes rather than a small subset of them, it is preferred and will be used in the remainder of this chapter.

3. CENSORING

As mentioned, the handling of censoring is what makes TTE analyses unique. This section will explore censoring in more detail. The following list gives several examples of censoring.

1. A subject does not incur the outcome of interest during the study period. In a study with death as the outcome, a subject still alive at the study's end would be considered censored.
2. A subject voluntarily withdraws from the study before the study protocol indicates follow-up should end; this is commonly known as "lost to follow-up."
3. In a retrospective database analysis, information on the outcome of interest is missing for a subject.
4. Information on when a subject entered a study is not available, but information on outcome of interest *is* available—an example of this is a study that wants to include all patients who had a particular surgical technique and has death as the outcome of interest; if a subject is known to have had the surgical technique but the exact date is not available, this subject is considered censored, even if that subject's date of death is known.
5. The exact date of the outcome of interest is not known, but a date when the outcome had not occurred yet is known and a date when the outcome had already occurred is known—this is fairly common in studies that use periodic scans or blood draws to measure the outcome of interest. In a study using bone scans to measure disease progression, with the bone scans taken every 6 mo, if a subject had not progressed at one reading and had progressed at the next reading, the subject is considered to be censored.

Censoring can take a variety of forms. The form of censoring that occurs most commonly is called *right censoring*. Examples 1, 2, and 3 above are examples of right-censoring. Right-censoring occurs when there are known time points at which the subject has not reached the event of interest, and hence the true interval of time between the start of the study and the outcome of interest occurred is not known. There are several reasons why this could happen. The outcome of interest could have happened, but the study investigators never learned about it. Or the outcome of interest may happen in the future, which is common when death is the outcome. In a longitudinal study, the outcome of interest may have occurred but never been properly recorded in the database. Whatever the reason, with right-censoring we do know that the patient made it until a certain point in time without having the outcome of interest and may or may not have had the event after that point. This partial information (i.e., the interval of time a subject "survived" without reaching the study end point of interest) can be used in TTE analyses.

Left censoring occurs less often and is when the time the subject entered a trial is unknown, but the subject has known follow-up times at which the status of the outcome of interest is known. Example 4 is an example of left-censoring. A common occurrence of left censoring will be when a study is focusing on the first time a surgical technique is performed, but the dataset containing information on a patient's medical history is incomplete and only contains information on the past several years. For example with low-stage bladder cancer, a transurethral resection of the bladder tumor (TURBT) is performed to obtain a tissue diagnosis and may be therapeutic. If a study wished to evaluate the duration of time from first TURBT until death from bladder cancer, the ideal dataset would include all TURBTs that the patient has ever had. However, it is often the case that the first TURBT may have been performed at another institution and that information was not available for analyses; hence, the first TURBT data point in the dataset may not actually be the first TURBT that the patient received. These incomplete data at the earliest initiation of followup would be considered left-censored.

Another form of censoring that is not seen commonly in the literature and may not always be adequately dealt with in the analysis is called *interval censoring*. Example 5 is an example of interval censoring. Interval censoring occurs when a date is known at which the subject had not reached the outcome of interest and a separate date is known at which the subject had reached the outcome of interest, but there is a large gap between the dates. This type of censoring essentially happens all the time when death is not the outcome of interest. For example, if PSA recurrence is the outcome of interest after prostate cancer surgery, the date of the latest PSA test taken that had a value considered to be not detectable (e.g., 0.2 ng/mL or less) is the left side of the interval and the date of the first PSA test taken that had a value considered to be detectable (e.g., >0.2 ng/mL) is the right side of the interval. The investigators have no idea whether the outcome of interest actually occurred the day after the first PSA, the day of the last PSA, or anywhere in between. Even if PSA values are taken frequently (e.g., once per week), there is a small interval in which the investigator is unsure of the PSA recurrence status. However, if the interval is small enough, ignoring the interval censoring will have minimal effects on the analysis. Interval censoring may become analytically problematic when the interval between assessments is large.

4. KAPLAN–MEIER PLOTS AND LOG-RANK TESTS

By far the most common way to illustrate the distribution of TTE data is to display the data using a Kaplan–Meier plot, often abbreviated as KM plot. These plots are recognized by their distinctive "step" design. Interestingly, the plots derive their name from the estimates that are displayed in the plots and not from the graphical presentation of the data. Kaplan and Meier, working independently, each arrived at the maximum likelihood estimate for the survival function. This estimate became known as the Kaplan–Meier estimate, and it is widely recognized as the best estimate of the survival function.

KM plots have a distinctive "step" design, as illustrated in Figure 1, which displays the survival rate after radical cystectomy for patients who had a lower pathologic stage after surgery (i.e., downstaged), had the same stage, or had a higher stage after surgery (i.e., upstaged). Typically, the risk at study onset is the same (e.g., all subjects are alive). As events (e.g., deaths) occur, the proportion of patients remaining in the study and still at risk decreases and thereby lowers the survival rate. KM plots are interpreted by the dispersion of the lines over time. Generally, lines that separate over the duration of follow up suggest a difference in survival between the groups. Overlapping lines suggest a lack of significant difference.

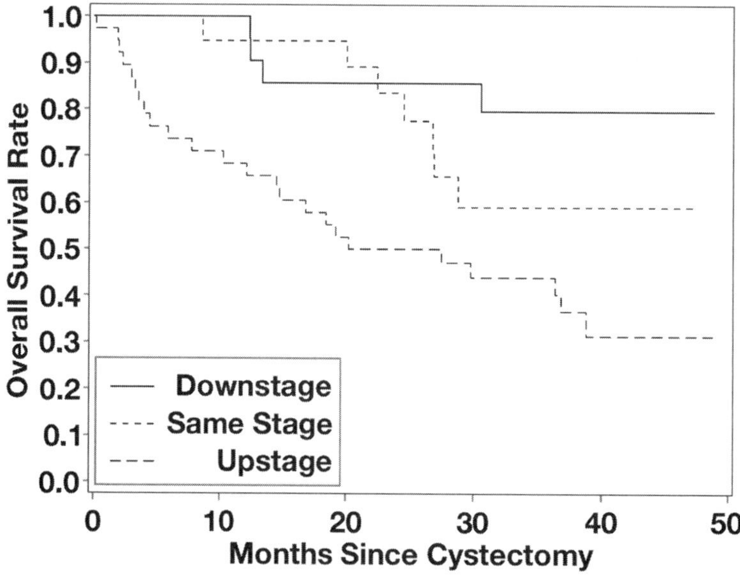

Figure 1: Kaplan-Meier plot.

Normally, line graphs for a statistical analysis will directly connect two points. However, the reason that KM plots have their distinctive steps is a direct result of the manner in which the KM estimates are calculated. The actual formula for the KM estimate is:

$$\hat{S}(t) = \prod_{j=1}^{k} \left(\frac{n_j - d_j}{n_j} \right)$$

In plain language, this formula has the following meanings:

- $\hat{S}(t)$ means that the survival estimate is a function of t, or time.
- $\prod_{j=1}^{k}$ means that the survival estimate is a product of the survival estimates calculated at each unique event time and not at time points at which only censored values occur. For example, in a study of a new drug for HIV/AIDS with an outcome of death, the 10 study subjects incurred events and censored times as indicated in Example 1 that follows, with their follow-up times ordered from shortest to longest. The survival estimate would be calculated only at 1, 4, 5, 7, and 10 mo. Thus the survival estimate is not calculated at time points at which only a censored event occurs (i.e., months 2, 3, and 8) and is only calculated once at time points at which a study end point occurs, even if multiple events occur at the same time.

Example 1

Study subject #	*001*	*005*	*006*	*008*	*007*	*002*	*003*	*004*	*009*	*010*
Duration of follow-up until death (mo)	1	2*	3*	4	4	5	5*	7	8*	10

* Indicates censored value.

- $\left(\frac{n_j - d_j}{n_j}\right)$ means that the survival estimate at each event is calculated as the number of subjects who "survived" that time point, divided by the number of subjects at risk before that time point.

Other important points to note about the survival estimate include:

- The survival estimate will always range from 1 (no events) to 0 (every subject has an event [e.g., all subjects have died]). From baseline until the first event time, the survival estimate will equal 1. Often, the range of the survival estimate will be reported as 0% to 100%.
- The survival estimate for a particular event time remains the survival estimate for all time points until the next event time. For example, if a study has event times of 10 and 14 days and the survival estimate at day 10 is estimated to be 60%, the survival estimate at days 11, 12, and 13 is also assumed to be 60%, even if a censored event occurs at day 12.
- The information from censored observations is used by reducing the number of subjects at risk at each event time larger than the censored time.

To help illustrate this further, the survival estimates based on Example 1 will be calculated. When doing this by hand, it is easiest to use a simple table. The table should initially look like the following.

Time Interval	n_j	d_j	c_j	$(n_j - d_j)/n_j$	$\hat{S}(t)$
0–<1					
1–<4					
4–<5					
5 <7					
7–<10					
10					

The table contains six rows because there are five unique event times (in a clinical trial, these are typically the scheduled study follow-up periods or windows), plus one row for the baseline survival estimate. The time intervals are defined from each event time until the moment before the next event time. The table contains columns for the number at risk during each time interval (n_j), the number of deaths during each time interval (d_j), the number of censored values during each time window (c_j), the formula for the survival rate during each time interval ($(n_j - d_j)/n_j$), and the overall survival estimate during each interval ($\hat{S}(t)$).

Now, start by filling in the first row of the table. It should look like this:

Time Interval	n_j	d_j	c_j	$(n_j - d_j)/n_j$	$\hat{S}(t)$
0–<1	10	0	0	1.000	1.000

In this initial time period, which begins at baseline and continues until the moment before the first event time, there are 10 subjects at risk, because we begin with our entire study population. There were no deaths during this interval, because the interval concludes immediately before the first event time, and there also happens to be no censored values. The survival rate is thus (10–0)/10, which equals 10/10, or 1.000. And because this is the first survival estimate, the overall survival estimate also is equal to 1.000.

Moving on to the second row of the table, we have the following.

Time Interval	n_j	d_j	c_j	$(n_j - d_j)/n_j$	$\hat{S}(t)$
1–<4	10	1	2	0.900	0.900

The number at risk during each time interval is equal to the number at risk for the preceding time interval minus the number of deaths and the number of censored values during the previous time window. Because there were no deaths or censored values in our first time interval, there are still 10 subjects at risk for the time interval (1-<4 mo). There was one death during this window (at 1 month), and there were two censored values (at times 2 and 3 mo). The survival rate for the interval is (10 – 1)/10, or 9/10, which equals 0.900. We then multiply this survival rate by the overall survival rate from the previous time window, which results in (0.900)*(1.000), which equals 0.900. Then we proceed to the next study period, which is filled in as follows.

Time Interval	n_j	d_j	c_j	$(n_j - d_j)/n_j$	$\hat{S}(t)$
4–<5	7	2	0	0.714	0.643

In this time interval, we subtract the 1 death and 2 censored values (from the previous time period) from the 10 subjects who were still at risk during the previous time period and get 10 – 1 – 2 = 7 subjects at risk (n_j) during this interval. During this interval, there are 2 deaths (each at time 4) and no censored cases. The survival rate is (7 – 2)/7, or 5/7 = 0.714. Finally, we multiply the new survival rate by the previous overall survival estimate, for (0.714)*(0.900) = 0.643.

Time Interval	n_j	d_j	c_j	$(n_j - d_j)/n_j$	$\hat{S}(t)$
5–<7	5	1	1	0.800	0.514

In this time interval, we subtract the 2 deaths from the 7 subjects who were still at risk during the previous time period and get 7 – 2 = 5 subjects at risk (n_j) during this interval. During this interval, there is a single death and 1 censored case (both occurring at time 5 mo). The survival rate is (5 – 1)/5, or 4/5 = 0.8. Finally, we multiply the new survival rate by the previous overall survival estimate, for (0.8)*(0.643) = 0.514.

Using the same methods, we can fill in the rest of the table, which ultimately looks like the following. It is highly recommended that you take the time to calculate these numbers to fully comprehend how they are derived.

Time Interval	n_j	d_j	c_j	$(n_j - d_j)/n_j$	$\hat{S}(t)$
0–<1	10	0	0	1.000	1.000
1–<4	10	1	2	0.900	0.900
4–<5	7	2	0	0.714	0.643
5–<7	5	1	1	0.800	0.514
7–<10	3	1	1	0.667	0.343
10	1	1	0	0.000	0.000

After completion of this particular study, all subjects have died and hence the survival function reaches zero. The completed table above is often referred to as an actuarial table.

Although this section is about how to create KM plots, we have already taken a great deal of space to describe the creation of the survival estimates and spent time to calculate the tables above on what is a very small and simple dataset without talking about the actual plots at all. Fortunately, with many statistical software packages, such as SAS, S-plus, or SPSS, simple code can be written in minutes that will create the necessary calculations of the survival estimates for you. For example, in SAS, if the data from Example 1 were already created in a dataset called KM, with SURVTIME representing the time to event and CENSOR indicating whether the value was censored (with 1 representing censored and 0 representing noncensored), the following code could be used to create the table above:

```
proc lifetest data = km;
  time survtime*censor(1);
run;
```

In SAS, the data will look like the output in Table 1, in which "survtime" is equivalent to the start of the Time Interval in our table, Survival is equivalent to the $\hat{S}(t)$ column in our table, and Number Left is equivalent to the n_j column. In addition, Failure is equal to 1-Survival and Number Failed is a cumulative version of the d_j column from our table. Finally, the Survival Standard Error is a measure of the amount of variance in the data, which is useful when calculating other survival statistics not covered in this chapter.

After the survival estimates are created, the plotting of these estimates to create a KM plot is fairly straightforward, especially if you remember that the survival rate ($\hat{S}(t)$) remains constant for each time interval, then for the next time interval, the survival rate ($\hat{S}(t)$) again remains constant, and so forth. It is this fact which leads to the "steps" in the plotted lines. To create the plot, use the following steps:

1. The survival estimate will go on the vertical (or *y*-) axis, and the time to event will go on the horizontal (or *x*-) axis.
2. Simply place points on the graph for each pairing of survival estimate ($\hat{S}(t)$) and beginning of each Time Interval.
3. Begin by starting at the left upper corner of the plot where the survival rate ($\hat{S}(t)$) = 1.0 and draw a flat line to the right for that entire Time Interval (in our example, that would be time 0 through 1 mo).

Table 1
SAS Output

The LIFETEST Procedure

Product-Limit Survival Estimates

survtime	Survival	Failure	Survival Standard Error	Number Failed	Number Left
0.0000	1.0000	0	0	0	10
1.0000	0.9000	0.1000	0.0949	1	9
2.0000*	.	.	.	1	8
3.0000*	.	.	.	1	7
4.0000	.	.	.	2	6
4.0000	0.6429	0.3571	0.1679	3	5
5.0000	0.5143	0.4857	0.1769	4	4
5.0000*	.	.	.	4	3
7.0000	0.3429	0.6571	0.1830	5	2
8.0000*	.	.	.	5	1
10.0000	0	1.0000	0	6	0

4. At the end of that Time Interval, draw a vertical line down to the survival rate ($\hat{S}(t)$) for the next Time Interval (i.e., 0.90). This line is again carried to the right for the entire second Time Interval (in our example, that would be time 1 mo through 4 mo).
5. Then draw a vertical line down to the next survival rate ($\hat{S}(t)$), and so forth.
6. If the greatest time to event in your data is an actual event (i.e., noncensored value), the line will end at the last dot. However, if the greatest time to event is censored, the line will continue to the right of the last dot until you reach the time of the last censored event.
7. Often, tick marks or some other symbol are used to indicate censored values.

As an example, the KM plot of the data presented in Example 1 will look like the graph in Figure 2.

Frequently, KM plots are used to illustrate the TTE distributions of distinct groups of subjects (also called strata). In the medical literature, examples of these strata could be different surgical techniques or other treatment differences, different patient characteristics (such as age groups), or different baseline disease severity characteristics (such as cancer stage or cancer grade). The key point is that they have to be mutually exclusive.

When multiple strata are involved, the time to event and censoring data are first grouped by strata, then the survival estimates are calculated separately for each stratum using identical techniques to those described above. When using SAS to create the survival estimates for each stratum, the following code can be used (based on the earlier set of SAS code and assuming the strata of interest is called STAGE).

```
proc lifetest data = km;
  strata stage;
  time survtime*censor(1);
run;
```

When creating a KM plot based on the estimates, lines are created for each strata, usually on the same plot (especially when comparing the distributions is of interest), with

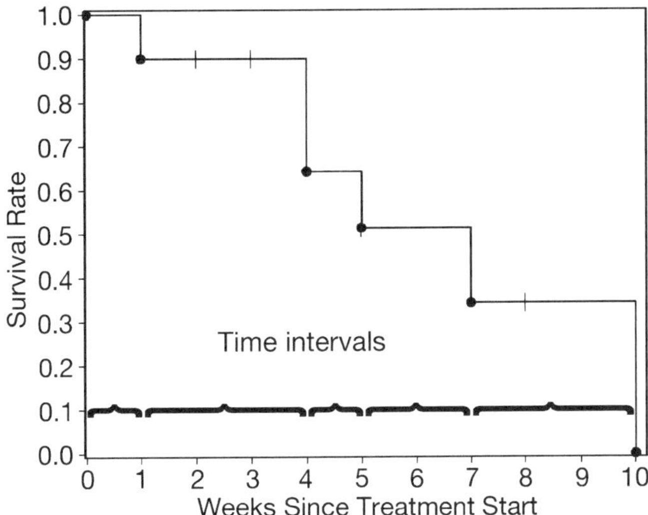

Figure 2: Kaplan-Meier plot of data from Example 1.

different color or line styles used to distinguish between the strata. An example of this can be seen in Figure 1 at the beginning of this section.

As with anything else involving statistics, estimates are more reliable when based on larger sample sizes. For this reason, what we call the tail of the plot, or the right hand side of each line, can fluctuate wildly between one data point to the next because of the small number of subjects at risk in the tail portion of the plot. Because of this, visual differences between the strata (or the lack thereof) in the left-most portion of KM plots are usually more meaningful (and more in agreement with the log-rank test p value) than treatment differences in the tail region. To make it easier to determine whether a tail region is based on sufficient information, some KM plots put the number at risk at each horizontal axis tick mark below the axis.

The KM plots alone can not indicate whether a statistically significant difference in treatment arms exist, even when the lines appear to be wildly divergent. This will be true more often for plots based on a small number of subjects or a small number of noncensored events.

Fortunately, there is a statistical test that addresses this issue. This test, called the log-rank test, measures the probability that the survival estimates between the strata would be as different as they are if the strata truly had identical chances of survival. In other words, are the differences we see in the lines of the KM plot from a real difference or chance alone? As with most statistical tests, the log-rank test produces a probability value, or p value, that can be compared with a cut point to determine statistical significance. We usually consider a log-rank test with a p value of 0.05 or less to be statistically significant.

KM plots are useful tools for visually presenting the survival distribution of an entire sample or of various subgroups of a sample or study arms in a randomized clinical trial. These plots help to simplify the presentation of the results from a TTE analysis and allow medical investigators to be able to quickly assess how a certain population is performing with regard to the outcome of interest. Used in conjunction with the log-rank test, KM plots are a useful tool to determine whether TTE differences exist between various strata.

5. COX PROPORTIONAL HAZARDS MODEL

The last section covered KM plots, which are a wonderful way to graphically illustrate TTE data. However, the KM approach does have some drawbacks. Even though KM estimates can be produced for strata of interest, they can not be easily adjusted for the impact of other, "secondary," variables.

For example, say that a KM plot was produced with two survival curves, one for Drug A and the other for Drug B. The two curves on this plot do not overlap at all, with Drug A having much higher survival estimates than Drug B at nearly all time points. A log-rank test confirmed that the two drugs had significantly different survival estimates, with a p value of 0.002. It seems obvious from this scenario that Drug A is better than Drug B, right?

Now assume that you found out that the mean age for those receiving Drug A was 30 and that the mean age for those receiving Drug B was 80. This additional information puts the survival differences between the two drugs in a new light. Were the survival differences noted earlier the result of better efficacy for Drug A compared with Drug B, or were the survival differences attributable to the age differences between the two populations? Or were the survival differences from a combination of the drug and age factors? The earlier KM plot does not address this issue at all, and it would be very difficult, if not impossible, to use a KM plot and a log-rank test to answer this question.

Fortunately, statistical tools exist to deal with this issue as well. The most commonly used approach is to use what is called a proportional hazards model. Because this model was initially proposed by David Cox in 1972, it is often referred to as the Cox regression model.

Before we cover the Cox proportional hazards regression models in detail, we should review some basic principles of mathematics and statistics. First, you may recall from a geometry class that the basic mathematical form to represent a line in a graph is:

$$Y = a + bX$$

In this case, a represents the intercept, or the value on the vertical (or y-) axis when the line crosses the axis (when $x = 0$). b represents what is called the slope of the line. You may have also learned slope as "rise over run." Let's say that we had a value with a slope of 3. That means that for every increase in X by 1, we have an increase in Y by 3.

For example, say we had the line represented as $Y = 2 + 4X$. We could make a table of some of the values that would fall on this line.

X	Y
0	2
1	6
2	10
3	14
4	18

Now, you may be wondering what the form of a line has to do with survival analysis. This is because all regression models are actually representations of a line. In fact, the term *regression* is used with these statistical models because we are "regressing" all of the information contained in our data into a single line.

For example, we might have a linear regression model that is fitting the effect of age on a biomarker level. After fitting the model using statistical software, we may find the best model is:

$$\text{Biomarker level} = 42.934 + 2.349 * \text{age} + \text{error}$$

Notice the similarity in the format to the form of a line presented earlier. Except for the error term, this could be rewritten as $Y = 42.934 + 2.349X$. The error term merely indicates how far off the line the actual data point falls. The model is chosen in such a way as to minimize the error values across all the data.

Cox models are regression models, so they have a similar representation. However, rather than an untransformed outcome variable as the "Y" variable, as in the linear regression model, Cox models use the natural log of the hazard function as the "Y" variable, so the form of the Cox model is $Log(H(t)) = H_0(t) + b X$, where $H(t)$ represents the hazard function and $H_0(t)$ represents the baseline hazard function.

There are two things that we should know about the hazard function. First, it represents the probability that an individual has the event at time t, provided that the individual did not have the event before time t. Second, the hazard function is a direct function of the survival function. The survival function is what is plotted in KM plots and is generally what we are most interested in when using survival analyses. Knowing that the hazard and survival functions are direct functions of each others means that if you know one of them, you know the other. The formula relating these functions is:

$$S(t) = exp\{-H(t)\}, \text{ or}$$

$$H(t) = -ln\ (S(t))$$

Earlier in this section, we stated that Cox models are also called *proportional* hazards models. As we have shown, the *hazards* part of the name comes from the fact that the dependent variable is a function of the hazard function. However, what about the proportional part?

As with all statistical models and tests, the Cox model is based on certain assumptions about the data. They are based on the assumption that the hazard functions for each stratum are proportional to the hazards for every other stratum. For example, let's say that we are fitting a Cox model for overall survival to a set of data with two strata, perhaps an indicator for Drug A vs Drug B. Let's say that at 1 yr, the risk of dying is 25% higher for the Drug A group than for the Drug B group. Say that the actual 1-yr risk is 0.20 for Drug B and 0.25 for Drug B. For the proportional hazards assumption to be met, the risk of dying at every point in time must be 25% higher for Drug A than for Drug B. The *actual* risk of dying can change in each group over time, but the *proportion* must remain constant. So, at 2 yr, if the risk of dying in group B drops to 0.10, the risk of dying for group A must also drop to 0.125 for the assumption to be met.

A common misconception is that for the proportional hazards assumption to be met, the lines in a KM plot need to be proportional to each other. However, as explained earlier, the lines in a KM plot illustrate the survival function, which is different than the hazard function that serves as the basis of the outcome in Cox models. So, KM plots can not be used to determine whether the proportional hazards assumption is met.

There are methods that can be used to test whether the proportional hazards assumption is met, but they are all fairly complex and beyond the scope of this book. A biostatistician should be able to run these tests. Fortunately, the Cox model is fairly robust to the proportional hazards assumption, so this assumption does not need to be strictly met for a Cox model to be fit and for its results to be accurate.

Now that we have covered the model assumptions and the form of the model, let's look at the details of Cox models. As with the KM plots, a unique aspect of Cox regression models is the handling of censored data. Other types of regression models have only a single outcome variable on the left hand side of the model equation. Cox regression models also have a single outcome variable (the variable representing time to the event), but also have a variable on the left-hand side indicating whether the TTE variable is censored or not. For example, in SAS, the left side of the equation looks like the following, if the TTE variable is called SURVTIME and the censor variable is called CENSOR, with a value of 1 if censored and 0 if not censored:

```
survtime*censor(1)
```

As with other regression models, Cox models can be divided into simple and multivariable versions. A simple Cox model merely means that only a single independent factor is included in the model. If the independent factor is a categorical variable, the results of the Cox model will be nearly identical to those from the log-rank test for the same analysis, in terms of the p value measuring statistical significance. A categorical variable is a variable that has only a finite number of values. For example, common categorical independent variables used in Cox models are:

1. An indicator of treatment, such as a variable that has value 1 for subjects receiving Drug A and value 0 for subjects receiving Drug B
2. Numerical categories (also called ordinal categorical variables), such as age groups
3. Nonnumerical categories (also called nominal categorical variables), such as racial groups.

Simple Cox regression models can also contain a noncategorical (or continuous) independent variable. A common example of this is if the study was designed to determine the effect of age on TTE, but the investigator was not interested in a grouped age effect. Groupings are often arbitrary or may not accurately indicate the effect of a factor on TTE. When the independent factor is a continuous variable, a log-rank test can not be used; therefore, the Cox model is the only common analysis method that can be used for this situation.

Many different software titles can perform Cox regression models. In SAS, the code to run a simple Cox model on a dataset called SURVDATA, where the TTE variable is called SURVTIME, the censor variable is called CENSOR (with value 1 representing a censored value), and the independent factor is called AGE, would be:

```
proc phreg data = survdata;
  model survtime*censor(1) = age/ties = exact;
run;
```

The underlying formula for determining the fit of a Cox regression model does pairwise comparisons between data points and compares the TTE for each member of the pair. If the two data points being compared have the exact same TTE, it is called a tie. There are three main methods (Exact, Breslow, and Effron) for dealing with tied data, and each handles ties in a slightly different manner. The details of these methods are beyond the

Table 2
Analysis of Maximum Likelihood Estimates

Analysis of Maximum Likelihood Estimates								
Variable	DF	Parameter Estimate	Standard Error	Chi-Square	PR > ChiSq	Hazard Ratio	95% Hazard Ratio Confidence Limits	
age	1	0.03272	0.00783	17.4656	<.0001	1.033	1.018	1.049

scope of this book. The Exact method is the most accurate choice, but is much more computationally intensive than the other two methods. However, with the increase in computing power in the last couple of years, the Exact method will be the best choice for the handling of ties in all cases except when the dataset is extremely large. The output from this procedure is shown in Table 2. The part of the output that is most interesting is found in the rightmost three columns. The column labeled Pr > ChiSq is the p value for the model. In this case, the model has a p value of <0.0001, so age has a statistically significant impact on the TTE.

The column labeled Hazard Ratio is what can be used to show the impact of an independent factor. This statistic is as its name implies—a ratio of the hazard functions for two strata. A value greater than 1 indicates the group of interest has a higher hazard (or risk of the event at any point in time) than members of the reference group. A value less than 1 indicates the opposite, that the group of interest has a smaller risk of having the event than the reference group. "Groups" are easier to understand when the independent factor is a categorical variable. For example, when the independent factor is an indicator with value 1 for the group receiving Drug A and value 0 for the group receiving Drug B, Drug B is the reference group (because it has value 0) and we interpret the hazard of Drug A in comparison to the hazard for Drug B. For example, in this case, if we had a hazard ratio of 1.50, we would say that the risk of having the event for the group receiving Drug A is 50% higher than the risk of the group receiving Drug B.

With a continuous independent factor, as in the output above with age, the interpretation is a bit harder. In that case, the hazard ratio represents the risk for each unit of increase. Specifically from our example, the hazard ratio of 1.033 indicates that the risk of dying is 3.3% greater for every 1 yr increase in age.

Often, a 1-unit increase in the independent factor may not be all that interesting—this is especially true with age. In the clinic, there is not a large difference between a person who is 62 and one who is 63. Thus we may want to interpret a larger increase in the independent factor. With age, a 5-yr or 10-yr increase often makes more sense to evaluate. Fortunately, calculating the hazard ratio for an alternative unit increase is straightforward with the help of a calculator. The hazard ratio is equal to the exponentiated value of the parameter estimate (found in the third column from the left in the previous example). In formula, we would write it as:

$$HR = \exp(\text{parameter estimate})$$

If we wish to look at an increase different than 1 unit, the formula becomes:

$$HR = \exp(\text{number of units} * \text{parameter estimate})$$

Say that we are interested in the hazard ratio for every 10-yr increase in age for the data in the example above. We would calculate this as:

$$HR = \exp(10 * 0.03272) = 1.387$$

This indicates that we see a 38.7% increase in the risk of dying for every 10-yr increase in age.

The last important column from the output above is the rightmost column, which is split into two numbers. This column is the 95% confidence interval around the hazard ratio and indicates the level of uncertainty in the estimate of the hazard ratio derived from our data. Specifically, this provides a lower and an upper value for a range that will contain the true value of the hazard ratio with 95% probability. Generally, the more data and less variability contained in an analysis, the tighter this range will be. This interval will also be in sync with the *p* value for the analysis. If the *p* value is greater than 0.05, the 95% confidence interval will contain the value 1 (which indicates no differences between the strata). If the *p* value is lower than 0.05, the interval will be entirely above or entirely below 1.

The other type of Cox model is the multivariable Cox model. These models are identical to simple Cox models except that they have more than one independent factor. Most of the survival models reported in the literature are multivariable models. For example, a multivariable Cox model may include both age and race as independent factors when looking at overall survival. Or a model looking at the effect of a biomarker on survival may be age-adjusted. There are no equivalent log-rank tests for multivariable models because multivariable models determine the *independent* effect of each of the factors being considered. For example, say that we were using a Cox model for overall survival that contained both age and race (African American vs Caucasian). A simple Cox model containing only race as an independent factor would be looking at whether survival differences existed between the two racial groups. However, the age-adjusted multivariable model with age and race would be looking at whether there was an independent effect of race. Stated another way, the multivariable model would be looking for the race effect if there were no age differences between the racial groups. So, it determines the race effect *independent* of age.

In terms of the output, multivariable Cox models are nearly identical to simple Cox models—they merely have more rows to account for each of the independent factors. For example, if we added an indicator for Group A (vs the reference Group B) to the model from our example above, the output would look like Table 3. The important columns of the *p* value, hazard ratio, and 95% confidence interval remain in the same locations of the table.

Occasionally, when there is a strong covariate effect in the Cox model, the Cox model will reach a different conclusion about strata differences from what is shown in the KM plot. This is especially problematic because KM plots are the only common way to graphically illustrate survival data and as such are the most user-friendly way of indicating strata differences with survival data. In this situation, sometimes KM plots stratified by the influential covariates will accurately indicate the strata effect, whereas in other situations, there are no clean or easy methods to graphically show the strata effect.

Cox models are very powerful and flexible and can handle many more complex forms of analyses than those shown earlier. These are beyond the scope of this book, but include the use of interactions (which model the situation where the effect of one factor depends on the value of another factor) and the use of time-dependent covariates (in which the effect of a factor is different at different points of follow-up).

Table 3
Analysis of Maximum Likelihood Estimates

Analysis of Maximum Likelihood Estimates								
Variable	DF	Parameter Estimate	Standard Error	Chi-Square	PR > ChiSq	Hazard Ratio	95% Hazard Ratio Confidence Limits	
age	1	0.03538	0.00802	19.4621	<.0001	1.036	1.020	1.052
Group A	1	0.27239	0.08994	9.1720	0.0025	1.313	1.101	1.566

Cox regression models allow us to determine the independent impact of a main variable or several main variables on the time it takes to reach a defined event, adjusting for the impact of one or more confounding factors. Because of this adjustment, the results of these models may be more accurate (and will be at least as accurate) as the results of a log-rank test. In addition, Cox models can also properly analyze TTE data when the only covariate of interest is not a categorical variable, such as age. With many powerful options, Cox regression models are a very important tool in the analysis of TTE data.

CONCLUSION

In summary, survival analyses are powerful methods for evaluating data where a TTE is involved. These methods should not be limited to studies in which the outcome is alive or dead *per se*. In fact, their application may be more innovative when applied to a broader range of condition specific outcomes. A clear understanding of the concepts behind survival analyses covered in this chapter will aid in demystifying the literature and allow the investigator to increase his or her statistical armamentarium.

REFERENCES

Collett D. Modelling survival data in medical research. Chapman and Hall, London; 1994.

Cox D. Regression models and life tables. J Roy Stat Soc B 1972:74;187–220.

Kaplan E and Meier P. Nonparametric estimation from incomplete observations. J Am Stat Assoc 1958:53;457–481.

10 Assessing the Performance and Validity of Diagnostic Tests and Screening Programs

David C. Miller, MD, MPH, *Rodney L. Dunn,* MS, *and John T. Wei,* MD, MS

CONTENTS

Although surgery is primarily a therapeutic intervention, surgeons also play a pivotal role in the initial evaluation and diagnosis of surgical disease. Indeed, recent scientific and technologic advances (e.g., molecular markers of disease) have considerably expanded the catalog of diagnostic tests available to contemporary surgeons. At the same time, many established (and widespread) screening programs (e.g., mammography, colonoscopy, prostate-specific antigen [PSA]) are designed to detect conditions that are treated primarily with surgical interventions. Moreover, given the substantial morbidity that may accompany surgical intervention, it is imperative that surgeons critically assess the value of a diagnostic test before using its results as the basis for intervention.

In this context, it is essential for surgeons to understand fundamental concepts related to the evaluation of clinical test performance, and for surgical investigators to be skilled in the interpretation of measures of test validity. Whether the test in question is from the patient history, physical examination, a laboratory test, or an imaging study, surgeons must be able to answer the question: How useful is this test for distinguishing diseased from disease-free individuals? *(1)*.

In this chapter, we will cover basic concepts related to the assessment of clinical test performance. We will introduce several statistical methods for assessing the validity of diagnostic tests including sensitivity, specificity, positive and negative predictive values, likelihood ratios and receiver operating characteristic curves (Appendix 1). To highlight their appropriate clinical application, the various measures of validity will be covered separately for tests with categorical (dichotomous) versus continuous results.

From: *Clinical Research for Surgeons*
Edited by: D. F. Penson and J. T. Wei © Humana Press Inc., Totowa, NJ

In addition, this chapter will address some of the most salient issues related to the selection, implementation, and evaluation of screening tests and programs. The rationale for disease screening efforts, as well as various risks and benefits associated with such programs, will be discussed. Finally, several sources of potential bias associated with screening programs, including lead-time bias and length-bias sampling, will be covered to provide a comprehensive framework for assessing the value and validity of a screening program.

1. ASSESSING THE VALIDITY OF DIAGNOSTIC TESTS

1.1. Sensitivity, Specificity, and Accuracy

The validity of a test refers to its ability to measure what it is purported to measure; in most clinical situations, this involves the ability of a diagnostic test to distinguish between individuals with and without a particular disease. Two principal measures of test validity are sensitivity and specificity. In general terms, sensitivity may be characterized as the degree to which a particular test correctly identifies diseased individuals; in contrast, specificity reflects the capacity of the test to distinguish individuals that are free of disease *(1)*. In statistics, sensitivity is defined as the proportion of diseased individuals with a positive test result; specificity, on the other hand, is the proportion of disease-free individuals with a negative test result. A complementary measure of the validity of a given test is its accuracy, which can be defined as the proportion of all tests results (both positive and negative) that are concordant with true health status.

An important caveat with regard to assessing the validity of a diagnostic is that, to assess the performance of a particular test, there must be a "gold standard" test available for comparison. In other words, a different and established test must be available that reliably and precisely differentiates individuals with and without a given disease. In many cases the gold standard may be the pathologic findings from an invasive procedure such as tissue biopsy or extirpative surgery. Alternatively, the gold standard may be based on an objective or subjective set of clinical findings, such as the National Institutes of Health/National Institute of Diabetes and Digestive and Kidney criteria for the diagnosis of interstitial cystitis *(2–4)*. Thus, to properly assess the validity (sensitivity and specificity) of a diagnostic test, the investigator should identify and make use of an existing gold standard. Without a widely accepted gold standard for comparison, evaluations of test performance may be difficult.

1.2. How to Evaluate Tests With Categorical (Dichotomous) Results

A useful way to conceptualize the concepts of sensitivity and specificity is to start by examining a 2×2 table for a scenario involving a dichotomous disease state (i.e., disease present or disease absent) and a dichotomous test outcome (i.e. test positive or test negative) (Table 1). It should be mentioned that an ideal test would have both a sensitivity and specificity of 100%. Examining Table 1, such a test would classify subjects into only two outcome groups: individuals with the disease that have a positive test result (*true positives*, the upper left cell [a]) and individuals without the disease that have a negative test result (*true negatives*, the lower right cell [d]). In the clinical setting, there are no tests that perform at this ideal level. In fact, the outcomes of most tests include positive results in disease-free individuals (*false positives*, the upper right cell [b]) and negative results in people with that actually have the disease (*false negatives*, the lower left cell [c]). Based

Table 1
Standard Table for Comparison of Test Results With Actual Disease Status

	Disease Present	*Disease Absent*	
Test Positive	**a** (true positives)	**b** (false positives)	**a+b**
Test Negative	**c** (false negatives)	**d** (true negatives)	**c+d**
	a+c	**b+d**	**a+b+c+d**

Sensitivity = **a**/(**a+c**)
Specificity = **d**/(**b+d**)
Accuracy = **a+c**/(**a+b+c+d**)
Positive predictive value (PPV) = **a**/(**a+b**)
Negative predictive value (NPV) = **d**/(**c+d**)

on these four possible outcomes, this standardized 2 × 2 table can be used to further illustrate the calculation of sensitivity and specificity.

Recall that sensitivity is defined as the proportion of individuals with a disease that have a positive test result. From Table 1, the total number of diseased individuals is represented by the sum of cells a and c; the number of positive test results for this group is represented in cell a. Thus, for this standard 2 × 2 table, sensitivity is defined as:

$$\text{Sensitivity} = a/(a+c) \quad (1)$$

Similarly, specificity refers to the proportion of disease-free individuals (b+d) that have a negative test result (d) and is, therefore, represented by the following formula:

$$\text{Specificity} = d/(b+d) \quad (2)$$

It should also be noted that for a test with dichotomous results, the accuracy of the test is calculated based on the following formula:

$$\text{Accuracy} = (a+d)\ (a+b+c+d) \quad (3)$$

In a recent publication, Staib and associates used these calculations to evaluate the validity of a newly available diagnostic imaging modality. Specifically, the authors examined the ability of ^{18}F-fluorodeoxyglucose positron emission tomography (FDG-PET) to detect recurrent colorectal cancer in patients who had previously undergone surgical resection with curative intent. In this study, the diagnostic gold standard for recurrent cancer was either histologic confirmation via tissue biopsy or clinical progression of the presumably malignant site identified by FDG-PET *(5)*. The relevant results from this study are summarized in Table 2. Among the 58 patients with recurrent colorectal cancer, as documented by the gold standard described previously, 57 had increased tracer uptake on an FDG-PET scan (interpreted as a positive result). Therefore, the sensitivity of the FDG-PET scan was reported as 57/58 = 98.2%. In terms of specificity, negative FDG-PET results were observed in 38/42 men without recurrent cancer, indicating a specificity for this test of 90.5% (Table 2). The accuracy of FDG-PET imaging for detecting a recurrence was (57+38)/(57+4+1+38) = 95%. Based on these results, the authors concluded that FDG-PET had reasonable validity and may be a useful adjunct to conventional imaging studies in patients with colorectal cancer *(5)*.

Table 2
Validity of ^{18}F-Fluorodeoxyglucose Positron Emission Tomography (FDG-PET) for Detecting Recurrent Colorectal Cancer

	Recurrent Colorectal Cancer	*No Recurrent Colorectal Cancer*	
FDG-PET positive	57	4	61
FDG-PET negative	1	38	39
	58	42	100

Sensitivity = 57/58 = 98.2%
Specificity = 38/42 = 90.5%
Accuracy = (57+38)/(57+4+1+38) = 95%
PPV = 57/61 = 93.4%
NPV = 38/39 = 97.4%
Data from Staib et al. *(5)*.

1.2.1. Positive Predictive Value and Negative Predictive Value

Although sensitivity and specificity are useful measures for evaluating test validity, they are less helpful from a clinical standpoint where disease status is typically unknown and surgeons are faced with assessing the likelihood of disease given a particular test result. It is in this clinical context that understanding and applying the concepts of the positive predictive value (PPV) and negative predictive value (NPV) of a diagnostic test is essential. In general, the PPV (or NPV) helps clinicians answer the following question: "Given that this test is positive (or negative), what is the probability that this patient actually has (or does not have) the disease?" Similar to sensitivity and specificity, an ideal test would have both a PPV and NPV of 100%; however, tests with such optimal performance characteristics are exceedingly rare in clinical practice.

Turning our attention back to Table 1, the PPV of a test is defined as the proportion of individuals with positive tests that actually have the disease:

$$\text{PPV} = a/(a+b) \tag{4}$$

Correspondingly, the NPV is defined as the proportion of individuals with a negative test result that are actually disease-free:

$$\text{NPV} = d/(c+d) \tag{5}$$

In more general terms, the PPV is the probability that someone with a positive test result actually has the disease. The NPV describes how likely it is that a patient with a negative test result is truly unaffected. Based on these definitions, a general principle is that the number of false-positive and false-negative tests will affect the PPV and NPV, respectively. The study from Staib and colleagues (Table 2) can also serve as a useful example for calculating PPV and NPV. Specifically, the PPV of FDG-PET for detecting recurrent cancer was 57/61 = 93.4%; the corresponding NPV was 38/39 = 97.4% *(5)*.

An important caveat with regard to PPV and NPV is that the predictive value of a test may vary based on several factors, including disease prevalence in the community or study sample and the specificity and sensitivity of a particular test *(1)*. An example from the literature is useful to illustrate this concept *(6)*. Lachs and colleagues evaluated the

Table 3
Urine Dipstick Example Illustrating the Relationship Between Disease Prevalence and Predictive Value Data from Lachs et al. *(6)*

A: UTI Prevalence 7% (Low Prior Probability)

	Urine Culture Positive	*Urine Culture Negative*	
Dipstick positive	10	53	63
Dipstick negative	8	188	196
	18	241	259

Positive predictive value = 10/63 = 16%
Negative predictive value = 188/196 = 96%

B: UTI Prevalence 52% (High Prior Probability)

	Urine Culture Positive	*Urine Culture Negative*	
Dipstick positive	49	29	78
Dipstick negative	4	21	25
	53	50	103

PPV = 49/78 = 63%
NPV = 21/25 = 84%

performance of the rapid dipstick test for urinary tract infections (UTI) in two groups of patients that differed in their prior probability of UTI. The investigators defined patients at high-risk for UTI as those with a high proportion of symptoms (dysuria, urgency, frequency, hematuria, fever) and signs (abdominal and costovertebral angle tenderness) consistent with UTI. Conversely, the same signs and symptoms were significantly less frequent among patients classified as having a low prior probability of infection. As expected, the actual prevalence of UTI, based on urine culture as the diagnostic gold standard, was different for the two groups, with 52% (53/103) of the high-risk patients having a culture-proven UTI vs only 7% (18/259) of low-risk patients *(6)* (Table 3). Based on Table 3, in the sample with a prevalence of 7%, 18 women are affected with a UTI and 241 women are disease-free. However, 63 women in this sample have a positive result on their urine dipstick test, and only 10 of these were true positives. Therefore, in this low prevalence sample, the PPV of a urine dipstick test is only 10/(10+53) = 16% *(6)*.

Using the same urine dipstick test in the sample of women with a higher prevalence of UTI (52%) (Table 3), we see that among the 78 women with positive dipstick tests, 49 are true positives and 29 are false positives; the resulting PPV is 49/78 = 63% *(6)*. Therefore, as the prevalence of disease in the sample being tested increases, the PPV of the test increases as well. Likewise, as the prevalence of a particular disease *decreases*, the NPV *increases* (although, given the rarity of many diseases, this tends to be less dramatic than the association between prevalence and PPV). This correlation between prevalence and predictive value is an important and consistent principle that should be kept in mind when considering the potential applications for a clinical test. Furthermore,

Table 4
Urine Dipstick Example Illustrating the Relationship Between Test Specificity and Predictive Value

A: UTI Prevalence 7%

	Urine Culture Positive	*Urine Culture Negative*	
Dipstick positive	10	53	63
Dipstick negative	8	188	196
	18	241	259

Sensitivity = 56%
Specificity = 78%
PPV = 10/63 = 16%
NPV = 188/196 = 96%
Data from Lachs et al *(6)*.

B: UTI Prevalence 7%

	Urine Culture Positive	*Urine Culture Negative*	
Dipstick positive	10	12	22
Dipstick negative	8	229	237
	18	241	259

Sensitivity = 56%
Specificity = 95%
PPV = 10/22 = 45%
NPV = 229/237 = 97%
Data based on the results for a hypothetical urine dipstick test applied to the sample for **A** (*see* text) *(6)*.

this relationship provides the rationale for selective implementation of screening tests in populations that are at increased risk for a particular disease *(1,4)*.

Independent of the effect of disease prevalence, changes in the specificity, and, to a lesser degree, the sensitivity, of a particular test will also affect its predictive value. This principle is illustrated with a hypothetical example based on the study from Lachs and associates (Table 4). Suppose that a new rapid urine dipstick test was developed and found to have an improved specificity (but identical sensitivity) when compared with available tests. Suppose also that a subsequent study was undertaken to compare the predictive value of this new urine dipstick with the "conventional" dipstick test employed by Lachs et al. To control for the effect of disease prevalence on predictive value, the two dipstick tests were applied only in low-risk sample of patients (UTI prevalence = 7%). As determined by Lachs et al, the specificity of the "conventional" dipstick test in this sample is 78%; in contrast, the (hypothetical) specificity of the newly available dipstick in the same population is 95% (Table 4). The sensitivity of both tests is 56%. From Table 4, we see that a change in the specificity from 78% to 95% substantially decreases the number of false-positive test results (53 with the "conventional" dipstick vs 12 with the "improved" dipstick). Consequent to this improved specificity, there is a simultaneous improvement in the PPV of the rapid dipstick test from 16% to 45% (Table 4). The key principle in this example is that changes in the specificity of a diagnostic test tend to have a dramatic effect on the predictive values of the test, with increases in specificity increasing the PPV and vice versa. The PPV and NPV of a test will also increase concurrently with increases in the sensitivity of a particular test; however, the effect of sensitivity on predictive value is modest for low prevalence conditions.

Although their derivations are beyond the scope of this introductory chapter, the previously described relationships between predictive value, prevalence, sensitivity and specificity may also be summarized by the following equations (based on Bayes theorem):

$$\text{PPV} = \frac{(\text{sensitivity})(\text{prevalence})}{[(\text{sensitivity})(\text{prevalence}) + (1 - \text{specificity})(1 - \text{prevalence})]} \quad (6)$$

$$\text{NPV} = \frac{(\text{sensitivity})(\text{prevalence})}{[(1 - \text{sensitivity})(\text{prevalence}) + (\text{specificity})(1 - \text{prevalence})]} \quad (7)$$

Based on Equation 6, it is clear that as sensitivity, specificity or prevalence increase, PPV will increase correspondingly. Similar to PPV, increases in NPV will occur in concert with increases in specificity and sensitivity; however, increases in disease prevalence will actually be associated with a lower NPV (Table 4).

1.2.2. Likelihood Ratios

Another method for describing the performance of a diagnostic test is the likelihood ratio (LR). The use of LRs is increasingly common in the medical literature, and a basic understanding of their derivation is useful for clinical researchers in the surgical disciplines. In general, the LR indicates how much a particular test result raises (or lowers) the pretest probability of the disease of interest and provides an alternative method for determining the PPV and NPV. Furthermore, an important advantage of LRs is that, to determine the PPV and NPV, a clinician must only remember one number for a particular test (the LR) rather than having to recall both the sensitivity and specificity. Furthermore, the availability of validated nomograms has greatly enhanced the clinical value and application of this measure of test performance.

A positive LR is defined quantitatively as the probability of a positive test result in patients with the disease of interest divided by the probability of that test result in disease-free individuals *(7)*. Conversely, a negative LR is derived from the probability of a negative test result among healthy individuals divided by the probability of the same result among those affected with the disease of interest. To illustrate this point further, consider the following equations:

$$\text{LR for a positive test} = \frac{\text{Probability (+ test) among diseased individuals}}{\text{probability (+ test) among disease-free individuals}} \quad (8)$$

$$\text{LR for a negative test} = \frac{\text{Probability (– test) among disease-free individuals}}{\text{probability (– test) among diseased individuals}} \quad (9)$$

Recalling our definitions of sensitivity and specificity, equivalent equations for the LR of a positive and negative test, respectively, are:

$$\text{LR for a positive test} = \frac{\text{Sensitivity (true-positive “rate”)}}{1 - \text{specificity (false-positive “rate”)}} \quad (10)$$

$$\text{LR for a negative test} = \frac{\text{Specificity (true-negative "rate")}}{1 - \text{sensitivity (false-negative "rate")}} \quad (11)$$

As previously mentioned, the clinical value of a LR is based on the fact that this information can be combined with pre-test assessment of disease probability to calculate the posttest probability of disease (PPVs or NPVs) *(7)*. Indeed, the LR specifies how much a particular test result increases or decreases the pretest probability of the disease of interest. In practice, the pretest probability of disease is typically estimated by the clinician based on the patient's history and physical examination, as well as adjunctive epidemiologic data and personal experience.

In general, LRs greater than 1 indicate that the test result increases the probability that a patient has the disease of interest. Conversely, LRs less than 1 decrease the probability of the target disorder *(8)*. A LR equal to 1 indicates that the pretest and posttest probabilities of disease are equivalent. Some authorities define likelihood ratios ≥5 or ≤0.2 as being associated with moderate to large shifts in pretest to posttest probability (and therefore having a greater impact on clinical decision making).

In a recent article, McCormick and colleagues applied this concept to the diagnostic evaluation of orthopedic trauma patients *(9)*. In this study, the authors evaluated the accuracy of four different physical exam maneuvers for diagnosing posterior pelvic ring injuries in patients with traumatic pelvic fractures. For each physical examination modality, sensitivity and specificity for the detection of posterior ring injury was determined based on comparison with computed tomography findings (considered the diagnostic gold standard) *(9)*. One of the examination modalities assessed was posterior pelvic palpation, which involves careful palpation of the sacrum and bilateral sacroiliac joints; this diagnostic maneuver was considered positive when local tenderness was noted on examination. When compared with computed tomography scan results, the sensitivity and specificity of posterior pelvic palpation were 98% and 94%, respectively *(9)*. Based on Equation 10, the authors determined that the positive LR for posterior pelvic palpation (for the diagnosis of posterior ring injuries) was 16.3, indicating that this physical examination finding is 16 times more likely to be present in a patient with a posterior ring injury than one without such a lesion. Based on these results, the authors concluded that the positive findings on posterior palpation provide strong evidence in favor of a posterior ring injury and that this test can, therefore, be used to refine and guide the subsequent radiologic evaluation of patients with traumatic pelvic injuries *(9)*. Indeed, applying this concept further, a LR of 16.3 for pain on posterior palpation means that even if the pre-examination probability of a posterior ring fracture is fairly low (based, perhaps, on patient history and mechanism of injury), the presence of this physical exam finding generates a large, and potentially conclusive, change from pre-test to post-test probability of a posterior ring injury *(8, 9)*.

The mechanics by which LRs are used to translate from pretest to posttest disease probability are fairly complex and require a brief review of the concept of the odds of a disease. Statistically, the odds of an event (such as the presence of a disease) may be defined as follows:

$$\text{Disease odds} = \text{disease probability}/1 - \text{disease probability} \quad (12)$$

After calculating the pretest odds, this statistic may be combined with the LR to calculate the posttest odds of disease (which are much more useful to a clinician than the pretest odds). For a positive test result, the following equation illustrates this point:

$$\text{Posttest disease odds} = \text{pretest disease odds} * \text{positive LR} \tag{13}$$

The posttest disease probability (PPV) may then be determined as follows:

$$\text{Posttest disease probability (PPV)} = \frac{\text{posttest disease odds}}{1 + \text{posttest disease odds}} \tag{14}$$

It should also be noted that the posttest disease probability is mathematically equivalent to the positive predictive value for the diagnostic test. Similar calculations can be performed for negative test results, based on the corresponding negative LR. Recognizing the relative complexity and time requirements of such calculations, sophisticated nomograms have been developed that allow clinicians to move rapidly from pretest (based on clinical data and disease prevalence) to posttest disease probability, thereby facilitating clinical decision making and broadening the applicability of this measure of test performance *(8, 10)*.

2. HOW TO EVALUATE TESTS WITH CONTINUOUS RESULTS

Until now, we have focused on tests with only two possible outcomes (positive or negative). In surgical practice, however, clinicians frequently order and interpret diagnostic tests (e.g., PSA, carcinoembryonic antigen) that have continuous outcomes. In this context, there is no concrete positive or negative test result; rather, a threshold level must be established for the test such that values above this threshold are considered positive and those below the threshold are considered negative. In truth, the choice of cutoff levels can have important implications with regard to the performance of tests with continuous outcome values.

PSA, an important tumor marker for patients with prostate cancer, is an example of a test with continuous outcomes that is widely used in clinical practice. Indeed, the application of PSA as a diagnostic test for prostate cancer serves as a useful illustration of the effects of changes in cutoff levels on the performance of a diagnostic test. Consider, for example, the data in the attached PSA screening dataset, which summarizes serum PSA levels and cancer status for 100 men undergoing screening for adenocarcinoma of the prostate (Table 5). Overall, 40 men have biopsy-confirmed prostate cancer, whereas 60 patients had no evidence of cancer in their biopsy specimen. However, there is no precise PSA threshold that unequivocally separates men with and without prostate cancer; instead, there is overlap of diseased and nondiseased individuals at most levels of PSA. Nonetheless, in clinical practice, a PSA cutoff must be defined such that individuals with values above this level can be referred for additional testing (i.e., transrectal ultrasound-guided prostate biopsy), whereas those with PSA values below the threshold are spared further workup.

The most widely accepted cutoff for a normal PSA level is 4.0 ng/mL *(11)*. Based on this threshold, the PSA screening dataset (combined with Table 1 as a reference) can be used to estimate the sensitivity and specificity of PSA (as a diagnostic test for prostate cancer). In this example, the calculated sensitivity is 87.5% (35/40 cancers detected) and the specificity is 25% (PSA <4.0 for 15/60 men without prostate cancer). Some urologists contend that a PSA cutoff of 4.0 has an unacceptably low sensitivity and, therefore, application of this threshold fails to detect a significant number of men with important prostate cancers (in other words, this cutoff is associated with an unacceptably high false-negative rate) *(12, 13)*. As a result, some authorities have advocated a lowering of the

Table 5
Summary of Prostate-Specific Antigen Screening Dataset Format

Patient Number	*Prostate-Specific Antigen Level (mg/dL)*	*Cancer Status (0 = No cancer, 1 = Cancer)*
1	7.2	1
2	6.7	0
3	1.4	0
4	8.2	0
5	0.7	0
6	10	1
7	5.5	0
8	2.5	1
9	5.7	1
10	8.5	0
...	...	...
91	2	0
92	5.1	0
93	5.4	0
94	4.8	0
95	6.9	0
96	4.6	0
97	7.2	0
98	9.7	1
99	4.1	1
100	11.3	0

threshold for a positive result to 2.5 ng/mL *(12)*. In the PSA screening dataset, lowering the PSA threshold to 2.5 ng/mL would increase the sensitivity of this test to 95.0%; however, the specificity would decrease to 21.7% because of an increased number of false-positive test results. In this setting, we see that very few men with prostate cancer would be undiagnosed (2/40); however, a concurrent effect of changing this threshold is that a large number of men without prostate cancer (47/60) will now be, unnecessarily, subjected to additional invasive diagnostic tests (i.e., a prostate biopsy).

In contrast, an inverse effect is seen when a higher threshold is applied. For instance, if clinical practice was changed such that a higher PSA cutoff level (i.e., 10 ng/mL) was implemented, many men that actually have prostate cancer would not be referred for additional workup, and their cancer would likely remain undiagnosed. At the same time, however, very few disease-free men would be subjected to needless additional testing. In the PSA screening dataset, the net effect of choosing 10 ng/mL as the PSA cut point is a decrease in the sensitivity of this test to 25.0% (10/40 cancers detected), with a simultaneous increase in the specificity to 85.0% (PSA <10.0 for 51/60 men without prostate cancer). In fact, sensitivity and specificity will always vary in an inverse fashion when the "normal" threshold changes for a diagnostic test with continuous results (Table 6).

As illustrated by this example, the choice of cutoff levels can dramatically affect the performance (sensitivity, specificity, and accuracy) of a diagnostic test with continuous outcome values. In general, lowering the cut point will increase the sensitivity, while simultaneously decreasing the specificity. Conversely, raising the cutoff level will gen-

Table 6
Summary of the Effect of Different PSA Cut Points on Its Performance as a Diagnostic Test for Prostate Cancer (Based on the PSA Screening Dataset)

PSA Cut Point (ng/mL)	*Sensitivity (True-Positive "Rate")*	*Specificity*	*1 – Specificity (False-Positive "Rate")*	*# True Positives*	*# False Positives*
2.5	95.0%	21.7%	78.3%	38	47
4.0	87.5%	25.0%	75.0%	35	45
10.0	25.0%	85.0%	15.0%	10	9

erally improve specificity at the expense of sensitivity (Table 6). Clinically, the most salient effect of this principle is that changes in cutoff levels will result in a variable number of false-negative or false-positive test results (Table 6). Accordingly, the choice of an optimal threshold depends on the relative balance between the adverse effects of false positive versus false negative test results. In the case of PSA testing, regardless of the specific threshold applied, two groups of patients of patients will be identified: (1) those with "positive" results that will be referred for biopsy and (2) those with "negative" results that will be spared further testing. In this example, if a low PSA threshold is chosen (resulting in excellent sensitivity but many false positives), then many men will be referred for additional testing that is not only expensive, but also carries a risk of unnecessary morbidity. On the other hand, if a high threshold is chosen, many men that actually have prostate cancer will be inappropriately reassured and their (potentially curable) cancer may remain undetected. Ultimately, for continuous tests, the choice of a clinical threshold depends on the relative significance (e.g., morbidity, cost, availability of effective treatment) of false-positive and false-negative results for the disease of interest.

2.1. Optimizing the Diagnostic Threshold for Continuous Tests Using Receiver Operating Characteristic Curves

As described in the previous section, when test values are measured on a continuum, the sensitivity and specificity of a test will vary based on the position of the cutoff between "positive" and "negative" values. An efficient method for displaying the effects of different cut points on test performance is a receiver operating characteristic (ROC) curve. ROC curves were first developed and used in the engineering and communication fields; currently, they are widely employed as a valid and reliable approach to assessing and comparing the accuracy of various diagnostic tests *(14)*.

In the most general sense, an ROC curve is a plot of the true-positive rate (sensitivity) vs the false-positive rate (1-specificity) for a range of diagnostic test thresholds. The PSA Screening Dataset used earlier in this chapter can be reformulated to determine the true positive and false-positive rates for each of the previously mentioned cutoffs (Table 6). Plotting the true-positive rate vs the false-positive rate (for each PSA threshold) generates an ROC curve for PSA as a diagnostic test (Figure 1); this plot graphically demonstrates the tradeoff between sensitivity and specificity that results from changing the cut point of a diagnostic test. Specifically, as the PSA cut point shifts from 2.5 to 4 and then from 4 to 10, you can see the concurrent decrease in sensitivity and increase in specificity. It

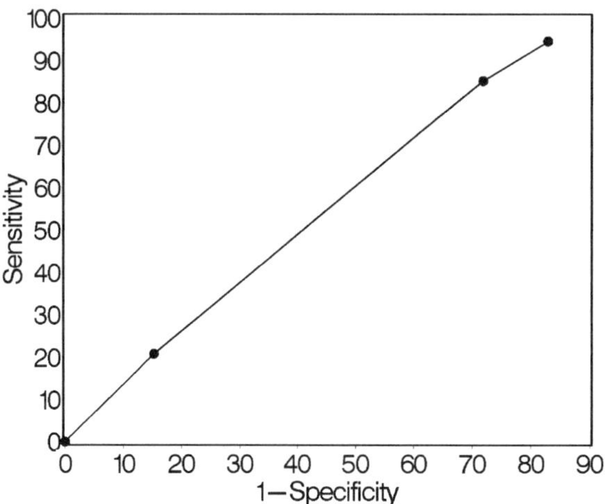

Figure 1: Receiver operating characteristic curve based on three prostate-specific antigen cut points (2.5, 4.0, 10.0 ng/mL) (from PSA Screening Dataset).

is important to recognize, however, that only three PSA cut points were used to generate the ROC curve in Figure 1; an idealized ROC curve for this example would be based on an infinite number of PSA thresholds and would have a (more typical) smoother appearance of the ROC curve in Figure 2.

There are several important caveats with regard to the interpretation of an ROC curve. First, the accuracy of diagnostic test can be assessed visually by examining the proximity of the ROC curve to the upper left-hand corner of the graph. An ROC curve for a "perfect" test would fill the entire area of the ROC space. Specifically, the closer the curve follows the upper left corner of the ROC space, the more accurate the test *(7)*. This makes sense because an ROC curve that approaches the upper left-hand corner of the graph reflects a test that achieves a high true-positive rate (sensitivity) while maintaining a low false-positive rate (1-specificity). Conversely, an ROC curve that approaches a 45° diagonal through the ROC space is a poorly performing test that does little to distinguish individuals with and without the disease of interest. In addition to visual inspection of an ROC curve, a more precise assessment of the accuracy of a test may be also obtained by measuring the area under the ROC curve.

As previously mentioned, the accuracy of a diagnostic test reflects how well the test distinguishes diseased from disease-free individuals. In the case of ROC curves, the most precise measurement of accuracy is the area under the curve; an area of 1 signifies a perfect test, while an area of 0.5 (represented by a 45° diagonal through the ROC space) indicates a poorly performing clinical test (e.g. the test performs no better than chance alone in terms of distinguishing between diseased and disease-free individuals). A useful way to conceptualize the meaning of this numeric value (area under an ROC curve) is to recognize that the area under the curve measures the discrimination of a particular test *(15)*. In other words, the area under the curve reflects the ability of a test to correctly classify individuals with and without the disease of interest. Continuing with our PSA example, consider a situation where the disease status is known for two different groups of men – one of the groups is comprised of men with prostate cancer (untreated) and the other group includes only men that are cancer-free. Suppose that one patient is randomly

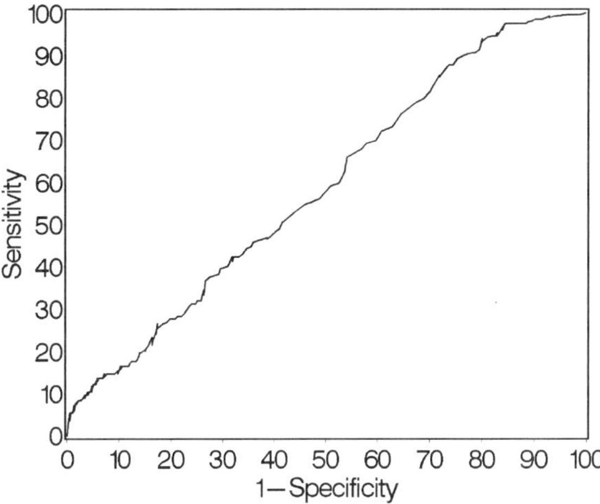

Figure 2: Idealized receiver operating characteristic curve–based PSA Screening Dataset.

selected from each group (e.g. one man with and one man without prostate cancer) and a PSA level is determined for each patient. If PSA is a useful diagnostic test, we presume that its value will be higher in the man with prostate cancer. Indeed, the area under the ROC curve (for PSA) is a numerical description of the percentage of times that this is true; more specifically, the area under the curve represents the percentage of randomly drawn pairs (cancer/cancer-free) for which the test of interest (i.e. PSA) correctly classifies the disease status of the two individuals in the random pair *(15)*.

Formal calculation of the area under an ROC curve is mathematically complex and almost exclusively performed by computer software. A comprehensive explanation of this methodology is beyond the scope of this chapter; however, suffice it to say that both non-parametric (trapezoidal rule) and parametric (maximum likelihood technique) techniques can be used to estimate both the area under the curve and its standard error *(15, 16)*. The point estimates for the area under the curve provide the basis for various statistical tests that assess whether or not two ROC curves are significantly different *(16)*. Although a detailed description is beyond the scope of this chapter, a common method for statistical comparison of ROC curves is to first calculate the area under each curve; the areas are then tested for statistically significant differences using a modification of the Wilcoxon rank-sum test *(7)*. A final caveat worth noting for ROC curves is that they are a function of disease prevalence like any other assessments of test performance such that using an identical assay, one can develop vastly different ROC curves in low prevalence and high prevalence populations.

3. SCREENING TESTS

No discussion of diagnostic test validity would be complete without considering the implications of test performance as they relate to the implementation and efficacy of disease screening programs. Screening tests (such a PSA, mammography and colonoscopy) are used to identify asymptomatic individuals with early-stage, potentially curable disease. In general, screening tests aim to classify individuals with regard to their probability of disease, rather than establishing a definitive diagnosis. The ultimate goal of screening is to alter the prognosis of a given condition by identifying patients in an early phase of the disease, thereby allowing the timely institution of effective therapy. For a screening program to be worthwhile and effective, the disease of interest (and screening test) must fulfill a number of criteria including: 1) the disease must be common and an important health problem; 2) the natural history of the disease should be well-defined and there should be an identifiable latent or presymptomatic stage; 3) if left untreated, the disease must be accompanied by significant morbidity or mortality; 4) there must be an accepted and effective treatment for patients with the disease and there must be some benefit, in terms of morbidity and/or mortality, when the disease is treated in the presymptomatic versus the symptomatic stage; 5) there must be a suitable screening test that is generally acceptable to the population; 6) the cost of screening (including diagnosis and treatment of diagnosed patients) must not be excessive relative to the overall costs of medical care; and 7) screening must be a continuous process and not a "one-time" event. For most widely available screening tests, including mammography, Pap smears and PSA testing, most, but not all, of these criteria are fulfilled *(1,4,17–25)*.

In cases where an available screening test fulfills most of the above criteria, there are several potential benefits to screening programs. For instance, effective screening programs (coupled with appropriate follow-up testing and intervention) may improve the prognosis for treated cases. In addition, by detecting disease in its earliest (and presumably most treatable) stage, there is a potential for a reduction in treatment-related morbidity among screen-detected cases. Furthermore, assuming that an accurate test is available, screening programs can provide reassurance to individuals with a negative test result. Finally, when appropriately implemented, screening programs can serve as a cost-effective use of health resources *(17,19–21,23,25,26)*.

However, there are also several potential disadvantages that must be considered when assessing the relative merits of a screening test. First, screening efforts that employ a test with limited accuracy can result in unnecessary morbidity and anxiety for individuals with false positive results, as well as false reassurance for diseased patients that test negative *(17,27)*. Furthermore, there is often concern that screening programs are implemented in the absence of data that supports their ability to alter disease prognosis *(18)*. Indeed, the true effectiveness of a screening test can only be established by expensive and time-consuming randomized, controlled trials that are designed to evaluate meaningful end points such as morbidity and mortality. In the absence of such data, interpretation of the effectiveness of screening programs can be obscured by bias and confounding and, in fact, the question of whether or not current screening programs (including PSA testing) have been successful in altering the natural history of the disease or improving outcomes for patients remains controversial *(18,24)*. Another potential limitation of screening programs may be a lack of consensus regarding the optimal treatment of patients diagnosed with early disease of uncertain prognosis. Finally, the relative economic and human resources devoted to screening programs may be excessive when considered in the context of widespread population based screening efforts.

As mentioned previously, assessments of the relative value of screening programs may be limited by several sources of bias that frequently plague such evaluations. One source of bias that must be considered is patient-selection bias. Specifically, the results of screening programs may be biased by the presence of systematic differences between individuals that voluntarily participate in a screening test or program and those that choose not to participate. Factors that may contribute to selection bias include significant differences (between participants and nonparticipants) in the following characteristics: baseline health status and sociodemographic characteristics, history of screening, and distribution of risk factors that predict future incidence and mortality from the disease of interest. Once again, systematic differences (between participants and nonparticipants) in one or more of these areas may irreparably bias the interpretation of screening test effectiveness.

Two other sources of bias that often occur in the context of screening programs are lead-time bias and length-time bias. Lead time is defined as the period of time between diagnosis with a screening test and the time when the disease would have been otherwise diagnosed based on various signs and symptoms that prompt medical attention. For a given disease and screening test, the duration of lead time depends on both the biology of the disease and the ability of the screening test to truly detect early disease. Lead-time bias occurs if early diagnosis (screen-detection) results in patients living longer with a disease without ultimately affecting mortality because of the disease. With lead-time bias, the apparent improvement in survival occurs only because of a shift in the date of diagnosis, and intervention produces no real prolongation of life. When evaluating a screening program, avoidance of lead-time bias can be achieved by random assignment of individuals to screening and control groups. Furthermore, rather than comparing survival rates from the time of diagnosis, the effects of lead-time bias can also be reduced by comparing age- and disease-specific mortality rates among screened and control individuals, which are independent of the time since detection.

Length-bias sampling (or length-time bias) refers to the tendency of screening programs to preferentially detect more slowly progressive disease. This occurs because aggressive conditions (such as highly malignant tumors) typically produce symptoms early in the course of the disease and are, therefore, primarily identified by routine diagnostic procedures rather than screening tests. Length-time bias occurs when there is an impression of improved survival because of screening, based solely on the preferential detection of slowly progressive disease. Analogous to lead-time bias, length-time bias may be reduced by repeated screening examinations as often occur in an randomized, controlled trials. In sum, it is crucial to consider the potential for selection, lead-time, and length-time bias when assessing the value of any screening program.

4. CONCLUSIONS

This chapter describes the most salient issues relating to the validity of diagnostic tests and their application to screening programs. It is important to recognize that sensitivity and specificity are generally fixed for a test with a dichotomous outcome; in contrast, sensitivity and specificity will vary based on different cutoff levels for tests with continuous outcomes. NPV and PPV are arguably the most useful measures for clinicians, given that disease status is generally unknown prior to performance of a particular test. The PPV and NPV of a test may vary based on disease prevalence in the sample being studied, as well as changes in the specificity and sensitivity of a particular test. ROC curves are a useful method for further assessing the validity of tests with continuous outcomes. By and large, these statistics are determined by straightforward calculations and should be established for all diagnostic tests. An appreciation of these measures of test performance will allow the surgeon to critically assess the value of both proposed and established disease screening programs.

REFERENCES

1. Gordis L. Epidemiology. 2nd ed. Philadelphia: W.B. Saunders Company, 2000.
2. Hanno PM, Landis JR, Matthews-Cook Y, Kusek J, Nyberg L Jr. The diagnosis of interstitial cystitis revisited: lessons learned from the National Institutes of Health Interstitial Cystitis Database study [comment]. J Urol 1999;161(2):553–557.
3. Kusek JW, Nyberg LM. The epidemiology of interstitial cystitis: is it time to expand our definition? [review]. Urology 2001;57(6:Suppl. 1):Suppl-9.
4. Hulley SB, Cummings SR, Browner WS, et al. Designing clinical research: an epidemiologic approach. 2nd ed. Baltimore: Lippincott Williams & Wilkins, 2001.
5. Staib L, Schirrmeister H, Reske SN, Beger HG. Is (18)F-fluorodeoxyglucose positron emission tomography in recurrent colorectal cancer a contribution to surgical decision making? Am J Surg 2000;180(1):1–5.
6. Lachs MS, Nachamkin I, Edelstein PH, et al. Spectrum bias in the evaluation of diagnostic tests: lessons from the rapid dipstick test for urinary tract infection [comment]. Ann Intern Med 1992;117(2):135–140.
7. Dawson-Sanders B. TRG. Basic and clinical biostatistics. 2nd ed. Norwalk, CT: Appleton & Lange, 1994.
8. Jaeschke R, Guyatt GH, Sackett DL. Users' guides to the medical literature. III. How to use an article about a diagnostic test. B. What are the results and will they help me in caring for my patients? The Evidence-Based Medicine Working Group. JAMA 1994;271(9):703–707.
9. McCormick JP, Morgan SJ, Smith WR. Clinical effectiveness of the physical examination in diagnosis of posterior pelvic ring injuries. J Orthopaed Trauma 2003;17(4):257–261.
10. Fagan TJ. Letter: nomogram for Bayes theorem. N Engl J Med 1975;293(5):257.
11. Arcangeli CG, Ornstein DK, Keetch DW, Andriole GL. Prostate-specific antigen as a screening test for prostate cancer. The United States experience [review]. Urol Clin N Am 1997;24(2):299–306,.
12. Punglia RS, D'Amico AV, Catalona WJ, et al. Effect of verification bias on screening for prostate cancer by measurement of prostate-specific antigen [comment]. N Engl J Med 2003;349(4):335–342.
13. Catalona WJ, Smith DS, Ornstein DK. Prostate cancer detection in men with serum PSA concentrations of 2.6 to 4.0 ng/mL and benign prostate examination. Enhancement of specificity with free PSA measurements [comment]. JAMA 1997;277(18):1452–1455.
14. Baker SG. The central role of receiver operating characteristic (ROC) curves in evaluating tests for the early detection of cancer [review]. J Natl Cancer Inst 2003;95(7):511–515.
15. Hanley JA, McNeil BJ. The meaning and use of the area under a receiver operating characteristic (ROC) curve. Radiology 1982;143(1):29–36.
16. McNeil BJ, Hanley JA. Statistical approaches to the analysis of receiver operating characteristic (ROC) curves. Med Decision Making 1984;4(2):137–150.
17. Goldstein MM, Messing EM. Prostate and bladder cancer screening [review]. J Am Coll Surg 1998;186(1):63–74.
18. Harris R, Lohr KN. Screening for prostate cancer: an update of the evidence for the U.S. Preventive Services Task Force [review]. Ann Int Med 2002;137(11):917–929.
19. Lindfors KK, Rosenquist CJ. The cost-effectiveness of mammographic screening strategies [comment] [erratum appears in JAMA 1996;Jan 10;275(2):112]. JAMA 1995;274(11):881–884.
20. Mahadevia PJ, Fleisher LA, Frick KD, et al. Lung cancer screening with helical computed tomography in older adult smokers: a decision and cost-effectiveness analysis [comment]. JAMA 2003;289(3): 313–322.
21. Marks D, Thorogood M, Neil HA, Wonderling D, Humphries SE. Comparing costs and benefits over a 10 year period of strategies for familial hypercholesterolaemia screening. J Public Health Med 2003;25(1):47–52.
22. McGrath JS, Ponich TP, Gregor JC. Screening for colorectal cancer: the cost to find an advanced adenoma. Am J Gastroenterol 2002;97(11):2902–2907.
23. Pignone M, Saha S, Hoerger T, Mandelblatt J. Cost-effectiveness analyses of colorectal cancer screening: a systematic review for the U.S. Preventive Services Task Force [summary for patients in Ann Intern Med 2002;Jul 16:137(2):I38; PMID 12118986] [review]. Ann Intern Med 2002;137(2):96–104.
24. Smith DS, Catalona WJ, Herschman JD. Longitudinal screening for prostate cancer with prostate-specific antigen [comment]. JAMA 1996;276(16):1309–1315.

25. van Valkengoed IG, Postma MJ, Morre SA, et al. Cost effectiveness analysis of a population based screening programme for asymptomatic Chlamydia trachomatis infections in women by means of home obtained urine specimens.[comment]. Sex Transm Infect 2001;77(4):276–282.
26. McGrath JS, Ponich TP, Gregor JC. Screening for colorectal cancer: the cost to find an advanced adenoma. Am J Gastroenterol 2002;97(11):2902–2907.
27. Harris R, Lohr KN. Screening for prostate cancer: an update of the evidence for the U.S. Preventive Services Task Force [review]. Ann Intern Med 2002;137(11):917–929.

Appendix 1
Equations for the Assessment of Clinical Test Performance

$$\text{Sensitivity} = \frac{\text{number true positive test results}}{\text{number diseased individuals}}$$

$$\text{Specificity} = \frac{\text{number false positive test results}}{\text{number disease-free individuals}}$$

$$\text{Accuracy} = \frac{(\text{number true positive test results} + \text{number true negative test results})}{\text{number disease-free individuals}}$$

$$\text{Positive predictive value} = \frac{\text{number true positives}}{\text{total number positive test results}}$$

$$\text{Negative predictive value} = \frac{\text{number true negatives}}{\text{total number negative test results}}$$

$$\text{LR for a positive test} = \frac{\text{probability (+ test) among diseased individuals}}{\text{probability (+ test) among disease-free individuals}}$$

or

$$\text{LR for a positive test} = \frac{\text{sensitivity (true-positive ``rate'')}}{1 - \text{specificity (false-positive ``rate'')}}$$

$$\text{LR for a negative test} = \frac{\text{probability (– test) among disease-free individuals}}{\text{probability (– test) among diseased individuals}}$$

or

$$\text{LR for a negative test} = \frac{\text{specificity (true-negative ``rate'')}}{1 - \text{specificity (false-negative ``rate'')}}$$

11 Secondary Data Analyses

Andrew L. Rosenberg, MD,
MaryLou V. H. Greenfield, MPH, MS,
and Justin B. Dimick, MD

Contents

INTRODUCTION

One of the most efficient and commonly used approaches to investigate a variety of clinical questions is performed by analyzing data that have been previously collected. This is known as secondary data analysis, and it has become a more common form of study methodology, in part because computerized datasets of information are more prevalent and the cost of data collection, storage, and data retrieval has decreased. It is also a popular method of research because the numerous databases available that have been collected by a wide variety of health and business organizations, research centers, hospitals, governments, and international agencies; data from these varied sources allow researchers to address a larger number of study questions. This chapter will discuss common components of these datasets, how they can be used to investigate clinical questions, how to initiate research using these datasets, pitfalls to avoid, and specific examples of datasets.

2. PRIMARY VS SECONDARY DATASETS

There are significant differences between primary and secondary datasets. Primary datasets contain information that has been prospectively collected for a specific purpose,

From: *Clinical Research for Surgeons*
Edited by: D. F. Penson and J. T. Wei © Humana Press Inc., Totowa, NJ

usually as part of a randomized clinical trial *(1)* or as part of prospective cohort studies that follow groups of patients. Examples of the latter are the Northern New England Cardiovascular Disease Study Group's dataset *(2)* and longitudinal epidemiologic studies, such as the Framingham heart study *(3)*. The number of patients included in the dataset, the variables evaluated, and the quality of data collection are determined by the original investigators. Alternatively, secondary datasets are used to investigate questions that may be different from the purpose for which the original data were collected. Typically, these secondary questions are investigated using existing data from a variety of sources such as a large medical center's Acute Physiology and Chronic Health Evaluation (APACHE) benchmarking and clinical information system *(4,5)* or state hospital association encounter datasets *(6)*. These databases are used primarily for risk-adjustment, predicting mortality, or administrative purposes. However, because of the vast amount of diagnostic, physiologic, procedural, and outcome data collected, they are useful for those who wish to explore clinical questions. A notable example of this was the Cleveland Health Quality Choice. This community-based outcomes assessment program was coordinated by a consortium of business, hospital, and medical leaders to provide comparative public data on hospital intensive care unit (ICU) performance *(7)*. The data from this study have been used for several secondary analyses of ICU benchmarking and hospital quality *(8–11)*. Similarly, the University HealthSystem Consortium maintains extensive clinical performance datasets from 212 academic medical centers that is available to investigators from member institutions *(12–16)*.

Secondary data analysis can also be carried out by combining existing databases into a new dataset. For example, data from a *primary study* can be combined with *administrative* data. An example of this is a study by Wennberg et al *(17)*, who used the hospital outcomes from centers that participated in a large, multicenter randomized controlled trials of carotid endarterectomy (the North American Symptomatic carotid Endarterectomy Trial [NASCET] and Asymptomatic Carotid Atherosclerosis Study [ACAS] studies) *(18,19)* with utilization and outcomes from the Medicare/Part A administrative dataset (Medicare Provider Analysis and Review [MEDPAR]). By combining these datasets, the investigators were able to demonstrate lower mortality rates for carotid endarterectomies among hospitals that participated in a clinical trial as opposed to hospitals that did not (1.4% vs 1.8% mortality, respectively) *(20)*. More commonly, secondary data analysis is performed with multiple clinical and administrative datasets that are combined, such as the study of Birkmeyer and colleagues who demonstrated reduced mortality rates after hospital discharge for pancreaticoduodenectomies in high- vs low-volume centers (12% vs 16%, respectively) *(21)*. These investigators analyzed surgical utilization data to determine the number of operations performed using the MEDPAR hospital discharge abstract file and combined them with the long-term outcome data available from other datasets. These datasets included the National Death Index maintained by the National Center for Health Statistics and the vital status information available from the Medicare enrollment dataset (the denominator) *(20)*. Another example of linking datasets to create a secondary research dataset with more clinically relevant information uses merged Medicare claims data and the National Cancer Institute's Surveillance, Epidemiology, and End Results (SEER) files and the Pennsylvania Health Care Cost Containment Council for coronary bypass surgery *(22)*. Several studies using these merged data have investigated the outcomes for patients with lung cancer based on insurance status *(23)*, complications of different cancer treatments *(24)*, and factors influencing the use of mammography *(25)*.

Table 1
Typical Information Sets in Administrative Vs Clinical Datasets

Demographics

- Identifier codes; Social Security number sometimes available, often these are removed
- Names, last, first; often removed
- Age
- Gender
- Race
- Zip code, state

Administrative

- Admission and discharge date
- Admission source; emergency department, clinic, hospital transfer
- Admission type; emergency, scheduled,
- Provider; name, medical service UPIN (universal provider identification number)
- Discharge location; home, rehab, nursing home
- Insurance/payer information
- Total/reimbursed charges

Diagnostic (case-mix)

- Diagnoses; free text, ICD-9-CM diagnosis codes
- Procedure; free text, ICD-9-CM Procedure codes
- Comorbidities; free text, ICD-9-CM diagnosis codes
- Other predictor variables

Outcomes

- Specific outcomes focused on by a clinical study
- Complications; free text, ICD-9-CM diagnosis
- ICU days
- Hospital days
- Discharge status; alive, dead

Examples are from a Medicare hospital discharge abstract and a clinical registry for cardiac surgery for the Society of Thoracic Surgery Dataset.

ICD-0-CM, International Classification of Diseases, Ninth Revision, Clinical Modification.

3. ORGANIZATION AND CONTENTS OF SECONDARY DATASETS

A typical secondary dataset used for clinical research will contain many variables (Table 1). It will often include identification and demographic information such as social security numbers, registration numbers, the dates of specific events such as admission, procedure, discharges, name, age, gender, and zip code. Diagnostic, procedural, and some comorbidity information are supplied with free text or using the International Classification of Diseases, Ninth Revision, Clinical Modification (ICD-9-CM) codes. Finally, specific variables used to evaluate associations or predict outcomes may be listed.

Table 2
Data Sources for Secondary Research

Database Type	*Example*	*Details*
	Clinical Databases	
Clinical Trials	—NASCET Study —ACAS Study	Source of precise clinical data for risk adjustment and outcomes. Limited by number and types of patients, data selected for study
Prospective Observational Cohorts	—APACHE —MPM	Prospective data collection, clinically precise data, limited by biases, variables not collected, correlation with underlying population
Registries	—SEER —UNOS —STS	Prospective collection of data of patients with specific conditions, rich clinical information, patient-specific or grouped data
	Administrative Databases	
Medical Record (electronic)	Hospital medical record departments Departmental quality assurance dataset	
Enrollment	—Medicare beneficiary file —State medical research file	Total number of eligible persons for medical services. Source of denominator for population-based rates
Encounter	—National Hospital Discharge Survey —VA EDR —State Hospital Discharge Databases	Track utilization and resource consumption, describe clinical events with ICD-9-CM (admissions, procedures), very large sample sizes. May be linked to enrollment to determine rates.
Performance/ survey	—HEDIS —UHC —PHC4 CABG	Encounter data enriched with specific clinical variables, often used for benchmarking, dataset patient-centered information such as satisfaction/ quality of life.

NASCET, North American Symptomatic Carotid Endarterectomy Trial; ACAS, Asymptomatic Carotid Atherosclerosis Study; APACHE, Acute Physiology and Chronic Health Evaluation; MPM, Mortality Probability Models; VA EDR Veteran's Administration Event Driven Reporting; HEDIS, Health Plan Employer Data and Information Set; UHC, University Health Systems Consortium; PHC4, Pennsylvania Health Care Cost Containment Council; CABG, Coronary Artery Bypass Graft surgery; for other abbreviations, *see* Figure 1, p. 185.

See text for description of dataset examples. Adapted from ref. *28*.

The datasets used for secondary analysis can be categorized in a number of ways. Perhaps the most fundamental is whether or not the data were collected primarily for clinical research or for administrative aims such as for billing or claims purposes (Table 2). Clinical datasets generally contain more detailed physiologic, diagnostic, and disease specific information than are found in administrative datasets. Also, the data are usually collected prospectively with more rigorous quality and accuracy standards and are often used for research where adjusting for case mix or evaluating risk factors or other predictors of outcome are required. For this reason, they may influence outcomes more directly and are more appropriate for benchmark performance of an individual, a hospital, or a medical center *(20,26,27)*. In clinical datasets, the individual patient is usually the unit of observation. Alternatively, registries may organized by groups of patients with certain disease conditions, or procedures. For example, the United Network of Organ Sharing registry contains data describing patients receiving organ transplantation, whereas the SEER registry contains both diagnostic and procedural information for many types of malignancies.

Administrative datasets, on the other hand, are frequently composed of a relatively limited amount of clinical, and no physiologic, information. In fact, many researchers feel that administrative data are too vague and define data too broadly. In response, a recent consensus conference of the American Thoracic Society suggests using a more accurate classification for secondary datasets based on their administrative purpose *(28)*. These authors propose describing administrative data as either encounter data, enrollment data, registry data, performance data, or survey data (Table 2). Enrollment data such as found in the Medicare Beneficiary File or in the Blue Cross Blue Shield datasets, are typically used to determine the number of patients in a population that are either eligible to receive a medical treatment or who are at risk of contracting a disease. When used in this way, enrollment databases supply the denominator used by researchers to calculate population-based incidence and rates.

The other types of files found in administrative databases are those for specific encounters or services for example when a patient is admitted to a hospital for an operation. Encounter databases are usually created by the payers of health care to keep track of utilization, reimbursement, and other financial outcomes. Because these databases include information on all admissions or procedures, they form the basis for calculating the numerator in population-based rates when used with enrollment data. Encounter databases include the MEDPAR database, statewide hospital discharge databases, private insurance claims (e.g., Blue Cross Blue Shield), and the Veteran's Affairs (VA) Event Driven Record datasets *(26)*. The Event Driven Record tracks all hospital admissions, clinic visits, and surgical procedures performed in one of the hundreds of VA centers in the United States. The majority of secondary data analyses for surgical patients involve encounter records because they track specific services performed such as surgical procedures *(29)*.

3.1. International Classification of Diseases, Ninth Revision, Clinical Modification (ICD-9-CM)

Most administrative datasets follow the format of the Uniform Hospital Discharge Data Set maintained by the National Center for Health Statistics (Table 3). A key feature of these and other administrative datasets is the use of a five-digit inpatient hospital services coding system known as ICD-9-CM *(30)*. There is an enormous amount of data available from ICD-9-CM codes. These codes achieve diagnostic and clinical detail by using up to five digits, three for the primary event and two for severity modifiers. The ICD-9-CM codes are used to indicate both primary diagnoses (e.g., 560.0 is intestinal obstruction) and secondary diagnoses that usually represent preexisting conditions such as emphysema or liver or renal insufficiency. The ICD-9-CM system also contains code modifiers in the last two digits to indicate additional severity of illness or complexity of primary diagnosis (e.g., 569.6 is colostomy and enterostomy complications; 569.61 is infection of colostomy or enterostomy cellulitis or abscess).

It should be noted that some administrative datasets use different coding systems for procedures. For example, the American Medical Association's Current Procedural

Table 3
Contents of the Uniform Hospital Discharge Dataset

Item	*Definition*
1. Personal Identification	Hospital assigned medical record number, Social Security number not recommended
2. Date of birth	More accurate than age
3. Gender	
4. Race and ethnicity	
5. Residence	Zip code
6. Hospital identification	Medicare provider
7. Admission and discharge dates	
8. Attending physician identification	Physician who is primarily responsible for the patient
9. Operating physician identification	Physician who is primarily responsible for the principal procedure
10. Diagnoses [a] (five)	The condition chiefly responsible for admission and other diagnoses associated with current hospitalization
11. Procedures [a] and dates	Principal procedure for definitive treatment or diagnosis
12. Disposition of patient	Home, short- or long-term rehabilitation, died
13. Expected principle source of payment	Medicare, Medicaid, private insurance, Blue Cross, other government payor, self pay, no charge

[a] Coded using ICD-9-CM.
Adapted from the National Committee on Vital and Health Statistics, US DHEW (1980).

Terminology (CPT) codes, which are used in the Medicare files, provide more clinical detail, and are different than the ICD-9-CM procedure codes. For example, the ICD-9-CM code for repair of a reducible, incisional hernia (CPT 49560) or a strangulated incisional hernia (CPT 49565) is the same (53.51). Other clinical information collected in ICD-9-CM codes include variables such as patient symptoms—for example, abdominal pain is 789.0, physical exam findings such as abdominal rigidity is 789.4 or benign hypertrophy of the prostate with or without urinary obstruction is 600.0/600.01, and various bone fractures are also coded 800.00 to 829.00. Laboratory and other test results such as myoglobinuria is 791.3, bacteremia is 790.7, and abnormal electrolytes are 276.0–276.9.

Two other supplemental classifications, V and E codes, exist to describe factors or events that affect a person's health but are neither diseases nor therapeutic interventions *(30)*. V codes generally indicate a variety of factors that influence health status. Examples include, long-term mechanical ventilator dependence is V 46.1, medication allergies such as penicillin allergy is V14.0, personal social circumstances such as homelessness is V.60.0, and wound care interventions such as dressing changes is V58.3. The E classifications generally are used for external causes of injury such as motor vehicle accident (E812.0), assault with a handgun (E965.0), and medical and surgical complications such as accidental perforation during a medical procedure (E870.0) and foreign object left in body (E871). The imprecision and vagaries of ICD-9-CM coding are exemplified by these supplemental codes. For example, some medical complications such as anesthetic overdose, 968.3, are found in the injury and poisoning section of the main ICD-9-CM codes, but not in the E-codes that designate other poisonings.

Despite the richness of detail provided by ICD-9-CM codes, several limitations exist, including a lack of clinical detail, nonuniform standards used to define some conditions, that codes may represent the interpretation of a medical chart by clerks, they may be retrospective, and they may be biased by "DRG [diagnostic-related group]-CREEP" *(31)*. Further limitations may include undercoding of secondary comorbidities, and restrictions of the numbers of categories available *(26)*.

Because of advances in diagnostic and medical/surgical procedures, a need to update the system arose. In 1994, under the auspices of the World Health Organization, many countries, including England, France, Japan, and Canada, began reporting mortality data using the revised ICD nomenclature, ICD-10. This updated version of the ICD codes was vastly expanded from 5000 to 8000 categories but unlike the ICD-9-CM, the ICD-10 version did not contain procedure codes. Instead, an entirely new procedure nomenclature will be introduced using the ICD-10 procedure coding system *(32)*. These new codes are alphanumeric with 7 characters, which will include16 possible types of basic procedures, and others representing organ system, basic operation, body part, approach, technique or device used, and a modifier for the procedure. The US ICD-10 implementation is waiting updates based on studies performed by the American Hospital Association and the American Health Information Management Association *(33)*.

Secondary datasets are generally composed of individualized data such as age, diagnosis, physiologic information, risk factors and outcomes, or aggregate data for groups of patients such as death rates for women with breast cancer among different age groups. Individualized data are usually found in datasets created for specific clinical trials, hospital datasets, many administrative datasets, and even several well-known national registries such as the National Death Index. Usually knowing two of three identifiers such as last name, birth date, and social security number is sufficient to access the individual data from many national registries *(20)*.

Aggregate datasets are less commonly used because the lack of individualized data increases the risk of confounding bias because groups of patients can frequently differ from each other in unpredictable ways. For example, the incidence of testicular cancer appears to be increased among subfertile men *(34)*. But these studies do not account for the confounding effect of increased testicular examinations that these men undergo by a urologist compared with the general male population *(35)*. The advantage of aggregate data is its availability and the ability to detect significant associations of a risk with an outcome. These types of datasets should be used, primarily, to generate hypotheses to better focus studies using individualized patient data.

A common source of data for secondary analysis that often contains both individualized and group information is from one of many health registries. These registries collect information about the incidence, effect, and extent of disease, as well as the types of treatment a patient with the disease may receive. These data may be reported to a central registry such as the SEER database that collects information from various medical facilities, hospitals, physicians' offices, therapeutic radiation facilities, freestanding surgical centers, and pathology laboratories. For example, US tumor registries report information on approx 26% of the US population to the SEER registry.

Data in registries are critical for programs focused on risk-related behaviors (e.g., tobacco use and exposure to the sun and their affect on cancer prevalence) or on environmental risk factors (e.g., radiation and chemical exposures and their affect on cancer prevalence). Such information is also essential for identifying when and where disease screening efforts should be enhanced and for monitoring the treatment provided to patients with the disease. In addition, reliable registry data are fundamental to a variety of research efforts, including those aimed at evaluating the effectiveness of disease prevention, control, and treatment programs. There are comprehensive population registries (SEER) and there are voluntary registries such as the Society of Thoracic Surgery cardiac surgery database. The latter clinical datasets consists of information sent in by each participating center performing cardiac surgery. There are many other registries of completed clinical trials that may be explored to answer or study other health questions such as those found in the various institutes of the National Institutes of Health.

Finally, other datasets may contain collections of clinically relevant scientific articles and images. The latter types of data are usually stored in some form of text oriented or natural language documents such as the National Library of Medicine's Medline and PubMed database of medical and scientific publications *(36)*, the Cochrane Database of Systematic Reviews *(37)*, or the National Clinical Guideline Clearinghouse *(38)* from the Agency for Healthcare Research and Quality.

4. META-ANALYSIS

Some forms of secondary data analysis use the data reported in published manuscripts. When a literature review incorporates transparent, uniform, and complete criteria for locating all relevant literature on a subject, and the articles are filtered through appropriate inclusion and exclusion criteria, such as a similar patient population that is exposed to similar treatments or risks within a logical time period, the result is known as a systematic review. These types of secondary data analysis are frequently carried out when clinical trials demonstrate negative results which are usually caused by small sample sizes. Occasionally, meta-analyses are also performed when the existing studies were from single hospitals or medical centers and thus their generalizability may be limited. Systematically identifying all relevant studies for a clinical problem and applying statistical analysis to the aggregate pool of subjects can help clarify whether an association exists when previously individual studies have had too few patients to demonstrate an effect *(36, 39)*. If the association is very strong, the meta-analysis may even support the causal relationship of a treatment or risk factor to an outcome *(40)*. Many datasets of medical literature including MEDLINE, OLDMEDLINE, Current Contents, CANCERLIT, EMBASE, AIDSLINE, and The Cochrane Registries of Controlled Trials and Systematic Reviews are the primary source for finding these articles of smaller clinical trials *(41,42)*. A major risk to validity for meta-analysis is being certain that all relevant studies have been included. Publication bias, in which articles with a positive outcome are published more often than negative studies, is the major risk to the validity of a meta-analysis *(36)*.

To perform a meta-analysis, the appropriate studies are first culled from a larger list of potential studies. The summary effect is essentially an average effect from each study weighted by its sample size *(43)*. When undertaking a meta-analysis, one must first determine whether the studies are clinically different from each other in terms of populations, treatments, and outcomes. If obvious clinical differences are not apparent, a test of heterogeneity using a form of the chi-square test to determine whether the results of

Table 4
Advantages and Disadvantages of Secondary Datasets

Advantages	*Disadvantages*
Data exist; readily available	Little control over content and quality of data
Large sample sizes	Accuracy of procedure classification
Populations of patients "real-world"	Accuracy of diagnosis classification
Ability to estimate population-based rates	Accuracy of comorbidities
Screening tool for rare events (registries)	Nonuniform disease modifier classification
Complete follow-up	Risk adjustment
Linkage between clinical and financial data	Limited number of diagnosis fields leads to "saturated" data
	Undercoding survivors

one study differ from another is performed and is reported as a Q-statistic with a p value *(44)*. If the test of heterogeneity suggests that the data within an individual study differs significantly from data between each of the studies, it is inappropriate to combine the studies. Many medical journals now require authors to use a random effects model because it is considered more conservative (i.e., random effects models are less likely to yield a statistically significant result because when heterogeneity exists, they tend to result in wider confidence intervals than fixed-effects models) *(36,39)*. A more in depth treatise for conducting a meta-analysis will be covered in Chapter 18.

5. ANCILLARY STUDIES

Occasionally, additional information is extremely useful when added to the existing data in a secondary dataset. New questions can be better addressed when these additional variables are added to the existing dataset (i.e., an ancillary study). Ancillary studies may be especially relevant to surgical patients if one pursues novel analyses of the stored serum, tissue, and DNA samples that are increasingly obtained as part of clinical trials *(43)*. As an example, stored serum was used to investigate the role of inflammatory cytokines in response to mechanical ventilator changes for patients with Adult Respiratory Distress Syndrome *(45)*. These samples exist at the data coordination centers of the primary study site and conceivably could be used to analyze novel molecular biologic questions as newer methods of analysis emerge. Similarly, tissue samples such as biopsy from other studies may also be available to use in combination with existing clinic data.

6. ADVANTAGES AND PITFALLS

Using data that have already been collected has the primary advantage of being immediately available for analysis, avoiding the months to years required to collect data prospectively (Table 4). For example, Rosenberg and colleagues evaluated the predictors of readmissions to the ICU by querying their institution's clinical data center, which had been collecting an enormous amount of clinical and diagnostic APACHE information for each patient admitted to the ICU during each day of their ICU stay *(46)*. Within 2 wk, data for more than 5000 consecutive admission to an ICU over a 4-yr period were obtained for data cleaning and analysis. Similarly, Rosenberg and colleagues have recently started to obtain specific patient population data from this institution's perioperative information system that contains all the physiologic data recorded during a patient's operation *(47)*. These data can be used to investigate a number of clinical questions related to anesthetic technique and patient outcome.

Many existing secondary datasets are less expensive to obtain, or to access, compared with performing a prospective clinical trial. The Healthcare Cost and Utilization Project, which provides state inpatient discharge datasets (starting at $20 per year) and the Nationwide Inpatient Sample ($200 per year), are very inexpensive. On the other hand, the Medicare files or proprietary datasets, such as the MediQual data files, may be expensive and difficult to obtain access and may require complex agreements for how the data will be used and presented. Finally, the complexity of Medicare and other large data files usually requires that the investigator have additional funding for analysis or statistician/programmer support.

Another advantage of secondary data analysis is the ability to investigate questions such as the associations of risk and outcomes, or patterns of disease, by using extremely large patient populations. These datasets incorporate data from wide geographic ranges such as the MEDPARS data, the National Death Index, or many state and national datasets based on the Uniform Hospital Discharge Dataset. Using these types of datasets, studies with large sample sizes from a variety of medical or health care centers offer real-world representative populations. Thus investigators avoid the problems of single-center studies, and the selection bias inherent when using data from only academic or referral centers *(20, 27)*. A large population-based dataset increases the statistical power and the generalizability of a study. This is especially true with administrative datasets that can often include several hundred thousand to several million patients. A recent example of this is a study of more than 750 million hospitalizations in the United States using the National Hospital Discharge Survey (Figure 1; Section III) to determine the causes and outcomes of sepsis among 10.3 million patients with that disorder *(48)*. Also large datasets, especially registries such as the SEER are excellent for studying rare conditions that would otherwise be difficult for an investigator to collect sufficient numbers of patients to analyze in a single (or even multiple) medical center study.

Another unique advantage of secondary databases is the ability to use encounter and enrollment datasets from the same population to calculate and analyze population-based rates of surgical utilization and outcome *(20)*. Trends in both surgical procedure and volume over time and across geographic areas, known as small-area analysis, can also be studied with large population-based datasets. Last, some secondary datasets are simple to obtain and use because they can be obtained less obtrusively and because the data are already collected and deidentified, informed consent is usually waived and no patient contact is required. An example of this is the Healthcare Cost and Utilization Project National Hospital Discharge Survey (NHDS) discussed below.

6.1. Pitfalls of Analyzing Secondary Datasets' Accuracy of Classifying Procedure, Comorbidities, Diagnosis, Complications Risk Adjustment–Saturated Diagnoses Fields

The primary disadvantage of using secondary datasets is that the data available may not contain the exact variables that would best answer a question. One does not have the

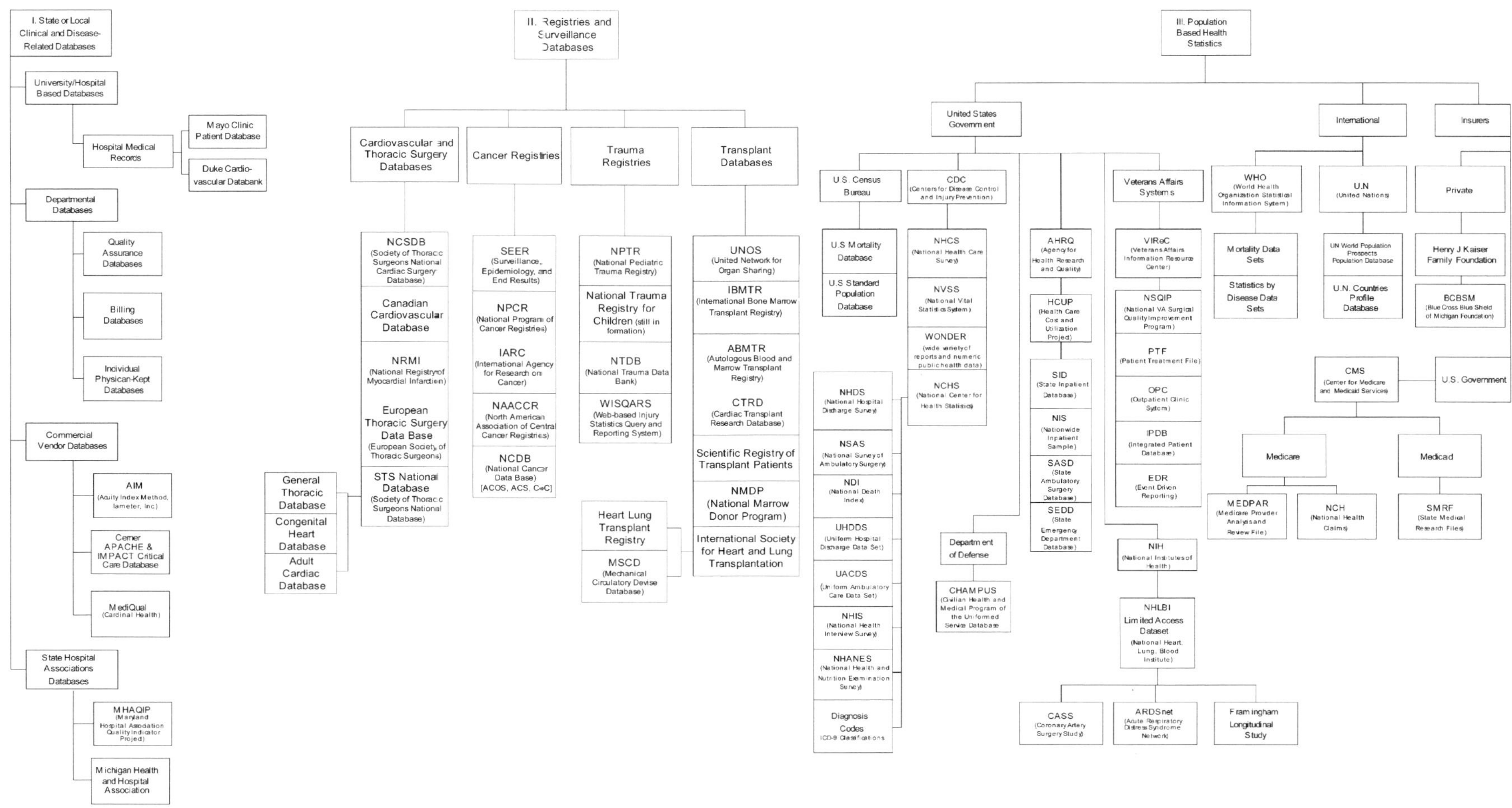

Figure 1: Overview of secondary datasets.

same control over what data have been collected or are available. Moreover, there are often missing or incorrect values, coding inaccuracies, and predictor variables or outcomes that were not measured *(20,43,49)*.

Perhaps the greatest threat to the validity of a clinical study using data that have already been collected is from the influence of bias and confounding *(50)*. The primary weakness of any nonrandomized prospective study is systematic errors in the design or conduct of a study, as well as the influence of unmeasured variables that are associated with the outcomes *(35)*. Common forms of bias in these studies include selection bias, measurement bias, misclassification bias, lead-time bias, recall bias, and publication bias. Selection bias occurs when different criteria are used to select a sample of patients or variables in creating a secondary dataset. For example, in a study of cognitive outcomes after coronary bypass surgery, it may be inappropriate to compare outcomes from a sample of patients who were enrolled in a clinical study of cerebral protection after bypass surgery to patients culled from a national institute on aging dataset. These two groups are likely different in other significant ways that may influence the incidence of antecedent cognitive decline. Similarly, if the methods to detect, measure, and classify cognitive decline in the two groups are significantly different, misclassification bias may also threaten the validity of the study. Complete populations may not be available in some datasets. MEDPARS, for example, only has information for inpatient admission. Thus patients admitted after a surgery are included and ambulatory patients, who go home on the same day of surgery, are not included *(20, 26)*. Also, studies using secondary data may be more susceptible to recall bias if new data are added retrospectively to the dataset. This is especially true if the data are from individuals surveyed after an event has occurred *(51)*. Other biases can result when the researchers must decide what variables to use when the outcomes are known.

These biases are especially prevalent with administrative datasets, especially when ICD-9-CM codes are used *(52)*. Not only are the codes entered by medical record clerks who try to pick appropriate codes from nonstandardized and often confusing medical records, but the numbers of codes allowed to describe a clinical situation are limited and may lead to bias if they are also picked in a nonstandardized fashion *(53)*. Moreover, Jencks et al *(54)* demonstrated that, by limiting the slots available for diagnostic codes in the ICD-9-CM system, patients with more severe comorbidities listed among their diagnoses had lower mortality rates. They found, paradoxically, that patients admitted with diabetes, angina, ischemic heart disease, or hypertension had lower mortality for a given primary diagnoses than patients admitted with the same diagnosis but none of these comorbidities. These investigators found that patients who died had a higher likelihood of having other, *acute* conditions listed on their ICD-9-CM codes. This would have the effect of making patient who died appear to have fewer chronic illnesses and comorbidities. These findings led to the modern UB-92 coding form having nine diagnostic codes (one principle and eight other diagnoses) and six procedure codes. Because there are only a fixed number of diagnostic slots available, there may be some degree of undercoding bias in these types of datasets *(26)* and the ability to develop comorbidity scores or case-mix adjust using these datasets is limited *(55,56)*.

Another potential source of error may occur when new codes are developed with numbers that are different from similar procedures or diagnoses. Coders and researchers may then have difficulty finding and combining similar codes such as laparoscopy hernia repair and laparoscopic lysis of adhesions. Another significant limitation of ICD-9-CM procedure codes is that they often fail to describe what was done in sufficient detail (e.g., nephrectomy [ICD-9-CM 55.53] does not indicate whether it was an open laparotomy or performed laparoscopically) *(26,30)*.

Another important risk of using secondary datasets is properly detecting and adjusting for confounding. Confounding occurs when a differential distribution of unmeasured variables exists among the samples of patients used to create a secondary dataset. Along with the known, measured variables, these confounding variables are also associated with the outcome and are linked with the risk factor of interest. They can cause associations to exist when in reality there are none and vice versa. This problem is often encountered in datasets where undercoding occurs. For example, in a study of outcome after high-risk operations such as pancreaticoduodenectomy, Birkmeyer and colleagues found that hospitals with high surgical volumes had almost fourfold lower mortality rates *(57)*. However, it is possible that some of the difference in mortality may be attributed to sicker patients having surgery at low-volume hospitals. As in this study, multivariate statistical models are often used to adjust for confounding by any measured variables. But even the most sophisticated modeling, techniques cannot adjust for unmeasured variables, and the likelihood of residual confounding exists.

The validity of secondary data has been the source of much debate, which centers on the accuracy of coding as compared to the gold standard of chart review *(58)* or prospectively collected data from a clinical trial. Some authors have suggested that, because encounter datasets are used for billing purposes, the financial incentives to code accurately may improve the validity of these datasets *(27,29)*. The Office of the Inspector General of the US Department of Health and Human Services has conducted several studies of the accuracy of coding for DRG assignment by sampling Medicare admissions data and reabstracting DRG information from the medical records *(26,59)*. These studies demonstrate coding errors, primarily from misclassifications, ranging from 21% in 1988 to 15% in 1992. An indication of how diagnostic information coded for DRGs may be biased is the finding that the Medicare case-mix indexes, which reflect the average effect of DRGs assigned to hospitalized patients, coding volume increased 32% in the first year after the case-mix indexes was introduced *(26)*. This increase in the use of more codes, reflecting higher severity or complexity, has been referred to as either DRG-Creep or "optimization." The increase in the case-mix indexes therefore may reflect purposeful financial manipulation of the coding or perhaps appropriate and accurate precision in coding practices. Regardless, when analyzing secondary datasets, one must be aware of differences and changes in coding practices over time, across geographic regions, and between institutions.

Variables that require special training or equipment to be accurately collected and measured may differ from site to site and therefore a larger registry may contain data of different quality even though they purportedly measure the same thing. Similarly, data that are subject to interpretation, or for which different descriptions are possible, such as diagnostic and procedural codes/descriptions, may be different between centers or geographic regions or have changed over time. Poor documentation of the secondary dataset is not uncommon and requires a great deal of data "cleaning." Electronic formats for how the data are stored may be incompatible with other data that one would like to merge into a secondary dataset. Occasionally, the raw data are no longer available and the existing data may have been modified (such as average blood pressure or highest or worse physiologic values recorded), limiting the secondary analysis. It is important that any variables that can identify a patient be removed from a dataset when it is ready for analysis. This is important for a secondary dataset researcher to ensure that it is done. However, it can cause a problem if the investigator wishes to merge separate datasets. Finally, though the regulatory hurdles that would otherwise increase the workload of the investigator using secondary datasets are less when dealing with data that have already been collected, there are still costs that can be encountered and sometimes extensive documentation may be required.

7. HOW TO GET STARTED

After an appropriate study question has been chosen, the researcher should identify potential sources of data. Figure 1 shows a number of sources from which one can find existing datasets that are readily available for research, and Table 5 provides Internet addresses to access these or obtain more information. The most commonly used datasets include state hospital discharge datasets, NHDS, MEDPAR, SEER, and the VA's patient treatment file (PTF) and National Surgical Quality Improvement Program (NSQIP). There are other large consortiums of research groups that have, and are conducting, large multicenter clinical trials that also may be good sources of data, especially after the primary studies have been published. A good starting point to determine what clinical studies are ongoing and may be the source of future data is through the National Institutes of Health's clinical trials summary site located at www.ClinicalTrials.gov. Perhaps the most direct route is to collaborate with local investigators who have conducted research in the areas of interest. One can also search the medical literature to find investigators who have published in the areas of interest, contact them, and determine if they have datasets that may be used in collaboration.

After a dataset has been located, it may be obtained for free or for a reasonable price. One exception is the data from private sources such as insurance firms and other payers (Kaiser, Blue Cross Blue Shield, Aetna), which is not generally available unless an investigator is working in collaboration with that group. When the data are available, they are often provided on CD-ROMs with "deidentified" data for all admissions to acute care hospitals for a specific year. Usually, the investigators must sign a data use agreement stating that they will not attempt to identify individual patients using the data. Such identification is generally not a problem unless studying a very rare medical condition.

Before using the data, one should obtain approval from the Institutional Review Board (IRB). Many times, with "deidentified" administrative datasets, the project may be exempt from review or will be expedited by an IRB because there are no risks to patients. But it is worth checking with the IRB and having a letter on file that documents the exemption. Also, it is important to note that for some administrative datasets such as the Medicare files, which are not "deidentified," the data use agreements limit how the data can be used and the IRB must review and approve the project. This additional level of approval is also necessary for many other secondary data sources.

7.1. Organizing and Analyzing the Data

Data can be collected and stored in a variety of methods, but ultimately, secondary data analysis relies on electronic data that are retrievable and in a digitized format. The data on CD-ROMS will usually come as a number of coded text files that may be difficult to access. The data will need to be imported, in part or in total, to a database management or statistical software package. The mainstay of secondary data analysis is a data matrix

Table 5
Selected Websites for Databases and Registries Listed in Figure 1

Section I. State or local clinical and disease-related databases

Acuity Index Method (AIM); Iameter, Inc.: http://www.iameter.com/iameterProducts.htm
APACHE/IMPACT (Cerner Corporation): http://www.cerner.com/products/products_4a.asp?id=2694
Duke Cardiovascular Databank: http://heartcenter.mc.duke.edu/
Maryland Hospital Association Quality Indicator Project: http://www.qiproject.org/
Mayo Clinic Patient Database: http://www.mayo.edu/
MediQual Dataset: http://www.mediqual.com/
Michigan Health and Hospital Association: http://www.mhaservicecorp.com/

Section II. Registries and surveillance databases

Autologous Blood and Marrow Transplant Registry (ABMTR): http://www.ibmtr.org/
Cardiac Transplant Research Database (CTRD): http://www.ctrd.org/
European Thoracic Surgery Database: http://www.ests.org.uk/
International Agency for Research on Cancer (IARC): http://www.iarc.fr/
International Bone Marrow Transplant Registry (IBMTR): http://www.ibmtr.org/
International Society for Heart and Lung Transplantation: http://www.ishlt.org/registries/
Mechanical Circulatory Device Database: http://www.ishlt.org/registries/mcsdDatabase.asp
National Cancer Database (NCDB): http://www.facs.org/cancer/ncdb/index.html
National Marrow Donor Program: http://www.marrow.org/
National Program of Cancer Registries (NPCR): http://www.cdc.gov/cancer/npcr/
National Registry of Myocardial Infarction: http://www.nrmi.org/index.html
National Trauma Data Bank (NTDB): http://www.facs.org/trauma/ntdb.html
National Trauma Registry for Children (NTRC): http://www.eapsa.org/outcomes/ntrc/index.htm
North American Association of Central Cancer Registries (NAACCR): http://www.naaccr.org/
Scientific Registry of Transplant Registrants: http://www.ustransplant.org/index.php
Society of Thoracic Surgeons National Database: http://search.ctsnet.org:8000/
Surveillance, Epidemiology, and End Results: http://seer.cancer.gov/
United Network for Organ Sharing: http://www.unos.org/data/
Web-based Injury Statistics Query and Reporting System (WISQARS): http://www.cdc.gov/ncipc/wisqars/

Section III. Population health-based statistics

ARDS Network: http://www.nhlbi.nih.gov/resources/deca/descriptions/ards.htm
Blue Cross Blue Shield of Michigan (BCBSM) Foundation: http://www.bcbsm.com/foundation/gp_iip.shtml
CASS (Coronary Artery Surgery Study): http://www.nhlbi.nih.gov/resources/deca/descriptions/cass.htm
Centers for Disease Control and Prevention CDC Wonder: http://wonder.cdc.gov/
Centers for Medicare and Medicaid Services (CMS): Medicaid http://www.cms.hhs.gov/medicaid/
CHAMPUS (Civilian Health and Medical Program of the Unformed Services) Database: http://www.tricare.osd.mil/training/tmart/index.cfm?fx=cmis
Diagnosis Codes: ICD-9 Classifications: http://www.cdc.gov/nchs/icd9.htm
Framingham Longitudinal Study: http://www.nhlbi.nih.gov/about/framingham/index.html
Healthcare Cost and Utilization Project: http://www.ahrq.gov/data/hcup/
Henry J Kaiser Family Foundation: http://www.statehealthfacts.org/

(continued)

Table 5 *(Continued)*

Medicare Provider Analysis and Review File (MEDPAR): http://www.cms.hhs.gov/data/purchase/directory.asp#ntl
National Center for Health Statistics: http://www.cdc.gov/nchs/
National Death Index: http://www.cdc.gov/nchs/r&d/ndi/ndi.htm
National Health and Nutrition Examination Survey: http://www.cdc.gov/nchs/nhanes.htm
National Health Care Survey: http://www.cdc.gov/nchs/nhcs.htm
National Health Interview Survey: http://www.cdc.gov/nchs/nhis.htm
National Hospital Discharge Survey: http://www.cdc.gov/nchs/about/major/hdasd/nhdsdes.htm
National Surgical Quality Improvement Project: http://www.nsqip.org/
National Survey of Ambulatory Surgery: http://www.cdc.gov/nchs/about/major/hdasd/nsasdes.htm
National Vital Statistics System: http://www.cdc.gov/nchs/nvss.htm
NHLBI Limited Access Data Sets: http://www.nhlbi.nih.gov/resources/deca/default.htm
US Census Bureau: http://www.census.gov/
Uniform Ambulatory Care Data Set: http://mchneighborhood.ichp.edu/eds/901027902.html
Uniform Hospital Discharge Data Set: http://mchneighborhood.ichp.edu/eds/901027520.html
United Nations Countries Profiles Database: http://unstats.un.org/unsd/cdb/cdb_country_prof_select.asp
United Nations Statistical Databases: http://millenniumindicators.un.org/unsd/databases.htm
United Nations World Population Prospects Population Database: http://esa.un.org/unpp/
Veterans Affairs Information Resource Center (VIReC): http://www.virec.research.med.va.gov/
World Health Organization European National Health Statistics: http://www.euro.who.int/InformationSources/Data/20010827_1
World Health Organization Mortality Database: http://www.ciesin.org/IC/who/MortalityDatabase.html
World Health Organization Statistical Information System: http://www3.who.int/whosis/menu.cfm

of variables listed in such a way that statistical software such as STATA, SPSS, or SAS can be used to "clean" the data, summarize them, and ultimately do statistical analyses. In some cases, the necessary code (directions for the computer) is included for the most commonly used statistical and data management software. For example, the Healthcare Cost and Utilization Project datasets are provided with the code necessary to import the text files into two statistical programs, SAS and SPSS. These programs will import the data, create variable names, and generate labels for the variables.

A common starting point is to select an appropriate study population. After you are able to query the database, the appropriate population should be selected. In a clinical or primary study dataset, the patients of interest are usually defined by variables with diagnoses or procedures. In an administrative datasets, patients are selected using ICD-9 diagnostic or procedure codes. To select patients who had an operation, the ICD-9 procedure codes should be determined. Because the codes chosen define the study population, it's important to use the right combination of codes. It is as easy to exclude a relevant code as it is to include an irrelevant code. One approach to aid code selection is to identify what codes were used in previous relevant publications, which have focused on similar predictors and outcomes as for one's own proposed study. Another valuable method to ensure the correct ICD-9-CM or CPT codes are being used from a larger data query is to refer to a local hospital billing database for a sample of patients who are known to have had the operation of interest. In any case, the final study sample size that is selected should be consistent with known incidence and prevalence data regarding the condition or procedure of interest.

The variables are then organized into a statistical software program beginning with identification and demographic variables, followed by other clinical or descriptive variables and finally outcome variables (Table 1). It is advisable from the planning and initiation phases of a study to have a statistician be part of the research team to discuss and advise on how the data will be delivered, stored, and analyzed to be sure that all variables are properly handled during data compiling, cleaning, and analysis. If the primary researcher does not have significant statistical methods training, analysts or programmers are essential to the success of the study. Many of these large administrative datasets will require a large amount of computer storage capacity and analytic time. It would not be unusual to commit an entire server to just one of these datasets.

8. EXAMPLES OF SPECIFIC DATASETS

For purposes of organization, available clinical datasets have been divided into three broad categories (Figure 1). The first consists of specific health care databases (Figure 1, Section I). These contain data from specific clinical trials or from individual physician datasets of patients they have cared for. This category also includes hospital datasets from medical records departments, specific clinical departmental databases, and hospital clinical benchmarking/quality datasets. An example of this type of dataset is the clinical and case-mix (diagnosis, demographics) data contained in an ICU benchmarking system such as the APACHE score *(4)* or the Mortality Prediction Model component of the IMPACT ICU model *(60)*. Another source of datasets is the multitude of registries and surveillance datasets established by medical societies such as the American Thoracic Society's cardiac surgery database, or larger registries of multiple datasets such as the SEER (Figure 1; Section II). Finally, there are numerous state and federal governmental and administrative datasets that are often used in secondary analysis such as the MEDPARS, NHDS, National Death Index, and NSQIP (Figure 1; Section III).

8.1. Prospectively Acquired Clinical and Outcome Datasets

A good example of a secondary dataset that contains prospectively acquired clinical and outcome data is the type found in a clinical information system such as the APACHE ICU risk-adjustment system. A hospital that has this, or a related type of system, will have among the most clinically valid data and precise outcome measures available. These datasets not only contain demographic and diagnostic information, they generally also contain a number of physiologic variables such as worst recorded values for vital signs, laboratory values, presence of vasopressor or inotropic support, fluid balances, and outcome data *(4)*. Because of the precise collection of variables in an intensive care unit, the evaluation of predictors of outcome is more likely to be valid. These datasets are limited to patients admitted to a particular area of the hospital such as the ICU where detailed information is collected. Furthermore, ICUs that collect these kinds of data are usually located in fairly large hospitals so that the information obtained may not be applicable to smaller hospitals, outpatient surgical patients, and procedures in which patients are usually not admitted to and ICU postoperatively.

8.2. State Administrative Databases

Data collected by states have resulted in several influential surgical studies, but the data have been limited by the inconsistencies in how states collect clinical information. For example, only 31 states report data on state Medicaid beneficiaries, and several studies using these datasets have encountered significant inconsistencies in how populations are defined, how variables are coded, and limited diagnoses fields available *(26,61)*. Most notably, state data have provided insights into the volume-outcome effect for high-risk procedures and allowed the study of geographic variations in the utilization of new surgical technologies *(6)*. Many statewide databases can be obtained, for a low price, directly from state hospital associations. Alternatively, the Agency for Healthcare Research and Quality makes available several state databases through its website.

One of the most highly cited studies of the volume-outcome effect was published using the Maryland state administrative database *(62,63)*. The investigators demonstrated a large difference in mortality rates between high- and low-volume hospitals for pancreatic resection. From 1988 to 1993, there were 502 patients who underwent Whipple procedures, with a mortality rate of 2.2% for the single high-volume provider compared with 13.5% at the 38 remaining Maryland hospitals. Several additional state and national datasets have confirmed these findings. As a result, there are several groups, most visibly the Leapfrog group, advocating for selective referral of patients requiring pancreatic resection to high-volume providers *(64)*.

Another useful attribute of state databases is the ability to study changes in hospital referral patterns within an area. For instance, in a second study in Maryland, the market share of all pancreatic resections in the state was found to increase from 21% to 59% at the only high-volume hospital over the study period. These studies highlight the ability of administrative data to compare outcomes across a broad range of institutions *(62)*.

Dissemination of new surgical technology can also be investigated using state administrative datasets. Often, less-invasive new technology can lower the threshold for intervention and increase utilization rates for the procedures. With this in mind, a study combining administrative datasets for Vermont, New Hampshire, and Massachusetts looked at the rate of surgery for gastroesophageal reflux disease *(65)*. The population-based rate of antireflux surgery more than doubled over 5 yr (1993–1998) from 4.8 to 11.7 per 100,000. During the same period, laparoscopic antireflux surgery was found to increase more than sixfold (1.2–8.9 per 100,000). In addition to the change over time, the rates of surgery varied across geographic regions. During the most recent 2 yr (1997–1998), rates of antireflux surgery varied nearly fivefold (5.4–24.5 per 100,000) across hospital referral regions.

The principal disadvantage of using state databases is the lack of external validity. States may have unique hospital referral patterns that make a comparison with other areas difficult. For instance, in the above studies on pancreatic resection, Maryland had only a single high-volume provider *(63)*. Other factors may also make it hard to generalize from one state to another. For instance, not every state has the same size population, demographic composition, or rates of disease.

8.3. National Administrative Databases

Datasets that are representative of all US hospitals are advantageous when a researcher wishes to generalize his or her findings to the broadest patient population possible. However, there is no dataset that contains 100% of all patient discharges in the United

States. The data that do exist are subsets of hospital data (e.g., Nationwide Inpatient Sample) or patients (e.g., Medicare inpatient database for acute care hospitalizations.). The datasets that are samples of hospitals are chosen so they represent all US hospitals by containing discharge "weights" that can be applied to their sample in order to extrapolate to the entire nation. For instance, the Nationwide Inpatient Sample (NIS) is a stratified, random sample of 20% of US hospitals. For the NIS, hospital and discharge "weights" are calculated so they represent the universe of hospitals in the American Hospital Association survey. The "weight" is inversely proportional to the probability of being selected in the sample. It is important to use these weights correctly when using the NIS, or other hospital samples, when studying national trends in utilization or outcomes. There are special "survey" commands that must be used with statistical software to obtain correct estimates of the standard error and other statistics when using datasets that are stratified samples.

With the emergence of new surgical techniques there is potential for precipitous changes in the utilization rates. For instance, recently, less-invasive laparoscopic technology for obesity surgery has been introduced. Using the NIS database, the rate of obesity surgery from 1990 to 1997 was found to more than double from 2.7 to 6.3 per 100,000 adults *(66)*. As previously mentioned, evaluating specific operations using administrative data can be limited by the use of ICD-9 codes. For instance, there is no specific code for the laparoscopic approach in obesity surgery *(30)*. Thus changes in the approach to surgery cannot be directly tested. It is important to consider the limitations of the ICD-9 coding system when interpreting the results of any study using this source of data.

8.4. SEER

The SEER Program of the National Cancer Institute *(24)* (Figure 1; Section II) contains cancer incidence and survival rates in the United States from 14 population-based cancer registries and 3 supplemental registries covering approx 26% of the US population. The SEER registry also includes information on more than 3 million *in situ* and invasive cancers with approx 170,000 new cases added each year. The SEER program routinely collects data on patient demographics, primary tumor site, morphology, cancer stage at diagnosis, first course of treatment, and follow-up for vital status. The SEER program is a comprehensive source of population-based information in the United States that includes stage of cancer at the time of diagnosis and also includes survival rates within each stage. The mortality data reported by SEER are provided by the NCHS.

Reportable records for a patient are sent from pathology laboratories, doctors' offices, and hospitals to the SEER where a patient set is either created or modified. Variables are collected from three sources: Health Records, Supplemental Records, and other non-health information sources. A health record contains medical information, such as an autopsy, cytology, hematology, oncology, pathology, radiology, and radiotherapy reports. Hospital discharge files and the disease index, surgery logs, death certificates, obituaries, and other similar sources are also included. Supplemental records are those in which related data are contained but which contain no health information. Supplemental data are used to verify personal data and follow-up information, and may include department of motor vehicle records, insurance demographic data, Internal Revenue Service records, state birth records, and voter registration information. Non-health information sources are limited to census tract information and a name list for race and ethnicity. When the patient set is finalized, it is submitted to the SEER database at the National Cancer Institute. The SEER database is updated annually. Registry data are submitted electronically without personal identifiers biannually.

SEER data are available for public use and are free of charge. The SEER 1973–2001 Public-Use data are available in the binary format required by the SEER*Stat software and in an ASCII text format which can be analyzed with an investigator's own statistical software. A signed SEER Public-Use Data Agreement is required to access the data. Any publications resulting from the use of SEER data requires a citation for each database provided by SEER and should include information about the data submission and release date.

SEER data and SEER*Stat software are available using three methods. The most common method for accessing the SEER Public-Use Data is to use the SEER*Stat to access the data through the investigator's Internet connection (SEER*Stat's client-server mode). Alternatively, an investigator may have CDs containing the data and SEER*Stat software shipped to them. These CDs include the binary and text versions of the data. Finally, investigators may download two compressed files containing the CD images. SEER*Stat, the binary data, as well as the text data. The data are constantly checked for quality and completeness of data reported. Populations covered are comparable to the general U.S. population with regard to measures of poverty and education. A disadvantage of the SEER may be the lack of applicability to some settings since the SEER populations tend to be somewhat more urban and have a higher proportion of foreign-born persons than the general U.S. population.

Recent uses of the SEER registry data can be found in articles discussing the treatment of ductal carcinoma *in situ* of the breast *(67)*, valvular dysfunction and carotid, subclavian, and coronary artery disease in survivors of Hodgkin's lymphoma treated with radiation therapy *(68)*, and morbidity after radical prostatectomy *(69)*.

8.5. Medicare Provider Analysis and Review (MEDPAR)

Perhaps the most frequently used secondary dataset in the surgical literature has been the Medicare Provider Analysis and Review (MEDPAR) File (Figure 1, Section III). MEDPAR is a limited dataset (LDS) which contains inpatient hospital and skilled nursing facilities (SNF) records for services provided to Medicare beneficiaries admitted to Medicare-certified hospitals and SNFs *(70)*. National MEDPAR Data Base consists of approx 14 million records representing Medicare beneficiaries. About 12 million records are from hospitals and two million records are from SNF. There is also a subset of the national MEDPAR LDS that has hospitals abstracted by the state of residence of the Medicare beneficiary *(1)*. Data are available in 500 character, fixed-length record format for fiscal and calendar years from 1987 to the present *(1)*. MEDPAR files have individual hospital stay records coded in ascending order by health insurance claim number, along with the admission date, and provider number so that the beneficiaries' records can be linked to inpatient histories. Specific information for each record includes the designated health insurance claim number, as well as the provider number. Other information comprise beneficiary demographic characteristics (age, gender, race), admission data, diagnosis and surgery information, hospital or SNF resources used, detailed charge data, days of care, and entitlement data (Table 3) *(1)*. The layout of a MEDPAR record including coding dictionary can be found at http://www.cms.hhs.gov/data/purchase/Medpar.

The MEDPAR datasets are a valuable tool to health care researchers who require beneficiary or facility-specific information. MEDPAR continues to be upgraded to permit an increasing level of detail on claims submitted. MEDPAR includes critical data on accommodations, services, and costs associated with each inpatient and SFN stay. These data allow researchers to follow patterns and outcomes of care provide to Medicare beneficiaries *(1)*. The collection of information on 14 million beneficiaries allows great generalizability using the MEDPAR data. In MEDPAR, comorbidity information is limited and other key determinants of outcomes, such as provider caseload, are not available *(26)*. Other limitations include those of any administrative database which has been previously discussed.

Access to MEDPAR is restricted because it contains patient identifiers. The Expanded Modified MEDPAR-Hospital File (National) is available to persons qualifying under the terms of the Routine Use Act as outlined in the December 24, 1984, Federal Register and amended by the July 2, 1985. This file consists of approximately seven 3490E cartridges and the cost to eligible investigators is $3,655 per year. The Expanded Modified MEDPAR-Hospital File (State), abstracted by provider state or beneficiary state of residence, consists of one 3940E cartridge and the cost to eligible investigators is $1,080 per year *(1)*. Recent articles in which the MEDPAR File has been used include studies of outcomes of surgeon volume and operative mortality rate,[70] coronary stent outcomes in a Medicare population *(71)*, and hospital cost of endovascular repair compared with open repair of aortic aneurysms *(72)*.

8.6. The National Hospital Discharge Survey (NHDS)

The National Hospital Discharge Survey (NHDS) (Figure 1, Section II) is an annual survey developed by the National Center for Health Statistics (NCHS) in 1965. It samples hospital and discharge records for medical and demographic data. In particular, institutions from which the records are sampled are non-federal short-stay hospitals that have average lengths of stay less than 30 d for all patients. The NHDS contains approx 270,000 inpatient records taken from a US sample of approx 500 hospitals. The NHDS is used for calculating statistics on hospital utilization and on the nature and treatment of illness among the hospitalized population *(72)*. "The database uses a national, stratified multistage probability sample in which patient records are selected at random and weighted to represent more than 30,000,000 hospitalizations that occur annually" *(73)*. NHDS data comprise age, sex, race, ethnicity, marital status, and expected sources of payment. Admission and discharge dates and discharge status are also collected. Coding using the ICD-9 classification permits collection of patient diagnoses and procedures *(72)*.

The system for data collection by NHDS began in 1965. It was redesigned in 1985 to improve efficiency and analytic capabilities of the survey by adding discharge data available in electronic forms and linking the NHDS to the design of the NCHS's National Health Interview Survey *(74)*. NHDS data are available in publications, on public use data tapes, data diskettes, CD-ROMs, and files that can be downloaded from the internet. Individual year public use data files can be downloaded directly from the NCHS at http://www.cdc.gov/nchs/about/major/hdasd/nhds.htm. Multiyear data files for analysis of trends may be obtained on CD-ROM at a current cost between $305 and $454, depending on the year. These and other products are available at the NCHS Data Dissemination Branch at 301-458-INFO or by e-mail at NCHSquery@cdc.gov.

Recent uses of the NHDS can be found in articles discussing rates of lower extremity amputation and arterial reconstruction *(75)*, health economic benefits in supplemental calcium for the prevention of hip fractures *(76)*, trends in prostatectomy among black and white men *(77)*, abdominal aortic aneurysm repair *(72)*, and patterns of inpatient surgeries for cancers of the lung, female breast, prostate, and the colon and rectum *(78)*.

8.7. National Surgical Quality Improvement Program

In 1986, the US Congress passed a law requiring VA hospitals to track the outcomes of surgical patients. The NSQIP (Figure 1; Section III) is a national, validated, outcome-based, risk-adjusted program for the measurement and enhancement of the quality of surgical care. The NSQIP incorporates 128 Veterans Affairs Medical Centers and 10 beta sites in the private sector. The program compares the quality of surgical care among all 132 VA hospitals in the United States *(79)*. Based on the results of the National Virginia Risk Study (NVASRS) and the VA Continuous Improvement in Cardiac Surgery Program (CICSP), the VA established the NSQIP in 1994 in all the medical centers performing major surgery *(80)*. In addition to reporting results, the National Virginia Surgical Risk Study (NSQIP) was designed to provide feedback to the individual institutions for quality improvement. Since the NSQIP began collecting data, there has been a 27% reduction in the 30-d mortality rate and a 45% reduction in the 30-d morbidity rate in VA hospitals *(81)*.

Besides providing reports on hospital performance and feedback for quality improvement, the NSQIP provides a rich resource for secondary data analysis. Unlike administrative databases, the NSQIP has detailed clinical data. Information on preoperative patient characteristics, intraoperative variables, and postoperative complications are available for each patient. To further ensure the quality of data collection, each center receives an audit with repeated data abstraction and estimates of interrater reliability.

With the large number of patients and rich clinical detail, many investigators have used this data source for research purposes *(82–85)*. Several published studies have focused on the validity of the NSQIP methods *(86, 87)*, whereas other reports have been focused on the risk of adverse outcomes for particular procedures *(88)*. For example, one recent study investigated the frequency of stroke after noncarotid vascular surgery *(89)*. Given the low stroke rate (<0.6%), single institution studies would not provide meaningful findings. However, using the NSQIP, there were 2551 abdominal aneurysm repairs, 2616 aortobifemoral bypass operations, 6866 lower extremity bypasses, and 7442 major lower extremity amputations *(79)*. Using this large dataset, the authors were able to demonstrate several risk factors for stroke (i.e., mechanical ventilation, previous stroke, and return to the operating room). Other recent uses of the NSQIP can be found in articles discussing a demonstration project of NSQIP in non-VA hospitals *(64)*, surgeon volume and operative mortality, outcomes in transthoracic versus transhiatal esophagectomy patients, pancreaticoduodenectomy for periampullary cancer *(64)*, and appendicitis *(80,81)*.

Despite the clear advantages of the NSQIP compared with administrative databases, there are some problems with it. First, the external validity is questionable. Patients in VA hospitals tend to be older males and do not represent the entire US population. Second, the data are not universally available to the public. The database may only be accessed by VA investigators and their institutional partners, such as non-federal academic centers (e.g., Emory University, the University of Kentucky, and the University of Michigan) as well as a recent private sector initiative with the American College of Surgeons involving 11 additional university hospitals *(90)*. The NSQIP provides an infrastructure for VA investigators to query the database and produce scientific presentations and publications. The NSQIP executive board must approve the use of the data through a formal review process. Forms can be requested from the NSQIP and a standard proposal is submitted. Both university and VA hospital IRB approval is also required for the release of data to investigators. Recently, the NSQIP approach has been expanded into the private sector, with data from more than 50,000 operations from more than a dozen hospitals now available *(90, 91)*. This additional data will be an added resource that may help to overcome the problems with external validity. So far, however, most of the medical centers are large academic centers and the data are therefore not yet applicable to all US hospitals.

9. CONCLUSIONS

Secondary data analysis is a commonly used study methodology in the surgical literature. Because of the expanding use and number of clinical and administrative datasets, it is likely that more research pertaining to surgical patients will be done in this manner. The strengths of large relatively easy to obtain datasets must be tempered by the limited clinical and diagnostic precision found in many of them. In the future, merging administrative data with more clinically granular data will likely improve prognostication, quality improvement efforts (such as the NSQIP is already accomplishing), and cost-effective analyses. It is also possible that in the future regulatory agencies and payers of health care will demand these kinds of data to support continued patient referrals or care. Medicare populations and patients belonging to a managed care organizations may be among the first to require these data.

REFERENCES

1. Lu-Yao GL, McLerran D, Wasson J, Wennberg JE. An assessment of radical prostatectomy. Time trends, geographic variation, and outcomes. The Prostate Patient Outcomes Research Team. JAMA 1993;269:2633–2636.
2. O'Connor GT, Plume SK, Olmstead EM, et al. A regional prospective study of in-hospital mortality associated with coronary artery bypass grafting. The Northern New England Cardiovascular Disease Study Group. JAMA 1991;266:803–809.
3. Framingham Heart Study. (Accessed , at http://www.nhlbi.nih.gov/about/framingham/bib-menu.htm).
4. Rosenberg AL, Zimmerman JE, Alzola C, Draper EA, Knaus WA. Intensive care unit length of stay: recent changes and future challenges. Crit Care Med 2000; 28: 3465–3473.
5. Rosenberg AL, Hofer TP, Strachan C, Watts CM, Hayward RA. Accepting critically ill transfer patients: adverse effect on a referral center's outcome and benchmark measures. Ann Intern Med 2003; 138:882–890.
6. Dimick JB, Cattaneo SM, Lipsett PA, Pronovost PJ, Heitmiller RF. Hospital volume is related to clinical and economic outcomes of esophageal resection in Maryland. Ann Thorac Surg 2001;72:334–9; discussion 339–341.
7. Sirio CA, Angus DC, Rosenthal GE. Cleveland Health Quality Choice (CHQC)—an ongoing collaborative, community-based outcomes assessment program. New Horiz 1994;2:321–325.
8. Baker DW, Einstadter D, Thomas CL, et al. Mortality trends during a program that publicly reported hospital performance [see comment]. Med Care 2002;40:879–890.
9. Baker DW, Einstadter D, Thomas C. et al. The effect of publicly reporting hospital performance on market share and risk-adjusted mortality at high-mortality hospitals. Med Care 2003;41:729–740.
10. Clough JD, Engler D, Snow R, Canuto PE. Lack of relationship between the Cleveland health quality choice project and decreased inpatient mortality in Cleveland. Am J Med Qual 2002;17:47–55.
11. Rosenthal GE, Hammar PJ, Way LE, et al. Using hospital performance data in quality improvement: the Cleveland health quality choice experience. Jt Comm J Qual Improve 1998 24:347–360.
12. University Health Consortium. (Accessed 2004, at http://www.uhc.edu/).
13. Adams SW, Schultz S 2nd, Elias A, et al. Using comparative clinical information to understand practice patterns and affect organizational change. Proc Annu Symp Comp Applic Med Care 1991;938–940.
14. Ozbolt JG. From minimum data to maximum impact: using clinical data to strengthen patient care. MD Comp 1997;14:295–301.
15. Schultz S. A new era dawns for the university hospital consortium. Interview by Carolyn Dunbar. Comp Healthcare 1992;13:32–37.

16. O'Neal PV, Ozcan YA, Ma Y. Benchmarking mechanical ventilation services in teaching hospitals. J Med Sys 2002;26:227–240.
17. Wennberg DE, Lucas FL, Birkmeyer JD, Bredenberg CE, Fisher ES. Variation in carotid endarterectomy mortality in the medicare population: trial hospitals, volume, and patient characteristics [see comment]. JAMA 1998;279:1278–1281.
18. Anonymous. Beneficial effect of carotid endarterectomy in symptomatic patients with high-grade carotid stenosis. North American symptomatic carotid endarterectomy trial collaborators [see comment]. N Engl J Med 1991;325:445–453.
19. Executive Committee for the Asymptomatic Carotid Atherosclerosis Study. Endarterectomy for asymptomatic carotid artery stenosis. JAMA 1995;273:1421–1428.
20. Birkmeyer JD. Using administrative data for clinical research. In: Souba WW, Wilmore DW, eds. Surgical research. San Diego, CA: Academic Press, 2000.
21. Birkmeyer JD, Warshaw AL, Finlayson SR, Grove MR, Tosteson AN. Relationship between hospital volume and late survival after pancreaticoduodenectomy. Surgery 1999;126:178–183.
22. PA Health Care Cost Containment Council (PHC4). A consumer guide to coronary artery bypass graft surgery, 1993 data. PA health care cost containment council (PHC4). Harrisburg, PA: PA Health Care Cost Containment Council, 1995.
23. Potosky AL, Merrill RM, Riley GF, et al. Breast cancer survival and treatment in health maintenance organization and fee-for-service settings. J Natl Cancer Inst 1997;89:1683–1691.
24. Potosky AL, Warren JL, Riedel ER, et al. Measuring complications of cancer treatment using the SEER-Medicare data. Med Care 2002;40:IV-62–IV-68.
25. McCarthy EP, Bruns KM, Freund KM, et al. Mammography use, breast cancer stage at diagnosis, and survival among older women. J Am Geriatr Soc 2000;48:1226–1233.
26. Iezzoni LI. Coded data from administrative sources. In: Iezzoni LI, ed. Risk adjustment. Chicago, IL: Health Administration Press, 2003; 83–138.
27. Pronovost P, Angus DC. Using large-scale databases to measure outcomes in critical care. Crit Care Clin 1999;15:615–631.
28. Rubenfeld GD, Angus DC, Pinsky MR, et al. Outcomes research in critical care: results of the American Thoracic Society Critical Care Assembly Workshop on Outcomes Research. The Members of the Outcomes Research Workshop. Am J Respir Crit Care Med 1999;160:358–367.
29. Mitchell JB. Using physician claims to supplement hospital data in medical effectiveness research data methods AHCPR #92-0056. In: Agency for healthcare policy and research, 1998; 77–83.
30. McMahon LFJ, Smits HL. Can Medicare prospective payment survive the ICD-9-CM disease classification system? Ann Intern Med 1986;104:562–566.
31. Simborg DW. DRG creep: a new hospital-acquired disease. N Engl J Med 1981;304:1602–1604.
32. Averill RF, Mullin RL, Steinbeck BA, Goldfield NI, Grant TM. Development of the ICD-10 Procedure Coding System (ICD-10-PCS). J AHIMA 1998;69:65–72.
33. About the International Classification of Diseases, Tenth Revision Clinical Modification and (ICD-10-CM). National Center for Health Statistics, Centers for Disease Control and Prevention, 2004.
34. Moller H, Skakkebaek NE. Risk of testicular cancer in subfertile men: case-control study. BMJ 1999;318:559–562.
35. Rosenberg AL, Wei JT. Clinical study designs in the urologic literature: a review for the practicing urologist. Urology 2000;55 468–476.
36. Hunt DL, McKibbon KA. Locating and appraising systematic reviews. Ann Intern Med 1997;126:532–8.
37. Cochrane. (Accessed December 1, 2004, at http:www.cochrane.org/index0.htm).
38. NGC—National Guideline Clearinghouse. (Accessed January 4, 2003, at http:www.guideline.gov/).
39. Hunter JE, Schmidt FL. Fixed effects vs. random effects meta-analysis models: implications for cumulative research knowledge. Int J Selection Assess 8:275–292.
40. Rodgers A, Walker N, Schug S, et al. Reduction of postoperative mortality and morbidity with epidural or spinal anaesthesia: results from overview of randomised trials. BMJ 2000;321:1493.
41. Haynes RB, Wilczynski N, McKibbon KA, Walker CJ, Sinclair JC. Developing optimal search strategies for detecting clinically sound studies in MEDLINE. J Am Med Inform Assoc 1994;1:447–458.
42. Huston P. Cochrane Collaboration helping unravel tangled web woven by international research. CMAJ 1996;154:1389–1392.
43. Hulley SB, Cummings SR, Browner WS, et al, eds. Research using existing data: secondary data analysis, ancillary studies, and systematic reviews. Philadelphia, PA: Lippincott Williams & Wilkins, 2001.
44. Hearst N, Grady D, Barron HV, Kerlikowske K. Research using existing data: secondary data analysis, ancillary studies, and systematic reviews. Philadelphia, PA: Lippincott Williams & Wilkins, 2001.

45. NHLBI—Dataset descriptions—ARDSNet. (Accessed December 2, 2004, at http:www.nhlbi.nih.gov/resources/deca/descriptions/ards.htm.).
46. Rosenberg AL, Hofer TP, Hayward RA, Strachan C, Watts CM. Who bounces back? Physiologic and other predictors of intensive care unit readmission. Crit Care Med 2001;29:511–518.
47. Tremper KK, O'Reilly M, Kazanjian P, Van der Spek A, Kheterpal S. A perioperative information system: design and implementation. In Katz, R, ed. Seminars in Anesthesia Perioperative Medicine and Pain; Perioperative Information Systems. 2004;23:72–85.
48. Martin GS, Mannino DM, Eaton S, Moss M. The epidemiology of sepsis in the United States from 1979 through 2000. N Engl J Med 2003;348:1546–1554.
49. Romano PS, Harold LS. Getting the most out of messy data: problems and approaches for dealing with large administrative sets. In: Grady ML , Schwartz HA, eds. Medical effectiveness research data. 1992.
50. Poses RM, Smith WR, McClish DK, Anthony M. Controlling for confounding by indication for treatment. Are administrative data equivalent to clinical data? Med Care 1995;33:AS36–AS46.
51. Sackett DL. Bias in analytic research. J Chronic Dis 1979;32:51–63.
52. Romano PS, Roos LL, Luft HS, Jollis JG, Doliszny K. A comparison of administrative versus clinical data: coronary artery bypass surgery as an example. Ischemic Heart Disease Patient Outcomes Research Team. J Clin Epidemiol 1994;47:249–260.
53. Iezzoni LI, Foley SM, Daley J, et al. Comorbidities, complications, and coding bias. Does the number of diagnosis codes matter in predicting in-hospital mortality? JAMA 1992;267:2197–2203.
54. Jencks SF, Williams DK, Kay TL. Assessing hospital-associated deaths from discharge data. The role of length of stay and comorbidities [see comment]. JAMA 1988;260:2240–2246.
55. Elixhauser A, Steiner C, Harris DR, Coffey RM. Comorbidity measures for use with administrative data. Med Care 1998;36:8–27.
56. Wray NP, Hollingsworth JC, Peterson NJ, Ashton CM. Case-mix adjustment using administrative databases: a paradigm to guide future research. Med Care Res Rev 1997;54:326–356.
57. Birkmeyer JD, Finlayson SR, Tosteson AN, et al. Effect of hospital volume on in-hospital mortality with pancreaticoduodenectomy. Surgery 1999;125:250–256.
58. Gibbs J, Clark K, Khuri S, et al. Validating risk-adjusted surgical outcomes: chart review of process of care. Int J Qual Health Care 2001;13:187–196.
59. Hsia DC, Ahern CA, Ritchie BP, Moscoe LM, Krushat WM. Medicare reimbursement accuracy under the prospective payment system. JAMA 1992;268:896–899.
60. Lemeshow S, Le Gall JR. Modeling the severity of illness of ICU patients. A systems update. JAMA 1994;272:1049–1055.
61. Ku L, Ellwood MR, Klemm J. Deciphering Medicaid data: issues and needs. Health Care Financ Rev 1990;Dec:35–45.
62. Dimick JB, Cowan JA Jr, et al. Surgeon specialty and provider volumes are related to outcome of intact abdominal aortic aneurysm repair in the United States. J Vasc Surg 2003;38:739–744.
63. Gordon TA, Burleyson GP, Tielsch JM, Cameron JL. The effects of regionalization on cost and outcome for one general high-risk surgical procedure. Ann Surg 1995;221:43–49.
64. Halm EA, Lee C, Chassin MR. Is volume related to outcome in health care? A systematic review and methodologic critique of the literature [summary for patients in Ann Intern Med. 2002;17:i52; PMID: 12230383] [review]. Ann Intern Med 2002;137:511–520.
65. Finlayson SR, Birkmeyer JD, Laycock WS. Trends in surgery for gastroesophageal reflux disease: the effect of laparoscopic surgery on utilization. Surgery 2003;133:147–153.
66. Pope GD, Birkmeyer JD, Finlayson SR. National trends in utilization and in-hospital outcomes of bariatric surgery. J Gastrointest Surg 2002;6:855–861.
67. Baxter NN, Virnig BA, Durham SB, Tuttle TM. Trends in the treatment of ductal carcinoma in situ of the breast. J Natl Cancer Inst 2004;96:443–448.
68. Hull MC, Morris CG, Pepine CJ, Mendenhall NP. Valvular dysfunction and carotid, subclavian, and coronary artery disease in survivors of hodgkin lymphoma treated with radiation therapy [see comment]. JAMA 2003;290:2831–2837.
69. Begg CB, Riedel ER, Bach PB, et al. Variations in morbidity after radical prostatectomy. N Engl J Med 2002;346:1138–1144.
70. Birkmeyer JD, Stukel TA, Siewers AE, et al. Surgeon volume and operative mortality in the united states. N Engl J Med 2003;349:2117–2127.
71. Ritchie JL, Maynard,C, Every NR, Chapko MK. Coronary artery stent outcomes in a medicare population: less emergency bypass surgery and lower mortality rates in patients with stents [see comment]. Am Heart J 1999;138:437–440.

72. Sternbergh WC III, Money SR. Hospital cost of endovascular versus open repair of abdominal aortic aneurysms: a multicenter study. J Vasc Surg 2000;31:237–244.
73. Graves EJ. National hospital discharge survey: annual summary, 1988. Vital & Health Statistics—Series 13: Data From the National Health Survey 1991;106:1–55.
74. Dennison C, Pokras R. Design and operation of the national hospital discharge survey: 1988 redesign. Vital & Health Statistics—Series 1: Programs & Collection Procedures 2000;39:1–42.
75. Feinglass J, Brown JL, LoSasso A, et al. Rates of lower-extremity amputation and arterial reconstruction in the United States, 1979 to 1996. Am J Public Health 1999;89:1222–1227.
76. Bendich A, Leader S, Muhuri, P. Supplemental calcium for the prevention of hip fracture: potential health-economic benefits. Clin Therap 1999;21:1058–1072.
77. Xia Z, Roberts RO, Schottenfeld D, Lieber MM, Jacobsen SJ. Trends in prostatectomy for benign prostatic hyperplasia among black and white men in the united states: 1980 to 1994. Urology 1999;53:1154–1159.
78. Wingo PA, Guest JL, McGinnis L, et al. Patterns of inpatient surgeries for the top four cancers in the united states, national hospital discharge survey, 1988–95. Cancer Causes Contr 2000;11:497–512.
79. Khuri SF, Daley J, Henderson W, et al. The national veterans administration surgical risk study: risk adjustment for the comparative assessment of the quality of surgical care [see comment]. J Am Coll Surg 1995;180:519–531.
80. Khuri SF, Daley J, Henderson W, et al. The department of Veterans Affairs' NSQIP: the first national, validated, outcome-based, risk-adjusted, and peer-controlled program for the measurement and enhancement of the quality of surgical care. National VA surgical quality improvement program [see comment]. Ann Surg 1998;228:491–507.
81. Khuri SF, Daley J, Henderson WG. The comparative assessment and improvement of quality of surgical care in the department of veterans affairs. Arch Surg 2002;137:20–27.
82. Webster C, Neumayer L, Smout R, et al. Prognostic models of abdominal wound dehiscence after laparotomy. J Surg Res 2003;109:130–7. Notes: CORPORATE NAME: National Veterans Affairs Surgical Quality Improvement Program.
83. Horner RD, Oddone EZ, Stechuchak KM, et al. Racial variations in postoperative outcomes of carotid endarterectomy: evidence from the Veterans Affairs National Surgical Quality Improvement Program. Med Care 2002;40:I35–I43.
84. Collins TC, Johnson M, Daley J, et al. Preoperative risk factors for 30-day mortality after elective surgery for vascular disease in Department of Veterans Affairs hospitals: is race important? J Vasc Surg 2001;34:634–640.
85. Arozullah AM, Khuri SF, Henderson WG, Daley J. Development and validation of a multifactorial risk index for predicting postoperative pneumonia after major noncardiac surgery. Ann Intern Med 2001;135:847–857. Notes: CORPORATE NAME: Participants in the National Veterans Affairs Surgical Quality Improvement Program.
86. Daley J, Forbes MG, Young GJ, et al. Validating risk-adjusted surgical outcomes: site visit assessment of process and structure. National VA Surgical Risk Study. J Am Coll Surg 1997;185:341–351.
87. Khuri SF, Daley J, Henderson W, et al. Risk adjustment of the postoperative mortality rate for the comparative assessment of the quality of surgical care: results of the National Veterans Affairs Surgical Risk Study. J Am Coll Surg 1997;185:315–327.
88. Grossmann EM, Longo WE, Virgo KS, et al. Morbidity and mortality of gastrectomy for cancer in Department of Veterans Affairs Medical Centers. Surgery 2002;131:484–490.
89. Axelrod DA, Stanley JC, Upchurch GR Jr, et al. Risk for stroke after elective noncarotid vascular surgery. J Vasc Surg 2004;39:67–72.
90. Fink AS, Campbell DA Jr, Mentzer RM Jr, et al. The national surgical quality improvement program in non-veterans administration hospitals: initial demonstration of feasibility [see comment]. Ann Surg 2002;236:344–354.
91. Daley J, Khuri SF, Henderson W, et al. Risk adjustment of the postoperative morbidity rate for the comparative assessment of the quality of surgical care: results of the National Veterans Affairs Surgical Risk Study. J Am Coll Surg 1997;185:328–340.

III Outcome Measurement

12 Traditional Outcome Measures

Aruna V. Sarma, PhD, MHA
and Julie C. McLaughlin, MPH, MS

Contents

A Dictionary of Epidemiology by John M. Last defines *outcomes research* as "research on outcomes of interventions." Outcomes research defined in this manner comprises much of the effort of clinical epidemiologists. However, the Institute of Medicine elaborates on this definition of outcomes research to include the "...inquiry, both basic and applied, that examines the use, costs, quality, accessibility, delivery, organization, financing, and outcomes of health care services to increase the knowledge and understanding" of the structure, processes, and effects of health services for individuals and populations *(1)*. As this description entails, the field of outcomes research has certainly extended beyond clinical epidemiologists and it has become increasingly important to define, examine, and evaluate the definition, use, and value of outcome measurements for all of those who participate in this type of research. In particular, surgical disciplines have long been interested in the outcomes of treatment to determine whether that treatment was, in fact, effective. The most commonly used types of outcomes in the surgical discipline are those that measure the disease process: *mortality* and *morbidity*. This is primarily a function of the notion that surgery is often directed toward ameliorating abnormalities of structure or function *(2)*. In this chapter, we discuss the traditional measures of outcomes such as mortality and morbidity as end points, specific types of these indicators, sources of these types of information, and the strengths and limitations associated with using them.

1. MORTALITY AS AN END POINT

The term *mortality* means "death" or describes death or related issues, and information on mortality is a central facet of vital statistics, epidemiology, and demographic data. In

From: *Clinical Research for Surgeons*
Edited by: D. F. Penson and J. T. Wei © Humana Press Inc., Totowa, NJ

LF ____________

CF ____________

TYPE/PRINT IN PERMANENT BLACK INK

STATE OF MICHIGAN
DEPARTMENT OF COMMUNITY HEALTH
CERTIFICATE OF DEATH

STATE FILE NUMBER
1646718

NAME OF DECEDENT
FOR USE BY PHYSICIAN OR INSTITUTION

DECEDENT

1. DECEDENT'S NAME *(First, Middle, Last)*
2. SEX
3. DATE OF DEATH *(Month, Day, Year)*
4a. AGE - Last Birthday *(Years)*
4b. UNDER 1 YEAR — MONTHS | DAYS
4c. UNDER 1 DAY — HOURS | MINUTES
5. DATE OF BIRTH *(Month, Day, Year)*
6. COUNTY OF DEATH
7a. LOCATION OF DEATH (Enter place officially pronounced dead in 7a, 7b, 7c.) HOSPITAL OR OTHER INSTITUTION - Name *(If not in either, give street and number)*
7b. IF HOSP. OR INST. Inpatient, Op./Emer. Room, DOA *(Specify)*
7c. CITY, VILLAGE, OR TOWNSHIP OF DEATH
8. SOCIAL SECURITY NUMBER
9a. USUAL OCCUPATION *(Give kind of work done during most of working life. Do not use retired)*
9b. KIND OF BUSINESS OR INDUSTRY
10a. CURRENT RESIDENCE - STATE
10b. COUNTY
10c. LOCALITY *(Check one box and specify)* ☐ INSIDE CITY OR VILLAGE OF ☐ TWP. OF
10d. STREET AND NUMBER
10e. ZIP CODE
11. BIRTHPLACE *(City and State or Foreign Country)*
12. MARITAL STATUS - Married, Never Married, Widowed, Divorced *(Specify)*
13. SURVIVING SPOUSE *(If wife, give name before first married)*
14. WAS DECEDENT EVER IN U.S. ARMED FORCES? *(Specify Yes or No)*
15. ANCESTRY – Mexican, Puerto Rican, Cuban, Central or South American, Chicano, other Hispanic, Afro-American, Arab, English, French, Finnish, etc. *(Specify below)*
16. RACE - American Indian, Black, White, etc. If Asian, give nationality i.e., Chinese, Filipino, Asian Indian, etc. *(Specify below)*
17. DECEDENT'S EDUCATION *(Specify only highest grade completed)* — Elementary/Secondary (0-12) | College (1-4 or 5+)

PARENTS

18. FATHER'S NAME *(First, Middle, Last)*
19. MOTHER'S NAME *(First, Middle, Surname before first married)*

INFORMANT

20a. INFORMANT'S NAME *(Type/Print)*
20b. MAILING ADDRESS *(Street and Number or Rural Route Number, City or Village, State, ZIP Code)*

DISPOSITION

21. METHOD OF DISPOSITION - Burial, Cremation, Removal, Donation, Other *(specify)*
22a. PLACE OF DISPOSITION *(Name of Cemetery, Crematory, or other place)*
22b. LOCATION - City or Village, State
23. SIGNATURE OF FUNERAL SERVICE LICENSEE
24. LICENSE NUMBER *(of Licensee)*
25. NAME AND ADDRESS OF FACILITY

SAMPLE COPY

CAUSE OF DEATH

26. PART I. Enter the diseases, injuries, or complications that caused the death. Do **NOT** enter the mode of dying, such as cardiac or respiratory arrest, shock, or heart failure. List only one cause on each line.

Approximate Interval Between Onset and Death

IMMEDIATE CAUSE (Final disease or condition resulting in death) → a.
DUE TO (OR AS A CONSEQUENCE OF):
b.
DUE TO (OR AS A CONSEQUENCE OF):
c.
DUE TO (OR AS A CONSEQUENCE OF):
d.

Sequentially list conditions, IF ANY, leading to immediate cause. Enter UNDERLYING CAUSE (Disease or injury that initiated events resulting in death) LAST

PART II. Other significant conditions contributing to death but not resulting in the underlying cause given in Part I
27a. WAS AN AUTOPSY PERFORMED? *(Yes or No)*
27b. WERE AUTOPSY FINDINGS AVAILABLE PRIOR TO COMPLETION OF CAUSE OF DEATH? *(Yes or No)*

CERTIFIER

28. ACTUAL PLACE OF DEATH (Home, Nursing Home, Hospital, Ambulance) *(Specify)*
29. WAS CASE REFERRED TO MEDICAL EXAMINER? *(Specify Yes or No)*

CERTIFYING PHYSICIAN
30a. To the best of my knowledge, death occurred at the time, date and place, and due to the cause(s) stated
(Signature and Title)
30b. DATE SIGNED *(Mo., Day, Yr.)*
30c. TIME OF DEATH M
30d. NAME OF ATTENDING PHYSICIAN IF OTHER THAN CERTIFIER *(Type or Print)*

MEDICAL EXAMINER
31a. *(Check one only)* ☐ The case reviewed and determined not to be a medical examiner's case. ☐ On the basis of examination and of investigation, in my opinion death occurred at the time, date and place and due to the cause(s) and manner stated.
(Signature and Title)
31b. DATE SIGNED *(Mo., Day, Yr.)*
31c. CASE NUMBER
31d. PRONOUNCED DEAD *(Mo., Day, Yr.)* ON
31e. TIME OF DEATH M

32a. NAME AND ADDRESS OF PERSON WHO COMPLETED CAUSE OF DEATH (ITEM 26) *(Type or Print)*
32b. LICENSE NUMBER

MEDICAL EXAMINER

33a. ACC. SUICIDE, HOM., NATURAL OR PENDING INVEST. *(Specify)*
33b. DATE OF INJURY *(Mo., Day, Yr.)*
33c. TIME OF INJURY M
33d. DESCRIBE HOW INJURY OCCURRED
33e. INJURY AT WORK *(Specify Yes or No)*
33f. PLACE OF INJURY - At home, farm, street, factory, office building, etc. *(Specify)*
33g. LOCATION - Street or R.F.D. No. City, Village or Twp. State

34a. REGISTRAR'S SIGNATURE
34b. DATE FILED *(Month, Day, Year)*

DCH - 0483 10/98
(Formerly B-36)

Figure 1: Example of a standard certificate of death for the state of Michigan.

fact, the accounting of deaths is one of the most highly developed reporting systems in the country. By 1933, the US Bureau of Census' Death Registration covered the entire United States *(3)*. Death certificates, originally considered legal documents, are the most widely used source of mortality data, and are considered the most representative of the general population (Figure 1). When a death occurs, the name, date, and place of death of the deceased are recorded, and cause of death certified by a physician, medical exam-

Table 1
Total Mortality, Crude Mortality Rates, and Age-Adjusted Mortality Rates for the 15 Leading Causes of Death for the Total Population of the United States in 2000*

Rank	*Cause of Death*	*Total Mortality (n)*	*Crude Mortality Rate*	*Age-Adjusted Mortality Rate*
	All causes	2,403,351	873.1	872.0
1	Diseases of the heart	710,760	258.2	257.9
2	Malignant neoplasms	553,091	200.9	201.0
3	Cerebrovascular diseases	167,661	60.9	60.8
4	Chronic lower respiratory diseases	122,009	44.3	44.3
5	Accidents (unintentional injuries)	97,900	35.6	35.5
6	Diabetes mellitus	69,301	25.2	25.2
7	Influenza and pneumonia	65,313	23.7	23.7
8	Alzheimer's disease	49,558	18.0	18.0
9	Nephritis, nephritic syndrome, and nephrosis	37,251	13.5	13.5
10	Septicemia	31,224	11.3	11.4
11	Intentional self-harm (suicide)	29,350	10.7	10.6
12	Chronic liver disease and cirrhosis	26,552	9.6	9.6
13	Essential (primary) hypertension and hypertensive renal disease	18,073	6.6	6.6
14	Assault (homicide)	16,765	6.1	6.1
15	Pneumonitis from solids and liquids	16,636	6.0	6.0
	All other causes	391,904	142.4	—

*Mortality rates calculated on annual basis per 100,000 persons; age-adjusted mortality rates standardized to the year 2000 US population *(46)*.

iner, or coroner. If an autopsy is performed, the results are also recorded. A funeral director is responsible for obtaining personal information about the decedent, completes the death certificate, and files the certificate with the local health department or state office of vital statistics.

Death certificates provide valuable information not only on the total number of deaths, but also useful demographic information and important facts about the deceased, such as date of birth and death, cause of death, place or residence, sex, occupation, and marital status. The main cause of death and any underlying causes of death in the United States and much of the industrialized world are classified using the World Health Organization–derived International Classification of Disease coding *(4,5)*. The information from death certificates is ultimately sent to the National Center for Health Statistics (NCHS). The NCHS publishes reports annually on the actual number of deaths and death rates in the United States by age, sex, race, geographic area, occupation, cause of death, and other demographic variables (Table 1). International comparisons of mortality rates are possible in part because of the requirement of certification of death and the use of standard death certificates and International Classification of Disease coding for causes of death in many countries *(4, 6)*. Additional sources of mortality data include: financial records (insurance, pension), hospital records, Medicare/Medicaid records, and occupational records *(7)*.

To accurately interpret mortality data, it is important to understand how mortality can be calculated. Mortality can be expressed as either a count, ratio, proportion, or rate. A *count* of mortality is simply a measure of the total number of deaths and can be limited to a specific time period, population, or cause. For example, there were 1424 deaths from coronary bypass surgery in the years 1994 to 1996 in New York state *(8)*. A primary limitation of count measures is that there is no reference denominator. For example, 10 deaths from prostate cancer would be interpreted very differently if they occurred among 15 men vs 1500 men. A *ratio*, on the other hand, includes both a numerator and a denominator. A ratio is used to compare two similar constructs, and is the value obtained by dividing one quantity by another quantity. A ratio can be expressed as a fraction. For example: there were two males to every female in the study, or the ratio of males to females was 2/1. In a ratio, the numerator is not necessarily included in the denominator. A *proportion* is a type of ratio in which the numerator *is* included in the denominator and is the result of one part being divided by a whole. A proportion can and is often expressed as a percentage. For example, if there are 50 females in a study containing 200 people, than the proportion of females in the study is 50/200, or 25%.

Often a simple proportion does not provide adequate information for clinical meaning or comparison. Consider a news report that states that 25% of patients in a recent study who underwent open-heart surgery died from heart failure. The reader cannot glean much import from this statement because we are not told what the numbers are being compared with: "25% of how many people?" The interpretation of these results would be quite different if there were 4 people in the study or 40,000. A small study population might suggest results are not representative of the entire population, whereas a large study population would indicate a great number of people are at risk. Furthermore, the inclusion of information regarding time in the data would provide even greater significance to the statistics: 25% of the 40,000 men dying within 1 mo would have very different risk implications to both patients and physicians than if 25% of the 40,000 men dying within 10 yr.

Knowing the number of people with an outcome, the study population size, and the study time period allows for the calculation of a *rate*. A rate defined as "a measure of frequency or occurrence of a phenomenon" is specifically, the number or frequency of an outcome per unit of population, in a *specified period (9)*. The use of rates rather than raw numbers is essential for the comparison of experience between populations at different times, different places, or among different populations. A rate formula includes a numerator (the number of disease events or individuals affected), a denominator (the midpoint population of the study), and a specified time period (*see* formula) *(5)*. Note that because the population changes over time, often the number of persons in the population at the midpoint of the specified period is generally used as an approximation of the total population in the denominator. Furthermore, the results are usually multiplied by a constant (1000; 10,000; 100,000) to enable a standardized unit of population for interpretation. Time periods are standardized as well (i.e., 1 yr). For example, in 1998, there were 153 inpatient surgical procedures in the United States per 1000 of the population *(10)*. The numerator of a rate is confined to a specific set of characteristics such as age, sex, race, occupation, or any variable we wish to evaluate (i.e., those who underwent inpatient surgery in the United States). The denominator is limited to the population of the study group (i.e., US population). The rate formula must be balanced. For example, if the *US*

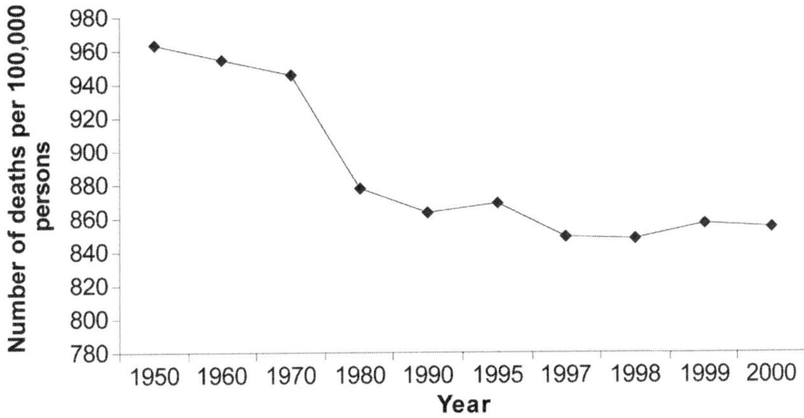

Figure 2: Crude mortality rates (per 100,000 persons) for all causes in the United States 1950–2000 *(15)*.

population is the study group, then the numerator must be restricted to inpatient surgical procedures *in the United States*. The denominator restricts the rate formula by determining the population at risk for an outcome, and the numerator includes only those affected by the outcome *(5)*.

$$\text{Rate (Per 1000 population)} = \frac{\text{Number of events in a specified time period}}{\text{Total population at midpoint of specified period}} \times 1000$$

2. MEASURES OF MORTALITY

2.1. Mortality Rates

A mortality rate is a measure of the frequency of occurrence of death in a defined population during a specified interval of time. There are several measures of mortality which can be defined as crude, adjusted, or specific. *Crude mortality rates* are based on the number of events that occur in a total population in a certain period. Although crude mortality rates relay vital event information and are useful for comparison of one country to another, they are only a summary rates derived from limited information. Crude mortality rates do not take into account the unique characteristics or behaviors of a population. Furthermore, they fail to show differences in and between population subgroups, because they do not take into account, for example, age, race, sex, socioeconomic status, or any other factor that could affect the probability of death (Figure 2). *Adjusted mortality rates*, therefore, represent mortality rates that are mathematically transformed to allow for comparisons among and between populations that differ in traits that may affect risk of death *(5)*. Using age-adjusted mortality rates, for example, we can more accurately compare prostate cancer mortality among various countries for the year 2000 (Figure 3) *(11)*. Finally, *specific mortality rates* provide detailed information, and express the rate of death for specific groups in the population defined, for example, by age, race, sex, marital status, religion, occupation, geography, or cause of death *(4)*. Specific measures of crude, adjusted, and specific mortality rates and their calculations and interpretations are discussed further in the next section.

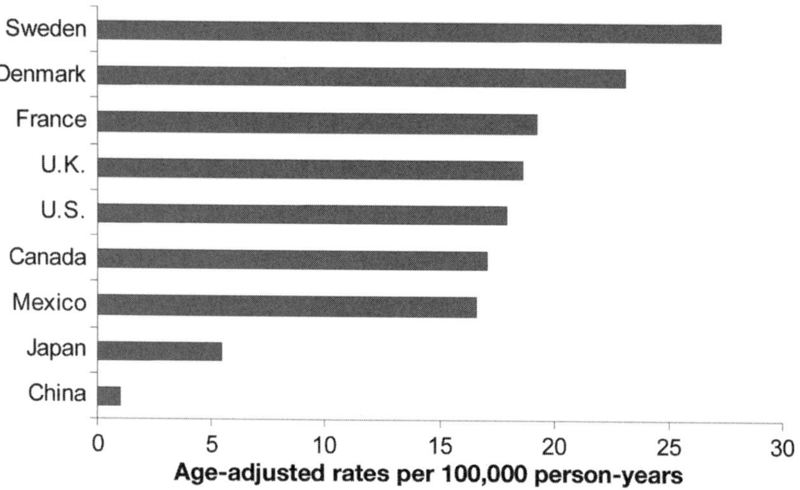

Figure 3: Age-adjusted mortality rates (per 100,000 person-years) for prostate cancer in selected countries for the year 2000. Mortality rates standardized to the year 2000 world population *(47)*.

2.2. Crude Mortality Rates

The most basic measure of mortality is a crude mortality rate, which measures death from all causes for a population during a specified time period. Crude death rate is calculated by the formula below *(5, 7)*.

$$\begin{array}{c}\text{Crude mortality rate}\\ \text{(per 100,000 population)}\end{array} = \frac{\text{Total number of deaths during specified period}}{\text{Total population at midpoint of specified period}} \times 100{,}000$$

A common measure of a crude mortality rate is an *annual mortality rate*, simply defined as the total number of deaths that occurred during a specific 12-mo period divided by the total population during that same 12-mo period (Figure 2) *(5, 7)*.

$$\begin{array}{c}\text{Annual mortality rate}\\ \text{(per 100,000 population)}\end{array} = \frac{\text{Total number of deaths during a specific 12-mo period}}{\text{Total population at midpoint of the 12-mo period}} \times 100{,}000$$

2.3. Specific Mortality Rates

Specific mortality rates are often generated from demographic components such as age, race, sex, or occupation, or to select for specific groups or subgroups within the population. A specific mortality rate selects for a particular component of the crude mortality rate and enables a more detailed view of a specific subgroup, providing more meaningful information than a crude mortality rate. A specific mortality rate is defined in the same manner as a crude mortality rate, except the numerator and the denominator are limited to a specific group *(5)*. For example, the age-specific mortality rate for adults ages 25–34 would be calculated as:

$$\begin{array}{c}\text{Age-specific mortality rate}\\ \text{(per 100,000 population)}\end{array} = \frac{\begin{array}{c}\text{Number of deaths in persons}\\ \text{ages 25–34 in a specified period}\end{array}}{\begin{array}{c}\text{Total persons ages 25–34 at}\\ \text{midpoint of specified period}\end{array}} \times 100{,}000$$

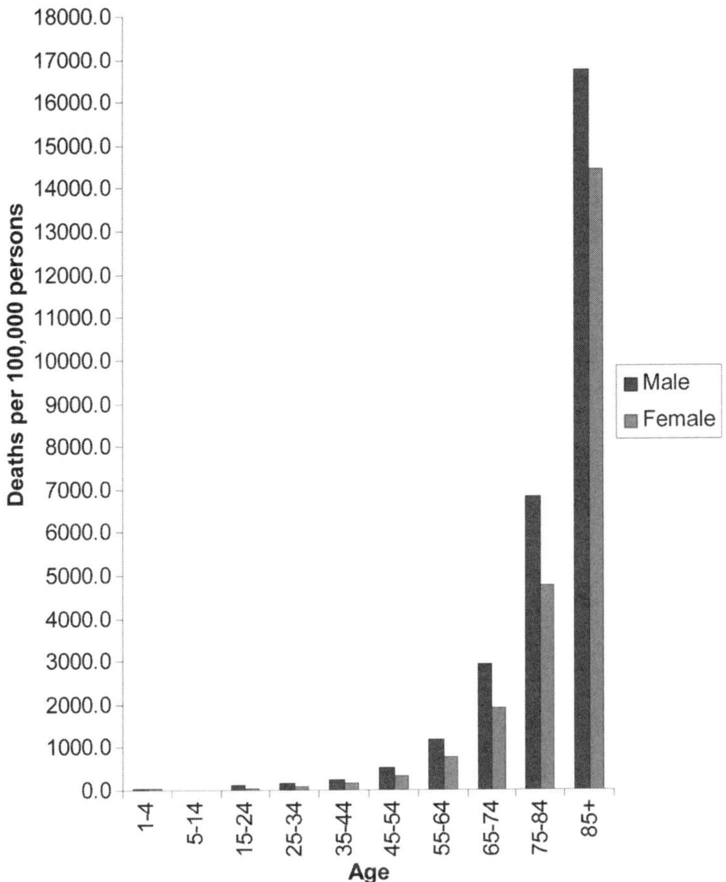

Figure 4: Age- and sex-specific mortality rates (per 100,000 persons) in the United States 2001 *(47)*.

Specific mortality rates for any other variable (i.e., sex) or subgroup are determined in the same fashion. The age- and sex-specific mortality rates for the United States in the year 2001 are shown in Figure 4.

2.4. Adjusted Mortality Rates

Often it is of interest to compare mortality between two different groups or populations. However, when comparisons are made between different groups or populations that inherently differ in their risks of death from factors or characteristics that are present in the two populations, an adjustment of the data is required. Adjustment or standardization of mortality rates allows for the comparison of mortality rates among groups that differ in risk from various factors, by controlling for differences in these select factors within the populations. For example, because age structures often elicit the greatest effect on death and morbidity rates in a population, age is the most common variable rates are adjusted for (Figure 5). Thus to compare risks of two or more populations at one point in time or one population at two or points in time, researchers typically use *age-standardized/age-adjusted rates*. An age-adjusted mortality rate is a summary measure of the mortality rate that a population would have if it had a standard age structure. Age

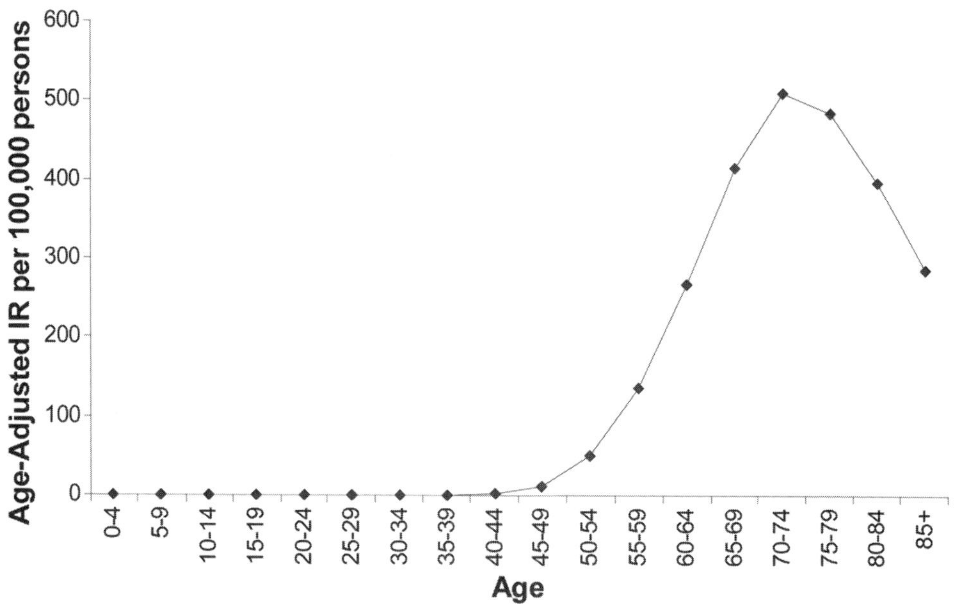

Figure 5: Age-adjusted Surveillance, Epidemiology and End Results program (SEER) average annual incidence rates (per 100,000 persons) of prostate cancer, United States 1992–1998, according to age. Incidence rates standardized to the year 2000 US population *(48)*.

adjustment eliminates the influence of different age distributions on the mortality rates of two populations being compared *(4,5)*. Age-adjusted rates are primarily useful for comparisons with other age-adjusted rates that are standardized to the same population only as their absolute value depends on the age distribution of the standard population chosen (Table 2). Adjustments can be made for many variables such as race, sex, or marital status. Using age as an example, mortality rates can be age adjusted in two ways: the direct method and the indirect method. Both methods are similar in that they consist of taking a weighted average of age-specific mortality rates. The difference between the two lies simply in the source of weights and rates.

2.4.1. Direct Method

In the direct method, the age-specific mortality rates of the two or more groups that one wishes to compare are applied to a population with a known age distribution, referred to as a standard population (i.e., US population). This eliminates differences in observed rates that result from age differences in population composition. The formula for age-adjusted mortality rate is shown below. In summary, multiplication of the age-specific rates by the standard population number within each age group generates the expected number of deaths that would have occurred in the standard population had the age-specific rates existed for that population *(5)*. The expected numbers of deaths for each age group in the respective populations are then summed, yielding the total number of expected deaths in each population. The total expected numbers of deaths divided by the

Table 2
Age-Adjusted Mortality Rates* by Race and Sex, United States 2000

Race	*Deaths per 100,000*	
	Male	*Female*
White	1029.4	715.3
Black or African American	1403.5	927.6
American Indian or Alaskan Native	841.5	604.5
Asian or Pacific Islander	624.2	416.8
Hispanic or Latino	818.1	546.0

*Mortality rates (per 100,000 persons) standardized to the year 2000 US population *(15)*.

Table 3
Calculation of Age-Adjusted Cancer Mortality Rates Using the Direct Method*

Age (yr)	*2000 Population (in Thousands) (a)*	*2000 Cancer Mortality Rates per 100,000 (b)*	*Expected Deaths in 2000 (a*b)*
<1	3795	2.4	91.08
1–4	15,192	2.7	410.18
5–14	39,977	2.5	999.43
15–24	38,077	4.4	16,75.39
25–34	37,233	9.8	36,48.83
35–44	44,659	36.6	16,345.19
45–54	37,030	127.5	47,213.25
55–64	23,961	366.7	87,864.99
65–74	18,136	816.3	148,044.17
75–84	12,315	1335.6	164,479.14
75–84	4259	1819.4	77,488.26
Total	274,634		548,259.90

Age adjusted cancer mortality rate = $\frac{548{,}259.90}{274{,}634{,}000} = 199.6/10^5$

*Mortality rates (per 100,000 persons) standardized to the year 2000 US population *(15)*.

total standard population produces age-standardized mortality rates for the individual populations that can then be compared (Table 3).

$$\text{Age standardized rate (direct method)} = \frac{\sum_{i=1}^{n} (r_i \times p_i)}{P}$$

where r_i = rate in age group i in the population of interest
p_i = standard population in age group i
n = total number of age groups in the population of interest
P = the total standard population

2.4.2. Indirect Method

The indirect method of age-adjustment is more frequently used, because it is preferential if there are small numbers of people in specific age groups. Because the death rates used are from the standard population, they tend to be more accurate *(5)*. To compare two populations with differing age distributions using the indirect method, we first need to establish the expected number of deaths in the populations. Death rates from a standard population are used to calculate the number of deaths that would have been expected in the study populations had people died at the same rate as the standard population. The expected numbers of deaths in each age stratum of the study populations are calculated by multiplying the age-specific rates of a standard population by the age-specific population number of the study groups. The total number of expected deaths for each study population is then calculated by summing the expected number of deaths for all age groups *(12)*. As with the direct method, the total expected numbers of deaths divided by the total standard population produces age-standardized rates for the individual populations that can then be compared (Table 4).

$$\text{Age standardized rate (indirect method)} = \frac{\Sigma\,(R_i \times p_i)\;\;{}^{n}_{i=1}}{P}$$

where R_i = rate in age group i in the standard population
p_i = population of interest in age group i
n = total number of age groups in the standard population
P = the total standard population

2.5. Standardized Mortality Ratio

In retrospective cohort or other studies measuring mortality, we often obtain information regarding the number of the number of deaths from a disease or condition that have been observed among the study population. To determine whether the number of deaths that have been observed among a study population are more or less than one might expect in the general population based on the age structure of the population, a *standardized mortality ratio* (SMR) is calculated. Rates from a standard population are used to calculate the number of cases that would have been expected in this group had they developed disease at the same rate as the general population. The indirect method is used to determine the expected numbers of deaths for calculating the SMR *(5)*. The expected number of deaths in each stratum (i.e., age) of the study population is calculated by multiplying the stratum-specific rates in the standard population by the number of person-years in each category. The sum of the expected number of deaths in each stratum yields the total number of expected deaths. The SMR, as shown in the formula below, is then calculated as the ratio of observed numbers of deaths in a selected group divided by the number of expected deaths in the same group, and multiplied by 100 to yield a rate. Table 4 presents hypothetical data from a retrospective cohort study of benzene plant workers from 1965 to 1975. Among these workers, 86 cancer deaths were observed. To calculate the expected number of cancer deaths, we multiply the number of person-years in each age-specific cohort of the study population by the category specific mortality rate among US white males during each study time interval (standard population). Adding the expected number of cancer deaths for each age-time stratum yields the total number of expected cancer deaths in the cohort. In the example provided in Table 4, the total number of expected

Table 4
Calculation of Age-Adjusted Cancer Mortality Rates Using the Indirect Method and Standardized Mortality Ratios (SMR)*

Age Group	*Person-Years in Cohort (a)*	*2000 Cancer Mortality Rates US White Males (per 100,000) (b)*	*Expected Cancer Deaths (a*b)*
25–34	1446	92.9	1.34
35–44	1332	30.9	0.41
45–54	1228	123.5	1.52
55–64	1252	401.9	5.03
65–74	1169	984.3	20.29
75–84	889	1736.0	15.43
85+	742	2693.7	19.99
Total	8058	64.01	

Age adjusted cancer mortality rate = $\frac{64.01}{8058}$ = 794.4/10^5 person-years

Example. In a hypothetical retrospective cohort study of benzene plant workers conducted from 1965 to 1975, 86 cancer deaths were observed.

Expected cancer deaths = 64.01

SMR = observed/expected × 100%
= 86/64.01 × 100%
= 134%

Thus the benzene plant workers had a risk of cancer mortality approx 34% greater than white men in the general population.

*Mortality rates (per 100,000 persons) standardized to the year 2000 US population.

cancer deaths was 64. The SMR is calculated as: 86/64 = 134%. This ratio can be interpreted in the following manner: the benzene plant workers had a risk of cancer mortality approx 34% greater than white men in the general population *(12)*. Thus the SMR indicates the excess risk of death due to a specified cause among a study population.

$$\text{Standardized Mortality Ratio} = \frac{\text{Observed deaths} \times 100}{\text{Expected deaths}}$$

2.6. Proportionate Mortality Ratio

To examine the impact of death from a specific cause on a population, the *proportionate mortality ratio* (PMR) is used. A PMR is a useful measure when one has information on the numbers and causes of death among an exposed group in a study population, but not the structure of the population from which the study group arose. The PMR is defined as the proportion of total deaths due to a specific cause in a given period per unit of deaths that occur in the same period and represents the proportion of deaths attributed to a specified cause among a study population. To determine the PMR, the proportion of deaths from a specified cause relative to all deaths among a study population is divided by the corresponding proportion in a comparison population, and multiplied by a constant

Table 5
Calculation of the Proportional Mortality Ratio (PMR)

Example. In a study of mortality among a cohort of male herbicide plant workers between 1970 and 1997, a total of 76 deaths were observed, 19 of which were attributable to cancer.

The proportional mortality for cancer among this cohort:
$= 19/76 \times 100\%$
$= 25.0\%$

The proportional mortality for cancer in the general population of US males of comparable age in 1980 was 21.0%.

Expected cancer deaths $= (76) \times (0.210)$
$= 16.0$

$$\text{PMR} = \frac{\text{observed number of deaths from specified cause}}{\text{expected number of deaths from specified cause}} \times 100\%$$

$= 19/16.0 \times 100\%$
$= 1.19 \times 100\%$
$= 119\%$

Thus the proportion of deaths attributable to cancer was almost 1.2-fold as great among the herbicide plant workers as among the US population.

*Data from MacLennan et al 2003 *(49)*.

unit of death (i.e., 100), as shown in Formula 1 *(12)*. The number of deaths among veterans from prostatectomy within 30 d of surgery, divided by the total number of deaths from all surgery in the Veterans Administration (VA) hospitals in that time period, is an example of a PMR. Alternatively, the PMR can be calculated as with the SMR by dividing the observed deaths from a specific cause by the expected number of deaths from the same cause, and multiplied by 100, as shown in Formula 2. The expected number of deaths is the number that would have occurred if the proportion of deaths from a specified cause relative to all deaths in the study population were the same as the corresponding proportion in the general population. Again, the expected number of deaths can be determined by either the direct or indirect method. The PMR can be useful in determining the extent to which a specific cause of death contributes to the overall mortality within a subgroup or population *(5)*. In addition, the PMR is often a more useful measure than the SMR, because the SMR requires knowledge of the age-specific death rates for a population, whereas the PMR only requires knowledge of the proportion of cause-specific deaths observed in each age stratum. Consequently, the PMR can be used when only death certificates are available. A limitation of the PMR is that underlying differences in lead causes of death among populations may make comparisons misleading *(12)*.

Table 5 presents data from a study of mortality among a cohort of nuclear shipyard workers. Between 1959 and 1977, 146 deaths were observed among the workers, 56 of which were attributable to cancer. The proportional mortality for cancer among the cohort is calculated as 56/146, or 38.4%. Among the general population of US white males in 1970, the proportional mortality for cancer was 21.5%. The expected number of deaths can then be calculated as the product of the total number of deaths observed in the cohort and the proportionate morality among the standard population. In the example provided in Table 5, the expected number of cancer deaths was 31.4. Thus the PMR can be calculated as: PMR = 56/31.4 = 178% and interpreted in the following manner: the shipyard workers had a risk of cancer mortality approx 78% greater than men in the general population *(12)*.

$$\text{Proportional Mortality Ratio (1)} = \frac{\text{Proportion of deaths from specified cause (exposed)}}{\text{Proportion of deaths from specified cause (comparison population)}} \times 100$$

$$\text{Proportional Mortality Ratio (2)} = \frac{\text{Observed deaths from specified cause}}{\text{Expected deaths from specified cause}} \times 100$$

2.7. Case Fatality Rate

The *case fatality rate* is the rate or proportion of persons dying from a certain disease/event within the same period, as shown in the formula below *(5)*. This rate is most often used in the setting of infectious disease to measure pathogenicity, severity, or virulence of an outbreak. However, the case fatality rate can be used to measure acute deaths elicited by other causes such as injury. In the context of a surgical procedure, a case fatality rate can be defined as the number of patients who undergo a specific surgical procedure and die, divided by the number of patients who undergo that surgery, and multiplied by 100.

$$\text{Case Fatality Rate} = \frac{\text{Number of deaths by a certain disease in a specified period}}{\text{Number of cases of disease in the specified time period}} \times 100$$

For example, in a National Surgical Quality Improvement Program (NSQIP) study of mortality among 16,994 patients after lower extremity amputation, 1318 postoperative deaths occurred within 30 d of amputation. This represents a case-fatality rate of 1318/16,994 = .078 or 7.8% (Figure 6) *(13)*. Because the case fatality rate is dimensionless (no units), it is limited in the information that it provides. Clinical medicine requires a time unit for fair comparison *(4, 14)*. A more useful measure is the cause-specific mortality rate, which measures the risk of death from a specific condition.

2.8. Cause-Specific Mortality Rate

The *cause-specific mortality rate* specifies the rate of death from a specific cause or source. Defined as the number of deaths from a cause in a year divided by the average population, and multiplied by a population constant (i.e., 100,000), the cause-specific mortality rate is often cited as the most important epidemiologic index available *(7)*. Mortality rates for any specific surgery or disease can be presented for the population or any subgroup specified by age, sex race or other variable (Table 1). Using a subgroup, the numerator of the cause-specific mortality rate includes deaths from a certain disease/surgery for the subgroup in a certain time period. The denominator is the total subgroup population for the same period, and the entire expression is multiplied by a constant unit of population (i.e., 100,000). Age-adjusted rates are often used in the calculation of cause-specific mortality rates, because age patterns of deaths from diseases such as cancer show distinct changes from one age category to the next *(5)*. The age-adjusted prostate cancer-specific mortality rates for various countries in the year 2000 are shown in Figure 3.

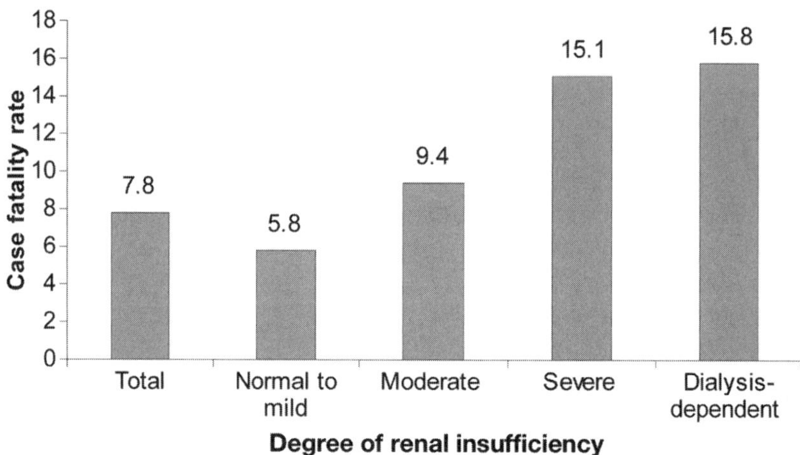

Figure 6: Case fatality rate among lower extremity amputation patients according to renal function. (Postoperative deaths within 30 d of amputation) *(13)*.

$$\text{Cause-specific mortality rate (per 100,000 population)} = \frac{\text{Number of deaths from a specific cause in a year}}{\text{Total persons at midpoint of year}} \times 100{,}000$$

2.9. Life Expectancy

Age is the most clear-cut predictor of differences in death or disease rates between various populations. *Life expectancy*, defined as the average number of years an individual of a given age is expected to live if current mortality rates persist, is a basic measure of the health status of a population and can be used to compare different races or countries *(4, 5)*. Shifts in life expectancy are often used to describe trends in mortality because life expectancy at birth is heavily influenced by infant and child mortality, whereas life expectancy later in life reflects death rates at or above a given age and is independent of the effect of mortality at younger ages (Table 6) *(15)*.

2.10. Years of Potential Life Lost

Another important measure of health status based on mortality data that has been increasingly used over recent years is *years of potential life lost* (YPLL). YPLL takes into account the value of human life and the economic implications of the loss of human productivity. YPPL is a measure used to distinguish the implications of deaths that occur early in life as compared with deaths that occur later in life. When death occurs at a young age, there is a greater loss to society in terms of cost of training, labor, tax dollars, and productivity than when death occurs at an older age.

The *YPLL rate* represents years of potential life lost per 1000 persons, assuming a healthy productive work life based on a retirement age of 65. YPLL for an individual is calculated by subtracting the age at death from an end point age (e.g., 65). The total YPLL for a population is calculated by summing the individual YPLLs. The YPLL rate is the total YPLL, divided by the population under the end point age, and multiplied by constant population unit (i.e., 1000) (Figure 7) *(5)*.

$$\text{YPLL rate} = \frac{\text{Total years of potential life lost}}{\text{Total population under end point age}} \times 1000$$

Table 6
Life Expectancy (yr) at Birth for Selected Countries According to Sex: 1980–1998.

	Male					*Female*				
Country	*1980*	*1990*	*1995*	*1997*	*1998*	*1980*	*1990*	*1995*	*1997*	*1998*
Australia	71.0	73.9	75.0	75.6	75.9	78.1	80.1	80.8	81.3	81.5
Canada	71.7	74.4	75.1	75.8	76.0	78.9	80.8	81.1	81.3	81.5
Cuba	72.2	74.6	75.4	75.7	75.8	—	76.9	77.7	78.0	78.2
France	70.2	72.7	73.9	74.6	74.8	78.4	81.0	81.9	82.3	82.4
Greece	72.2	74.6	75.0	75.6	75.5	76.8	79.5	80.3	80.8	80.6
Hong Kong	71.6	74.6	76.0	77.2	77.4	77.9	80.3	81.5	83.2	83.0
Israel	72.2	75.1	75.5	76.1	76.2	75.8	78.5	79.5	80.4	80.6
Italy	71.1	73.8	75.0	75.9	75.9	77.7	80.5	81.6	82.1	82.2
Japan	73.4	75.9	76.4	77.2	77.2	78.8	81.9	82.9	83.8	84.0
Norway	72.3	73.4	74.8	75.4	75.5	79.2	79.8	80.8	81.0	81.3
Spain	72.5	73.2	74.3	74.9	74.8	78.6	80.4	81.5	81.9	82.2
Sweden	72.8	74.8	75.9	76.7	76.9	78.8	80.4	81.3	81.8	81.9
Switzerland	72.8	74.0	75.3	76.3	76.3	79.6	80.7	81.7	82.1	82.4
United States	70.0	71.8	72.5	73.6	73.8	77.4	78.8	78.9	79.4	79.5

Source: Fried et al 2003 *(15)*.

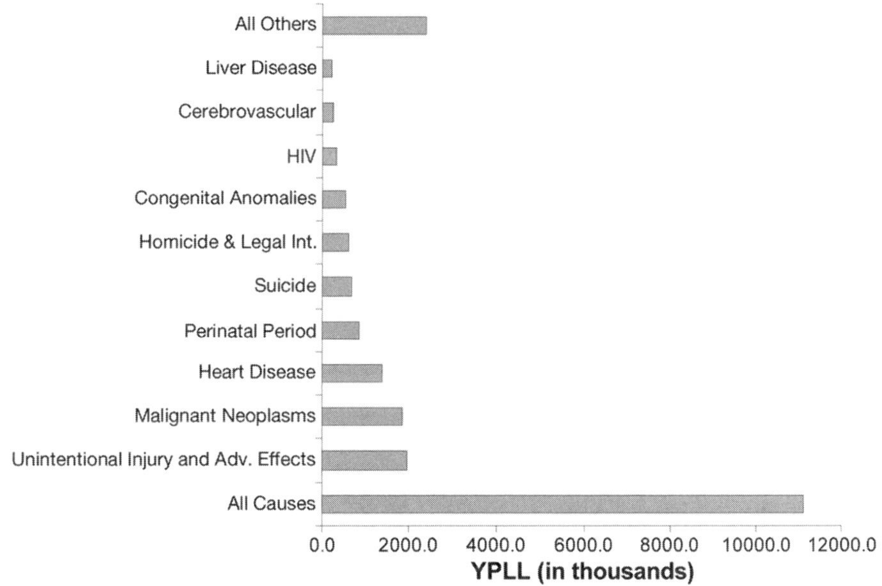

Figure 7: Years of potential life lost (YPLL) before age 65: United States 1998 *(50)*.

2.11. Survival Rate

When chronic diseases or events that take a long time to occur are evaluated, survival is often a preferred measure rather than the mortality rate. In clinical studies, the rates of development of unwanted outcomes including death are frequently measured in terms of survival or the proportion of the study group remaining free of outcome as time passes *(16)*.

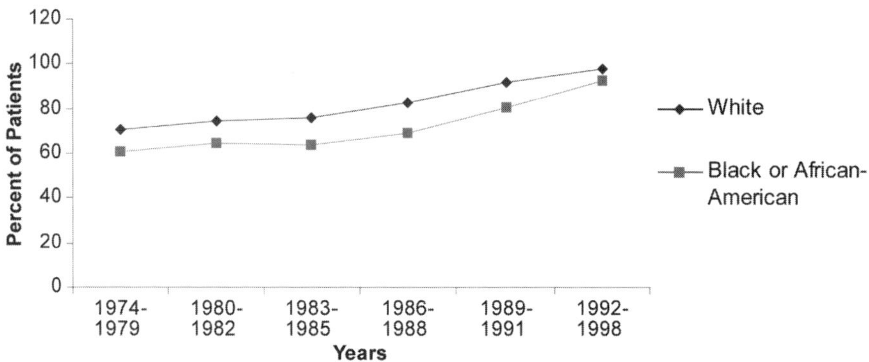

Figure 8: Five-year relative survival rates for prostate cancer, according to race: United States 1974–1998 *(48)*.

A *survival rate* is the cumulative probability of surviving a specified time period *(5)*. The survival rate, or the probability of surviving within a specific period after diagnosis, is of clinical and patient importance *(7)*. To calculate survival rates, data describing the time from entry into the study until death or withdrawal from the study are collected for each patient or subject. The period of reference is specified by the investigator. For example, a 5-yr survival rate enumerates the number of patients alive after 5 yr divided by the total number who underwent a particular treatment (Figure 8). Survival rates allow for the comparison of different populations even when the populations are observed for different lengths of time and can be used in any study that measures time to a particular event *(4)*.

3. SOURCES OF MORTALITY DATA

Vital statistics or events, refers to the process of "collecting, recording, and statistically analyzing data on morbidity, mortality, life expectancy, births, deaths, marriages, divorces, demographics, and census data" *(5)*. Data from local and state agencies are reported to national centers such as the US Public Health Service, Centers for Disease Control and Prevention, and NCHS that compile and distribute health data.

3.1. NCHS

The nation's principal health statistics agency, the NCHS, collects and publishes data on births, deaths, marriages, divorces, and other events in the United States *(17)*. The NCHS collects these data from birth and death certificates, medical records, interview surveys, and through direct physical exams and laboratory testing. Since 1985, all 50 states have participated in computerized reporting of vital statistics to the NCHS. A total of 99.3% of all births in the United States are registered, and it is widely assumed reporting of deaths is as complete *(5)*. Provisional deaths rates by cause, age, race, and sex are estimated from the Current Mortality Sample. The Current Mortality Sample is a 10% systematic sampling of death certificates received each month in the vital statistics office in the states, territories, and New York City.

Overall, the NCHS functions to document the health status of the population and important subgroups, identify disparities in health status and use of health care by race, ethnicity, socioeconomic status (SES), region, and other population gradients to describe experiences with the health care system, and to monitor trends in health status and health care delivery. NCHS publications include the National Vital Statistics Reports and *Health, United States,* an annual report on national trends in health statistics.

3.2. Health Insurance and Health Maintenance Organizations

Health insurance companies, health maintenance organizations, and medical clinics are other sources of vital events data. The usefulness of health insurance data is limited, in that the data represent only the clientele of the insurance company or health maintenance organization, and not the general population, so may be subject to selection bias *(5)*.

3.3. Surveillance, Epidemiology, and End Results Program

The Surveillance, Epidemiology, and End Results Program (SEER) in the National Cancer Institute collects and publishes cancer incidence and survival data from 11 population-based registries and 4 supplemental/expansion registries covering approx 26% of the US population *(18)*. The SEER registries routinely collect data on patient demographics, diagnosis/primary tumor site, morphology, stage at diagnosis, first course of treatment, and follow-up for survival status. The SEER Program is the only comprehensive source of population-based information in the United States that includes stage of cancer at the time of diagnosis and survival rates within each stage. Population estimates used to calculate incidence rates are obtained from the US Bureau of the Census, and mortality and life expectancy data is obtained from the National Center for Health Statistics (Figure 5).

3.4. The National Institute for Occupational Safety and Health

The National Institute for Occupational Safety and Health, created by the Occupational Safety and Health Act of 1970, is a useful source of data on injuries sustained at work, deaths from injuries at work, and deaths from work-related diseases in the United States *(19)*.

4. STRENGTHS AND LIMITATIONS OF MORTALITY DATA

Mortality rates are vital statistics that allow for understanding of how much a disease or medical event is regularly occurring in the population, and provide useful information on the overall effect of a disease in population *(5)*. Mortality rates present data/statistics in an easily understandable manner, and are comparable from one population group to another. Many studies of health outcomes look primarily at mortality, because it is easily defined, readily measured, widely available, and is a valuable measure to patients *(20)*. Despite its usefulness, the mortality rate is a summary rate developed from minimal data and limited information and reflects only one aspect of health *(5)*. Researchers may find it preferable to use survival rates rather than mortality rates to measure a particular outcome because trends in mortality data may be artifactual: reflecting changes in coding practices (International Classification of Disease revisions, training of physicians); diagnostic capabilities (new tests, procedures for assigning diagnosis); or the denominator population (changing geographic coverage or census inaccuracies). Furthermore, mortality rates tend to be more appropriate for short-term risks (i.e., when outcomes [mortality] occurs within relatively short periods). Survival rates are preferred for cases of chronic diseases when outcomes (mortality) may take longer to occur *(6)*.

However, several limitations exist in the use of survival rates as well. In most survival curves, the earlier follow-up periods usually include results from more patients than the later periods, and are therefore more precise. The study time interval must also be long enough to capture clinically meaningful events. In addition, results can be affected if too many people drop out of a study, and survival analysis can be subject to bias from competing risks or causes of death. If the study follow-up period is long, then the survival rate will reflect not only deaths from a specific cause, but also those from general causes of mortality in the population. Additionally, lead-time bias can occur if study patients are not all enrolled at similar, well-defined points in the course of their disease, and results in differences in outcome over time that may merely reflect differences in duration of illness. Lead time bias is an apparent increase in survival from earlier detection without any alteration in the natural history of disease.

Finally, the limitation of mortality data found in reports published by the government or other organizations is that the data may vary in their source, method of collection, definitions, method of reporting, and time period. Often, military personnel, institutionalized persons, or people living in nursing homes are excluded from studies, which may lead to under- or overreporting of a health condition. Furthermore, records may be inaccurate or incomplete, data collection systems are subject to error, and studies may be limited to only certain members of the population *(5)*. Selection bias occurs when different criteria are used to select study and control subjects and is often encountered when a selective or convenience sample rather than a true random sample is used. Furthermore, health maintenance organizations or other health insurance databases may have been set up for administrative purposes, thus limiting their clinical usefulness. Vital statistics have to be collected, compiled, and distributed in a standardized manner for correct interpretation and comparison.

5. MORBIDITY AS AN END POINT

Morbidity is defined as a measure of the amount of illness, disability, or injury in a defined population. Up to this point, we have described how mortality is defined, how information regarding mortality is gathered, and how mortality statistics are used as health indicators. Morbidity is also an important health indicator and morbidity data can be helpful in clarifying reasons for particular trends in mortality *(4)*. Morbidity can be expressed in terms of incidence, defined as the probability or risk of developing disease during a certain period, or prevalence, defined as the number of cases that are present during a certain period *(5)*.

5.1. Measures of Morbidity

5.1.1. Prevalence

Prevalence is defined as "the number of events, e.g. instances of a given disease or other condition, in a given population at a designated time" *(5)* and is often used when one is interested in the absolute number of cases present in a population at any given point in time. Prevalence measures are useful in determining the extent of disease or disease burden, projecting community health needs, and for monitoring disease control programs *(7)*.

There are two general measures of prevalence: *point prevalence* and *period prevalence*. Specifically, point prevalence is defined as "the number of persons with a disease or attribute at a specified point in time" *(9)* and can be calculated using the formula below (Table 7).

Table 7
Prevalence of Select Current Medical Conditions in a Cross-Sectional Study of 708 African-American Men

Medical Condition	*Number With Medical Condition*	*Point Prevalence per 100 Population*
Hypertension	407	407/708 = 54.1
Heart disease	77	77/708 = 759.4
Diabetes mellitus	139	139/708 = 16.4

Data from Joseph et al 2003 *(51)*.

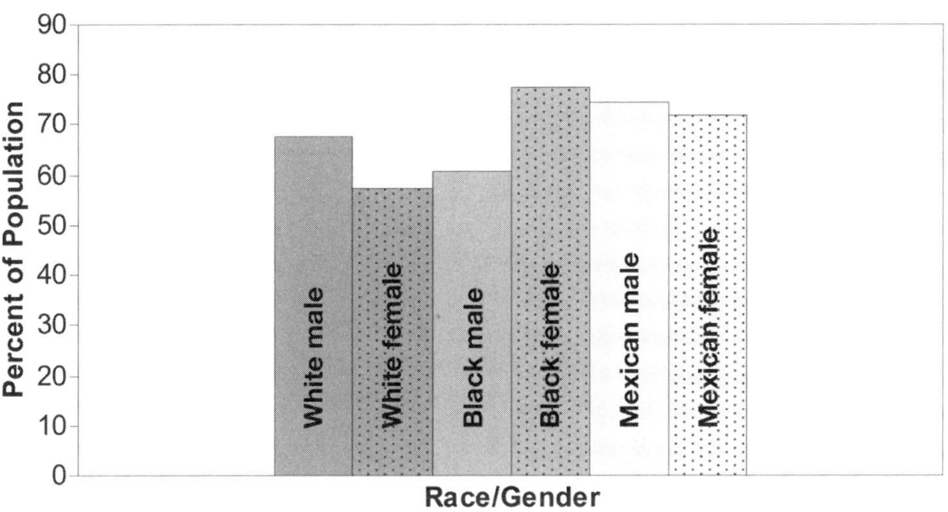

Figure 9: Age-adjusted prevalence of overweight among persons 20 yr of age and older, according to sex and race: United States 1999–2000. Mortality rates standardized to the year 2000 US population *(48)*.

$$\text{Point Prevalence} = \frac{\text{Number of } \textit{existing} \text{ cases of disease at a specified } \textit{point} \text{ in time}}{\text{Total population at specified } \textit{point} \text{ in time}} \times 1000$$

Period prevalence is defined as "the total number of persons known to have had the disease or attribute at any time during a specified period" (Figure 9) *(9)*. This measure is more complex than the point prevalence measure in that it includes all persons with the disease that have carried over from the previous time period (existing cases) as used in point prevalence, but also new cases occurring within the period of interest (incidence) *(5)*. The formula for period prevalence is as follows:

$$\text{Period Prevalence} = \frac{\text{Number of existing cases of disease within a specified period}}{\text{Total population at midpoint of specified period}} \times 1000$$

An average or midpoint population is used in the denominator to minimize the effects of in-migration of healthy people or out-migration of cases, and vice versa. Period prevalence is often used when ascertaining disease burden in a large population. For example, it would be very difficult to survey an entire city's population on a given day or specific

Table 8
Examples of Point and Period Prevalence and Cumulative Incidence in Interview Studies of Lower Urinary Tract Symptoms (LUTS)

Interview Question	*Type of Measure*
"Do you currently have LUTS ?"	Point prevalence
"Have you had LUTS during the last *(n)* years?"	Period prevalence
"Have you ever had LUTS?"	Cumulative or lifetime incidence

time point. Rather the enumeration of disease would take much longer and therefore period prevalence measures are more suitable. If the word *prevalence* is used without a specific reference to which type, it generally is assumed to mean point prevalence; for the remainder of the chapter, we will use the term *prevalence* to mean point prevalence.

5.1.2. Incidence

Incidence is defined as the number of new cases of disease which develop within a certain period in a specific population at risk for developing the disease. The critical element in the definition of incidence that sets it apart from prevalence is that incidence measures *new* cases of disease while prevalence measures *existing* cases of disease. Table 8 presents examples of survey questions used to delineate between prevalence and incidence measures. There are two types of incidence measures: cumulative incidence and incidence rate.

5.1.3. Cumulative Incidence

Cumulative incidence (CI) is a less strict measure of the occurrence of disease over a long period and is used to study a group of persons followed over the same period. The CI is defined as the number of people who get a disease during a specified period and divided by the at risk population at the beginning of the period, as shown in the formula below. The CI requires knowledge of the population size at the start of the study and is interpreted as the cumulative risk of individuals developing disease in a specified time period.

$$\text{Cumulative Incidence} = \frac{\text{Number of } \textit{new} \text{ cases of disease within a specified period}}{\text{Total population at risk at the beginning of specified period}} \times 100$$

For example, a prospective study was undertaken to compare the risk of repeat prostatectomy for benign prostatic hyperplasia in a population-based cohort of 19,598 men in Western Australia treated by transurethral resection of the prostate or open prostatectomy from 1980 to 1995 *(21)*. During the 16-yr follow-up period, 1095 repeat prostatectomies occurred, which results in a cumulative incidence of repeat prostatectomies among men treated with surgery for benign prostatic hyperplasia of 1095 per 19,598 or 5.59% during this 16-yr period. In a second example, a cohort of 208 patients who underwent surgical resection for primary gastric cancer were followed for 5 yr, and the cumulative incidence of cancer recurrence was calculated (Figure 10) *(22)*. During the 5-yr study interval, 109/208 patients developed a recurrence of gastric cancer, yielding a CI of 52.4%. Additionally, 28/208 patients developed liver metastases, yielding a CI of 13.5%.

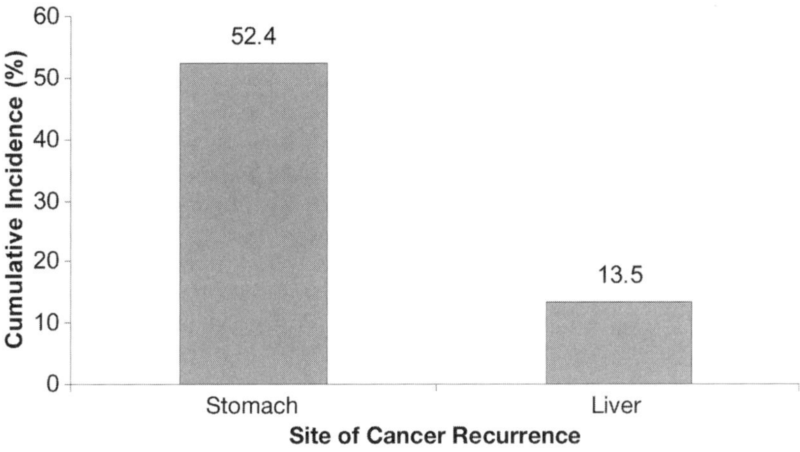

Figure 10: Five-year cumulative incidence of cancer recurrence after surgical resection for primary gastric cancer *(22)*.

5.1.4. Attack Rate

A specific type of cumulative incidence is an *attack rate*. Calculation of an attack rate is usually appropriate when the period of a disease under study is of only a short duration *(5)*. An attack rate measures the occurrence of a disease among a particular population at risk, which is observed for a limited period, often because of a very specific exposure *(12)*. This rate is calculated as the number of newly ill persons within the period, divided by the number of persons at risk within the period, and multiplied by constant unit of population (i.e., 100) as shown in the formula below. For example, in May 2000 a scout camp was held on an agricultural show ground in New Deer, Aberdeenshire, in the United Kingdom *(23)*. There were 337 campers at the event, of which 20 became ill between May 28 and June 3, and were confirmed as having *Escherichia coli* O157 infection, most likely from an environmental exposure at the camp. This results in an attack rate of 20/337 or 5.93%.

$$\text{Attack Rate} = \frac{\text{Number of new cases of disease within a specified period}}{\text{Total population at risk at beginning of specified period}} \times 100 \text{ (per 100 population)}$$

5.1.5. Incidence Rate

By comparing incidence rates of a disease among population groups that vary in one or more identified variables, researchers can determine if a factor affects the risk of acquiring the disease and by what magnitude *(7)*. The *incidence rate* (IR) is defined as the number of new cases of a disease within a population in a given period divided by the number of persons exposed or at risk of developing the disease in the same period. Incidence rates are often expressed as cases per unit of population, so the IR is multiplied by a constant (i.e., 1000) (Figure 3) *(5)*. As time passes, the number of people at risk in a population changes as people both enter and leave the population. There are two methods in calculating incidence rates that capture this dynamic including: incidence rates based on aggregate vs individual data *(24)*.

Incidence rates based on aggregate data are those typically calculated for geographic locations where the denominator includes the average population at risk (*see* formula

```
             Jan   July  Jan   July  Jan   July  Jan   July  Jan   July  Jan   Total
             1976  1976  1977  1977  1978  1978  1979  1979  1980  1980  1981  time
                                                                                at
                                                                               risk

Subject A   *------------------------                                          2.0
Subject B               *--------------------------------------x               3.0
Subject C         *-------------------------------------------                 3.5
Subject D               *------------------------------------------------      4.5
Subject E                            *-------------------------------x         3.0

Total years at risk                                                            16.0

*  = Initiation of follow-up
-- = Time followed
x  = Development of disease

ID  = 4 cases / 16.0 person-years
    = 25.0 / 100 person-years of observation
```

Figure 11: Calculation of person-years for incidence density.

below). To accurately represent the average population at risk, the population at the midpoint of the specified time period is often used. The calculation of incidence based on aggregate data is typically used to estimate mortality based on vital statistics information or incidence of newly diagnosed disease based on population-based registries. For example, when incidence needs to be estimated for an aggregate of individuals, defined by their residence in a given geographic area over some period as depicted by the examples of the various mortality rates described earlier in the chapter.

$$\text{Incidence rate based on aggregate data} = \frac{\text{Number of new cases of disease within a specified period}}{\text{Total population at midpoint of specified period}} \times 1000$$

When relatively precise information on the timing of events or losses are available for individuals from a define cohort, *incidence rates based on individual data* can be calculated. These types of incidence rates are frequently calculated using the unit "person-years" as the denominator, rather than the population, and are often referred to as incidence density (Figure 11). The unit person-year enables a more accurate method of calculating an IR, because it takes into account members of the study population that do not remain in the study for the entire period, either because of disease development, loss to follow-up, or entering or exiting the population. Each person in the study population contributes one person-year to the denominator for each year of observation the person is at risk (i.e., disease-free). Consequently, only the actual and complete period of time or cumulative years that the person was active in the study are used in the denominator *(5)*. Three conditions/assumptions must exist for the use of person-years to be valid. The probability of disease must be constant throughout the entire study period, those who drop out will have the same level of pathology as those who complete the study, and the disease may be so severe and advance so rapidly that some individuals are observed for less than the full period. When it is not possible to precisely measure disease-free periods, as in large population studies, person-years can be approximated by multiplying the average size of the study population by the length of the study. Other rates, including mortality rates, can also express risk in person-years.

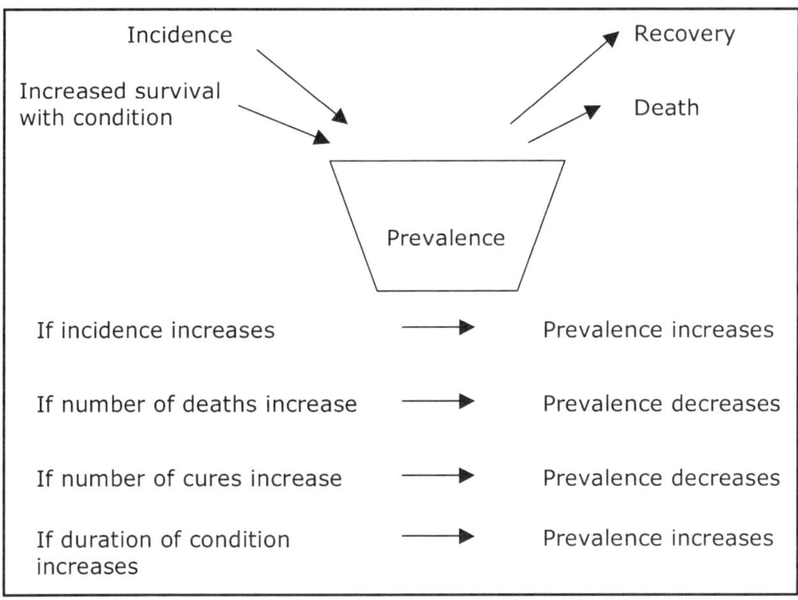

Figure 12: Factors influencing prevalence.

For example, in 1980, a prospective study of 85,118 female nurses was undertaken to examine the relation between vitamin C intake and risk of coronary heart disease (CHD) *(25)*. The women were followed from 1980 to 1994 and monitored for the development of incident CHD (nonfatal myocardial infarction and fatal CHD). During 16 yr of follow-up (1,240,566 person-years), 1356 incident cases of CHD were identified. Thus the IR can be calculated as 1356/1,240,566 person-years = 0.001093 or 109.3 cases of CHD per 100,000 person-years of follow-up.

$$\text{Incidence rate based on individual data} = \frac{\text{Number of new cases of disease during specified period}}{\text{Total person time}}$$

5.2. The Relationship Between Prevalence and Incidence

Prevalence measures can be influenced by many factors including the following: the severity of illness, the duration of illness, and the number of new cases. More specifically, if many people who develop the disease die quickly, its prevalence decreases, and if people live longer with the disease, prevalence at any given point in time increases. Additionally, if many people develop disease (increased incidence) its prevalence increases more than if few people develop disease (Figure 12; Table 9). As the example in Table 9 points out, prevalence is related to incidence and can be estimated using the formula below *(5)*.

$$\text{Prevalence} = \text{Incidence} \times \text{Duration of disease}$$

5.3. Morbidity Data Sources

Data on morbidity are less easily obtained than data on mortality. Morbidity data are often derived from sources such as: communicable disease reports, clinical and hospital

Table 9
Hypothetical Example of Tuberculin Testing: Prevalence, Incidence, and Duration

Screened Population	*Point Prevalence per 1,000*	*Incidence (Occurrences per Year)*	*Duration (yr)*
Hospital A	125	5	25
Hospital B	80	16	5

Prevalence = Incidence × Duration.

medical records, institutional and industrial records, health and disease surveys, disease control programs, and special research programs *(5)*.

5.3.1. Communicable Disease Reports and Sources of Data

Reports published by local and state public health departments, as well as federal agencies, are the main sources of communicable disease data *(5)*. The *Morbidity and Mortality Weekly Report (MMWR)*, published by the Centers for Disease Control and Prevention, is the most widely disseminated source of morbidity information in the United States. The Centers for Disease Control and Prevention also maintains a list of notifiable diseases, those that are of particular public health concern or may become epidemic, and are legally required to be reported by physicians to local public health departments. Notifiable disease reports are received by the Centers for Disease Control and Prevention from 52 areas in the United States and 5 territories. Completeness of reporting varies because not all cases receive medical care and not all treated conditions are reported. *Advance Data*, published by the National Center for Health Statistics, regularly reports chronic and acute disease data *(5)*.

5.3.2. Disease Registry

A disease registry contains information on all cases of an identified diagnosed disease in the population. The registry collects and registers health-related data. Cancer and tumor registries are the most common, although several different disease registries exist. Registries contain demographic and identifying data, disease diagnosis, frequency of occurrence, survival, follow-up or treatment, and other important information unique to a condition or disease *(5)*. Registries are available at the hospital, state, and national level, and vary in both accuracy and completeness.

5.3.3. Clinical and Hospital Medical Records

Clinical and hospital medical records, although readily available, are often biased by medical payment systems based on diagnosis or health insurance policy limitations *(5)*. Furthermore, morbidity data from hospitals may not reflect the level of disease in the community. Large or specialty hospitals can draw patients from large distances, and health insurance coverage issues can affect the reporting process. In addition, not all people who are experiencing a disease seek treatment, and treatments received in physician offices may not be reported or recorded in hospital records. However, hospitals do provide a defined set of study cases and a defined population in which to monitor the development of disease rates.

5.3.4. Managed Care

Managed care organizations are also a good source of disease rate data; however, they may not represent the entire community and may be subject to selection or referral biases *(5)*. Health maintenance organizations have contracts with large employers and the participants generally reflect a middle class population, rather than a cross-section of society.

5.3.5. Centers for Medicare and Medicaid Services

Formerly known as the Health Care Financing Administration, the Centers for Medicare and Medicaid Services (Current Mortality Sample) contains demographic information on participants, information on providers, types of services used, amounts paid for care, and diagnoses *(26)*. Although administrative data often lack clinical detail, Medicare/Medicaid data are useful for population studies because they are less prone to patient selection bias, incomplete follow-up, and generalizability difficulties than managed care or clinical and hospital records. In 2003, 41.0 million people were projected to be covered by Medicare and 41.4 million by Medicaid *(27)*.

5.3.6. Surveys

Surveys of specific diseases are often a better way to obtain a comprehensive view of health and factors that affect health in a population. The National Health Survey, established by the US Congress in 1956 and controlled by the NCHS, provides data on the heath status of the country by collecting morbidity information *(5, 28)*. Several national surveys providing morbidity data are conducted on a continuing basis.

The National Health Interview Survey, conducted annually since 1969, is a nationwide survey of about 36,000 households in the United States and a principal source of information on the health of the civilian noninstitutionalized population *(29)*. The National Health Interview Survey provides information on personal and demographic characteristics, illnesses, injuries, chronic conditions, disabilities, health behaviors, and health care access and utilization through continuous sampling and interviewing and by using core surveys and supplemental datasets.

The National Hospital Discharge Survey, conducted annually since 1965, is a national probability survey of non-federal short-stay hospitals in the United States, and reports patient information and data on all inpatient hospital discharges, excluding newborns *(30)*. The National Hospital Discharge Survey collects data including primary diagnoses, age, sex, and date of discharge, from a sample of approx 270,000 inpatient records acquired from a national sample of roughly 500 hospitals. Only hospitals with an average length of stay of less than 30 d for all patients, general hospitals, or children's general hospitals are included in the survey. Federal, military, and Department of Veterans Affairs hospitals, hospital units of institutions such as prisons, and hospitals with fewer than six beds staffed for patient use are excluded.

The National Survey of Ambulatory Surgery, initiated by the National Center for Health Statistics in 1994, is a national survey providing information about the use of surgical and nonsurgical procedures performed on an outpatient basis in a hospital setting *(31)*. Conducted annually from 1994 through 1996, the National Survey of Ambulatory Surgery includes data on approx 120,000 sampled visits per year in about 500 facilities. Data are available on patient characteristics including age and sex, administrative information including sources of payment and geographic region, and medical information including diagnoses and procedures performed.

The National Nursing Home Survey is a continuing series of national sample surveys of nursing homes, their residents, and their staff *(32)*. The National Nursing Home Survey collects and reports data on demographic characteristics, health status, medical services, and discharge information on patients, as well as size, certification, occupancy rate, number of days of care provided, and facility expenses, from a sampling of nursing and personal care homes and geriatric hospitals. Survey data are obtained through interviews with administrators and staff in a sample of approx 1500 facilities. The most recent National Nursing Home Survey was conducted in 1999.

The National Ambulatory Medical Care Survey is a continuing national probability sample survey of ambulatory care medical experiences conducted annually since 1989 *(33)*. The National Ambulatory Medical Care Survey is part of the ambulatory care component of the National Health Care Survey and measures health care utilization across various types of providers. Sample surveys are obtained from non-federally employed office-based physicians engaged in direct patient care, excluding physicians in the specialties of anesthesiology, pathology, and radiology. Data are obtained on patients' symptoms, physicians' diagnoses, and medications ordered or provided. The survey also provides statistics on the demographic characteristics of patients and services provided, including information on diagnostic procedures, patient management, and planned treatment.

The National Health and Nutrition Examination Survey is a continuing survey designed to collect information about the health and diet of people in the United States *(34)*. National Health and Nutrition Examination Survey provides current statistical data on the amount, distribution, and effects of illness and disability in the United States and has carried out national examination studies in the United States since 1960. The current National Health and Nutrition Examination Survey is the eighth in this series.

5.3.7. Surgery Vital Statistics

Clinicians and researchers are placing increasing emphasis on the use of outcome information to assess the effectiveness and quality of health care. Surgeons have recently moved to the forefront of this effort, involved in both regional and nationwide efforts to monitor and report patient outcomes and improve the quality of surgical care. The largest of these to efforts to date has been undertaken by the Veterans Health Administration, the largest single health care provider in the United States, and has already resulted in improved morbidity and mortality rates for surgical patients.

5.3.7.1. National Surgical Quality Improvement Program

In 1986, Congress passed a law mandating the VA report its surgical outcomes annually. To assess the quality of surgical care in the VA, the Department of Veterans Affairs conducted the first National VA Surgical Risk Study between 1991 and 1993 in 44 VA medical centers *(35–43)* (Chapter 7). This study was used to develop and validate models for 30-d morbidity and 30-d mortality after major surgery in eight noncardiac surgical specialties, adjusted for preoperative patient risk factors. Patients were evaluated postoperatively for mortality as a result of any cause inside or outside of the hospital occurring within 30 d of the surgical procedure and morbidity from 21 selected surgical complications within 30 d of the surgical procedure. Surgical risk study nurses at each medical center facilitated the collection of data and their electronic transmission for analysis at data coordination centers *(40, 41)*. Similar models have been developed for cardiac surgery by the VA's Continuous Improvement in Cardiac Surgery Program *(44, 45)*.

Based on the results of the National VA Surgical Risk Study, the VA established the NSQIP in 1994 to measure and enhance of the quality of surgical care in all the VA medical centers performing major surgery. The NSQIP maintains records on over 600,000 noncardiac surgical operations from 128 Veterans Affairs Medical Centers and 10 sites in the private sector *(37)*. The NSQIP program is validated *(36, 40)*, reliable *(40)*, outcome-based *(40,41)*, risk-adjusted *(35,39,40)*, peer-controlled *(40)*, and aimed at achieving continuous quality improvement by furnishing feedback to both providers and managers. Quality improvement is measured by changes in outcome rates over time. Since the inception of the NSQIP, 30-d postoperative mortality after major surgery in the VA has decreased by 27% and 30-d morbidity by 45% *(41)*.

The NSQIP database accrues prospectively collected presurgical patient risk factors, surgical process information, 30-d mortality and morbidity data, and length of hospital stay for approx 100,000 major surgeries annually. Veterans Health Administration surgeons and researchers can gain access to the database for research by submitting research proposals to the executive committee for peer review. Despite its clinical value, the NSQIP is limited in its generalizability to all population groups, because veterans tend to be an older and medically and socioeconomically disadvantaged population of men *(40)*. However, the eventual inclusion of non-VA institutions should minimize this concern.

6. SUMMARY

Measures of mortality and morbidity are important end points in outcomes research. Researchers must choose which measure most accurately reflects the data. Web-based databases for vital events information have greatly increased the access and usability of these data for research purposes. However, the strengths, limitations, and biases of surveys and databases must be recognized when interpreting results. Increasing concerns over the quality of health care will further the demand for national studies attempting to monitor and report patient outcomes and improve the quality of surgical care in the United States.

REFERENCES

1. Thaul S, Lohr KN, Tranquada RE. Health services research: opportunities for an expanding field of inquiry—an interim statement. Institute of Medicine. Washington, DC: National Academy Press, 1994.
2. Wright JG. Outcomes research: what to measure. World J Surg 1999,23(12):1224–1226.
3. Weed JA. Vital statistics for the United States: preparing for the next century. Population Index. Princeton, NJ: Princeton University, 1995.
4. Beaglehole R, Bonita R, Kjellstrom T. Basic epidemiology. Geneva: WHO, 1993.
5. Timmreck TC. An introduction to epidemiology. 3rd ed. Sudbury, MA: Jones and Bartlett Publishers, 2002.
6. Barker DJP, Cooper C, Rose G. Epidemiology in medical practice. 5th ed. New York: Churchill Livingstone, 1998.
7. Mausner JS, Bahn AK. Epidemiology: an introductory text. Philadelphia: W.B. Saunders Company, 1974.
8. Coronary bypass surgery in New York State 1994–1996. Albany, NY: New York State Department of Health, 1998
9. Last JM. A dictionary of epidemiology. 3rd ed. New York: Oxford University Press, 1995.
10. Hall MJ, Lawrence L. Ambulatory surgery in the United States, 1996. Advance data. Hyattsville, MD, National Center for Health Statistics, 2002. 300, 1–18.
11. Ferlay J, Bray F. GLOBOCAN 2000: cancer incidence, mortality, and prevalence worldwide. Lyon, France: International Agency for Research on Cancer, 2001.

12. Hennekens CH, Buring JE. Measures of disease frequency. In: Mayrent SL, ed. Epidemiology in medicine. Boston: Little, Brown and Company, 1987:54–98.
13. O'Hare AM, Feinglass J, Reiber GE, et al. Postoperative mortality after nontraumatic lower extremity amputation in patients with renal insufficiency. J Am Soc Nephrol 2004;15(2):427–434.
14. Wang J. Basic principles and practical applications in epidemiological research. Singapore: World Scientific, 2002.
15. Fried VM, Prager K, MacKay AP, Xia H. Chartbook on trends in health of Americans. 2003. Hyattsville, MD: National Center for Health Statistics, 2003.
16. Friedman GD. Primer of epidemiology. New York: McGraw-Hill, 1994.
17. National Center for Health Statistics (NCHS). Centers for Disease Control and Prevention. (Accessed November 11, 2003, at http://www.cdc.gov/nchs/
18. Surveillance, Epidemiology, and End Results (SEER). National Cancer Institute. (Accessed November 8, 2003, at http://www.seer.cancer.gov).
19. National Institute for Occupational Safety and Health (NIOSH). Centers for Disease Control and Prevention. (Accessed November 12, 2003, at http://www.cdc.gov/niosh/homepage.html).
20. Iezzoni LI. Using risk-adjusted outcomes to assess clinical practice: an overview of issues pertaining to risk adjustment. Ann Thorac Surg 1994;58(6):1822–1826.
21. Semmens JB, Wisniewski ZS, Bass AJ, Holman CD, Rouse IL. Trends in repeat prostatectomy after surgery for benign prostate disease: application of record linkage to healthcare outcomes. BJU Int 1999;84(9):972–975.
22. Marrelli D, Roviello F, De Stefano A, et al. Risk factors for liver metastases after curative surgical procedures for gastric cancer: a prospective study of 208 patients treated with surgical resection. J Am Coll Surg 2004;198(1):51–58.
23. Howie H, Mukerjee A, Cowden J, Leith J, Reid T. Investigation of an outbreak of Escherichia coli O157 infection caused by environmental exposure at a scout camp. Epidemiol Infect 2003;131(3):1063–1069.
24. Szklo M, Nieto FJ. Epidemiology: beyond the basics. Gaithersburg, MD: Aspen Publishers, Inc., 2000.
25. Osganian SK, Stampfer MJ, Rimm E, et al. Vitamin C and risk of coronary heart disease in women. J Am Coll Cardiol 2003;42(2):246–252.
26. Centers for Medicare and Medicaid Services (CMS). US Department of Health and Human Services. (Accessed November 12, 2003, at http://cms.hhs.gov/).
27. 2003 CMS Statistics. US Department of Health and Human Services. Centers for Medicare and Medicaid Services. Office of Research, Development and Information, 2003. Publication number 03445
28. National Center for Health Statistics (NCHS). Centers for Disease Control and Prevention. (Accessed November 12, 2003, at http://www.cdc.gov/nchs/).
29. National Health Interview Survey (NHIS). National Center for Health Statistics (NCHS). (Accessed January 14, 2004, at http://www.cdc.gov/nchs/nhis.htm).
30. National Hospital Discharge Survey (NHDS). National Center for Health Statistics (NCHS). (Accessed January 14, 2004, at http://www.cdc.gov/nchs/about/major/hdasd/nhds.htm).
31. National Survey of Ambulatory Surgery (NSAS). National Center for Health Statistics (NCHS). (Accessed January 14, 2004, at http://www.cdc.gov/nchs/about/major/hdasd/nsasdes.htm).
32. National Nursing Home Survey (NNHS). National Center for Health Statistics (NCHS). (Accessed January 14, 2004, at http://www.cdc.gov/nchs/about/major/nnhsd/nnhsd.htm).
33. National Ambulatory Medical Care Survey (NAMCS). National Center for Health Statistics (NCHS). (Accessed January 14, 2004, at http://www.cdc.gov/nchs/about/major/ahcd/nhamcsds.htm).
34. National Health and Nutrition Examination (NHANES). National Center for Health Statistics (NCHS). (Accessed January 14, 2004, at http://www.cdc.gov/nchs/nhanes.htm).
35. Daley J, Khuri SF, Henderson W, et al. Risk adjustment of the postoperative morbidity rate for the comparative assessment of the quality of surgical care: results of the National Veterans Affairs Surgical Risk Study. J Am Coll Surg 1997;185(4):328–340.
36. Daley J, Forbes MG, Young GJ, et al. Validating risk-adjusted surgical outcomes: site visit assessment of process and structure. National VA Surgical Risk Study. J Am Coll Surg 1997;185(4):341–351.
37. Feinglass J, Pearce WH, Martin GJ, et al. Postoperative and late survival outcomes after major amputation: findings from the Department of Veterans Affairs National Surgical Quality Improvement Program. Surgery 2001;130(1):21–29.
38. Khuri SF, Daley J, Henderson W, et al. The National Veterans Administration Surgical Risk Study: risk adjustment for the comparative assessment of the quality of surgical care. J Am Coll Surg 1995;180(5):519–531.

39. Khuri SF, Daley J, Henderson W, et al. Risk adjustment of the postoperative mortality rate for the comparative assessment of the quality of surgical care: results of the National Veterans Affairs Surgical Risk Study. J Am Coll Surg 1997;185(4):315–327.
40. Khuri SF, Daley J, Henderson W, et al. The Department of Veterans Affairs' NSQIP: the first national, validated, outcome-based, risk-adjusted, and peer-controlled program for the measurement and enhancement of the quality of surgical care. National VA Surgical Quality Improvement Program. Ann Surg 1998;228(4):491–507.
41. Khuri SF, Daley J, Henderson WG. The comparative assessment and improvement of quality of surgical care in the Department of Veterans Affairs. Arch Surg 2002;137(1):20–27.
42. National Quality Improvement Program. Veterans Health Administration. (Accessed November 8, 2003, at http://nsqip.org).
43. National Quality Improvement Program. Veterans Health Administration. (Accessed November 8, 2003, at http://nsqip.org).
44. Grover FL, Johnson RR, Shroyer AL, Marshall G, Hammermeister KE. The Veterans Affairs Continuous Improvement in Cardiac Surgery Study. Ann Thorac Surg 1994;58(6):1845–1851.
45. Hammermeister KE, Johnson R, Marshall G, Grover FL. Continuous assessment and improvement in quality of care. A model from the Department of Veterans Affairs Cardiac Surgery. Ann Surg 1994;219(3):281–290.
46. Minino AM, Arias E, Kochanek KD, Murphy SL, Smith BL. Death: final data for 2000. Hyattsville, MD: National Center for Health Statistics. National Vital Statistics Report. Vol. 50, number 15, 2002.
47. Ferlay J, Bray F. GLOBOCAN 2000: cancer incidence, mortality, and prevalence worldwide. Lyon, France: International Agency for Research on Cancer, 2001.
48. Ries LAG, Eisner MP, Kosary CL, et al. SEER Cancer Statistics Review, 1973–1999. Bethesda, MD: National Cancer Institute, 2002.
49. MacLennan PA, Delzell E, Sathiakumar N, Myers SL. Mortality among triazine herbicide manufacturing workers. J Toxicol Environ Health A 2003;66(6):501–517.
50. Years of Potential Life Lost (YPLL) before age 65. Office of Statistics and Programming, National Center for Injury Prevention and Control, CDC. National Center for Health Statistics (NCHS) Vital Statistics System. (Accessed November 8, 2003, at http://webapp.cdc.gov/sasweb/ncipc/ypll9.html).
51. Joseph MA, Harlow SD, Wei JT, et al. Risk factors for lower urinary tract symptoms in a population-based sample of African-American men. Am J Epidemiol 2003;157(10):906–914.

APPENDIX

The following data are from a hypothetical study of mortality among a cohort of 27,400 female plastic factory workers from 1990 to 2000 (all participants survived to midpoint of study or longer).

Table 1
Expected Breast Cancer Deaths for 25–34 Year Olds

Column 1	*Column 2*	*Column 3*	*Column 4*
Age (yr)	*2000 US Standard Population of Women (in Thousands) (a)*	*Age-Specific Breast Cancer Mortality Rates in Study (per 100,000) (b)*	*Expected Number Breast Cancer Deaths (in Thousands) (a*b)*
25–34	19,771	0.0014	27.68
35–44	22,701	0.0058	131.67
45–54	19,181	0.0076	145.78
55–64	12,629	0.0091	114.92
65–74	10,088	0.0062	62.55
75–84	7482	0.0066	49.38
85+	3013	0.0043	12.96
Total	94,865		544.94

Table 2
Breast Cancer Mortality Rates for 25–34 Year Olds

Column 1	*Column 2*	*Column 3*	*Column 4*	*Column 5*	*Column 6*	*Column 7*	*Column 8*
Age in Years	*N Persons (a)*	*N Person-Yrs (py)*	*N Cases of Breast Cancer*	*N Deaths From Breast Cancer*	*N Overall Cancer Beaths*	*U.S. 2000 Breast cancer Mortality Rates (per 100,000) (b)*	*Expected Deaths (py) (a*b)*
25-34	5000	46100	10	7	25	2.3	1.06
35-44	3250	31625	28	19	39	12.4	3.92
45-54	4200	29800	45	32	65	33.0	9.83
55-64	3750	26500	52	34	78	59.3	15.71
65-74	4700	33500	39	29	82	88.3	29.58
75-84	3500	20500	26	23	105	128.9	26.42
85+	3000	19400	18	13	132	205.7	39.91
Total	27,400	207,425	218	157	526		126.43

Table 3
YPLL From Breast Cancer for 25–34 Yr Olds (Based on End Point Age of 65)

Column 1	*Column 2*	*Column 3*
N Deaths	*YPLL (From Age 65)*	*Age at Death*
31	1	34
32	0	0
33	2	64
34	0	0
35	4	120
36	4	116
37	3	84
38	5	135
39	6	156
Total	25	709

Calculations

Crude cancer mortality rate= Total number of deaths during specified time period / Total population at midpoint of specified time period x 100,000
(per 100,000 population)

= 526 (Table 2, column 6) / 27400 (Table 2, column 2) = 1919.71 per 100,000 women in the population per 10 yr study period

Annual cancer mortality rate= total number of deaths per 12-month time period / total population at midpoint of 12-month time period x 100,00
(per 100,000 population)

= 526 / 10 years / = 27400 = 191.97 per 100,000 women in the population per yr

Age-specific cancer mortality rate= Number of deaths in specified time period / Total persons aged 25-34 at midpoint of time period x 100,000
for 25-34 year olds =
(per 100,000 population)

= 25 (Table 2, column 6) / 5000 (Table 2, column 2) = 500 per 100,000 women in the population per 10 yr study period

Age standardized mortality rate= Total expected number of deaths (Table 1, column 4) / Total 2000 U.S. population of women (Table 1, column 2)
(direct method)
(breast cancer)

= 544.94 / 94865 =574.44 per 100,000 women

Age standardized mortality rate= Total expected number of deaths (py) (Table 2, column 8) / Total person-years in cohort (Table 2, column 3)
(indirect method)
(breast cancer)

= 126.43 / 207,425 =60.95 per 100,000 person-years

Standardized Mortality Ratio= (breast cancer)

$$\frac{\text{Observed Deaths (Table 2, column 5)}}{\text{Expected Deaths (Table 2, column 8)}} \times 100$$

$$= \frac{157}{126.43} = 124\%$$

Thus, plastic factory workers had a risk of breast cancer mortality ~ 24% greater than women in the general population.

Proportional Mortality Ratio

Proportional mortality for breast cancer among this cohort:

$$\frac{157 \text{ (Table 2, column 5)}}{526 \text{ (Table 2, column 6)}} = 29.8\%$$

Proportional morality for breast cancer among U.S. population = 16.0%

Expected number of breast cancer deaths = 526 x 0.160 = 84.16

PMR =

$$\frac{\text{Observed Deaths from specified cause}}{\text{Expected Deaths from specified cause}} \times 100$$

$$= \frac{157}{84.16} = 186.5\%$$

Thus, the proportion of cancer deaths attributable to breast cancer was 1.87 fold as great among this cohort of plastic factory workers as among the general U.S. population.

Case Fatality Rate= (breast cancer)

$$\frac{\text{Number of deaths from disease in a specified time period}}{\text{Number of cases of disease in the specified time period}} \times 100$$

$$= \frac{157 \text{ (Table 2, column 5)}}{218 \text{ (Table 2, column 4)}} = 72.0\%$$

Cause-specific mortality rate= (breast cancer) (per 100,000 population)

$$\frac{\text{Number of deaths due to specific cause in time period}}{\text{Total persons at midpoint of time period}} \times 100{,}000$$

$$= \frac{157 \text{ (Table 2, column 5)}}{27{,}400 \text{ (Table 2, column 2)}}$$ = 573.0 per 100,000 women for the 10-yr study period

YPLL rate (among 25-34 yr olds)= (breast cancer)

$$\frac{\text{Total years of potential life lost}}{\text{Total population under endpoint age}} \times 1000$$

$$= \frac{709 \text{ (Table 3, column 4)}}{5000 \text{ (Table 2, column 2)}}$$ = 141.8 per 1000 women

Point Prevalence= (breast cancer)

$$\frac{\text{Number of } \textbf{\textit{existing}} \text{ cases of disease at a specified } \textbf{\textit{point}} \text{ in time}}{\text{Total population at specified } \textbf{\textit{point}} \text{ in time}} \times 100$$

Suppose at year 5 of the study, 124 breast cancers had developed.

Point Prevalence at year 5 of the study= $\frac{124}{27400}$ = 4.53 per 1000 women

Cumulative Incidence= (breast cancer)

$$\frac{\text{Number of } \textbf{\textit{new}} \text{ cases of disease within a specified time period}}{\text{Total population at risk at the beginning of specified time period}} \times 100$$

$$= \frac{218 \text{ (Table 2, column 4)}}{27{,}400 \text{ (Table 2, column 4)}}$$ = 79.6% during the 10-yr study period

Attack Rate= (per 100 population)

$$\text{Attack Rate} = \frac{\text{Number of new cases of disease within specified time period}}{\text{Total population at risk at beginning of specified time period}} \times 100$$

Suppose during the first months of the study, 2500 women developed influenza.

$$\text{Attack rate} = \frac{2500}{27{,}400} = 9.12\%$$

Incidence Rate = (breast cancer)

$$\text{Incidence Rate} = \frac{\text{Number of new cases of disease during specified time period}}{\text{Total person time}}$$

$$= \frac{218 \quad \text{(Table2, column 4)}}{207{,}425 \text{ py} \quad \text{(Table2, column 3)}} = 105.1 \text{ per } 100{,}000 \text{ person-years}$$

13 Health-Related Quality of Life

Mark S. Litwin, MD, MPH

Contents

Traditionally, the primary end points in evaluations of medical therapies have included improvement in clinical outcomes, cure, and survival; however, the advent of the medical outcomes movement and the worldwide effort to contain the rising costs of care have underscored the importance of patient-centered outcomes, such as health-related quality of life (HRQOL). This trend is especially relevant for individuals with chronic, nonfatal conditions who live for many years after diagnosis. If survival time is long, even modest changes in HRQOL may have a profound impact on the patient. Hence, in many cases, HRQOL may be just as important as survival, and treatment choices at various points in the chronic disease trajectory must constantly balance the dual goals of optimizing both quantity and quality of life. In light of evidence that survival and clinical outcomes may be similar across treatments for many conditions, quality of life considerations may be the critical factor in medical decision making for some.

HRQOL issues are even relevant for individuals with malignancies, particularly tumors known to behave in an indolent fashion. The impact of HRQOL on therapeutic decision making in oncology is now considered so important that some investigators consider a clinical cancer trial incomplete without HRQOL assessment *(1,2)*. HRQOL questionnaires may be successfully incorporated into large multicenter trials *(3–5)* if sufficient resources are available.

1. DEFINING HRQOL

HRQOL is one of several variables commonly studied in the field of medical outcomes research. It encompasses a wide range of human experience, including functioning and subjective responses to illness *(6, 7)*. Contemporary interpretations of HRQOL are based on the World Health Organization's definition of health as a state of complete physical,

From: *Clinical Research for Surgeons*
Edited by: D. F. Penson and J. T. Wei © Humana Press Inc., Totowa, NJ

World Health Organization *Definition of Health*
Health is not merely the absence of disease, but a state of complete physical, mental, and social well-being.

mental, and social well-being and not merely the absence of disease *(8)*. Because surgery can affect many aspects of quality of life, a wide spectrum of the components of well-being must be addressed when assessing outcomes in surgical patients, as shown in Table 1.

In broad terms, HRQOL may be conceived as the ratio of an individual's actual status over expected status. For example, to the degree that a prostate cancer patient's erectile dysfunction is expected, not bothersome, and not intrusive into his life or self-image, it does not affect his HRQOL. Conversely, a patient who is more focused on his expectations of good erectile function after therapy may perceive even the slightest decrement as having a potent effect on his quality of life *(9)*.

Unlike morbidity, which applies to the specific complications or consequences of an illness, HRQOL studies cast a broader net to include the bother associated with particular dysfunctions, any impact on normal functions or social roles, and a composite of other psychosocial domains. HRQOL is multidimensional and incorporates the impact of dysfunctions on the individual. This requires multidisciplinary research teams to measure and analyze the full effect of HRQOL disturbances *(10)*.

2. GOALS OF QUALITY OF LIFE RESEARCH

The ultimate goal of quality of life research must be to improve medical care and inform medical decision making. Individual patients who incorporate quality of life considerations into their decisions generally feel better about their treatment choices, are more satisfied overall with their care, and are less likely to experience regret *(11)*. Hence, patient education provides a strong impetus for studying and reporting quality of life. Through better education on the potential outcomes, quality of life research supports improved medical decision making for current and future patients. With accurate measurement of HRQOL outcomes, patients, clinicians, and researchers can better assess whether the goals of therapy have been met. This allows individuals and the public to balance the competing health care priorities of optimizing medical outcomes and resource utilization. Furthermore, the evaluation of quality of life, as perceived by the patient, allows for the assessment of subjective morbidity that, although not always life-threatening, may cause considerable distress. Such minor morbidity is often overlooked during the busy routines of clinical care. Finally, international HRQOL studies allow cross-cultural comparisons of the effects of the disease and its treatments (Table 1).

3. MEASURING HRQOL

During the past two decades, widespread interest in studying patient-centered outcomes has led to the development of a rigorous set of methods for HRQOL measurement. The unambiguous lesson from this work is that researchers and clinicians must ask about general and specific impairments in a standard manner. HRQOL outcomes are often complex, qualitative variables that are not easily simplified. For example, body image

Table 1
Quality of Life Research Objectives

- To assess overall treatment efficacy, including subjective morbidity
- To help determine whether the goals of treatment have been met
- To educate patients and clinicians about the full spectrum of treatment outcomes
- To facilitate medical decision making
- To provide the defining issue if treatments are otherwise equivalent
- To compare outcomes across treatments and populations

after open or laparoscopic surgery encompasses a wide range of feelings and activities that makes it difficult to dichotomize subjects for analysis.

3.1. Psychometric Test Theory

Although quantity of life is relatively easy to assess as overall or cause-specific survival, the measurement of quality of life presents more challenges, primarily because it is less familiar to most clinicians. To quantify these subjective phenomena, the principles of psychometric test theory are applied. This discipline provides the theoretical foundation for the field of survey research *(12–17)*. Data are collected with HRQOL surveys, called instruments. Instruments typically contain questions, or items, that are organized into scales. Each scale measures a different aspect, or domain, of HRQOL. For example, items of a particular instrument may address a patient's ability to have an erection and his satisfaction with ejaculation, both of which might be included in a sexual domain. Some scales comprise many items, whereas others may include only one or two items. Each item contains a stem (which may be a question or a statement) and a response set. Most response sets are one of the following types: (1) Likert scale, in which the respondent selects from a list of degrees of agreement or disagreement with the stem; (2) Likert-type scale, in which the respondent chooses from a list of text responses; (3) visual analog scale, in which the respondent marks a point on a line that is anchored on both ends by descriptors; or (4) numerical rating scale, in which the respondent chooses a number, usually between 0 and 10. Other response sets and approaches have been developed for children, people of low literacy, and various other populations *(18–20)*.

It is axiomatic that HRQOL assessments capture patients' own perceptions of their health and ability to function in life. Instruments are best when they are self-administered by the patient, but if interviewer assistance is required, it must be from a neutral third party in a standardized fashion. As an example, some studies have demonstrated that physicians typically underestimate the symptom burden experienced by prostate cancer patients, perhaps because their queries are not sensitive enough or because patients tend to understate their problems when speaking directly with the primary caregiver *(21–23)*. Other studies, however, suggest that physicians tend to overestimate the impact of the disease and its treatment on patients' psychosocial functioning and sense of well-being *(24–26)*. Conversely, spouses may overstate some domains and understate others when compared with patient assessments *(27)*. Kornblith *(28)* presented results from a large sample of patients and spouses, both administered several validated HRQOL measures. Spouses reported greater psychologic distress, but fewer sexual problems than did patients themselves. In a study of perspectives on HRQOL during antihypertensive therapy, Testa

(29) demonstrated that physicians were less sensitive to the impact of side effects, reporting less than 15% of the symptoms reported by patients. Spousal reports were more sensitive than patient self-assessments, particularly in the area of sexual functioning.

3.2. Comparison Groups

Prospective, longitudinal data collection is always best, because this approach may reveal time-dependent evolution of HRQOL domains *(30,31)*. Patients may then serve as their own controls. Assessing HRQOL at baseline before treatment allows for the inclusion of baseline age-related changes that should not be attributed to treatments. This approach facilitates the stratification of discriminants from determinants of HRQOL.

However, investigators often use methodologies in which HRQOL is assessed cross-sectionally, rather than longitudinally. In cross-sectional surveys, patients cannot serve as their own temporal controls, because it is well-established that patients' recall of pretreatment HRQOL is inaccurate *(32,33)*. Hence, studies must rely on appropriate comparison groups. Selecting the best normal comparison group is a critical step in conducting a meaningful analysis of HRQOL outcomes. If normal is defined as the absence of any dysfunction, then treatment groups may be held to too high a standard. If normal is determined by assessing age-matched subjects without the disease of interest, then HRQOL outcomes after treatment may be interpreted in a more valid context. Other factors, such as socioeconomic status, may also influence the care-seeking decisions of patients, and this may in turn affect how they perceive their HRQOL. In comparisons of treatment efficacy on HRQOL, longitudinal studies with concurrent controls provide the most valid results.

3.3. Reliability, Validity, and Responsiveness

The development and validation of new instruments and scales is a long and arduous process. It should not be undertaken lightly. Simply drawing up a list of questions that seems appropriate is fraught with potential traps and pitfalls. For this reason, it is always preferable to select instruments that have been validated and published. HRQOL instruments must be shown to have the fundamental properties of reliability, validity, and responsiveness *(34–36)*.

Reliability refers to how reproducible the scale is. Test–retest reliability is a measure of response stability over time. It is assessed by administering scales to subjects at two time points, with the time interval short enough to preclude the possibility that the domains being assessed will have been affected by the disease or its treatment during the intervening period. Correlation coefficients between the two scores reflect the stability of responses. Internal consistency reliability measures the similarity of an individual's responses across several items, indicating the homogeneity of a scale. The statistic used to quantify the internal consistency, or unidimensionality, of a scale is called Cronbach's coefficient alpha *(37)*. Generally accepted standards dictate that, for group comparisons, reliability statistics measured by these two methods should exceed 0.70 *(38)*. When used at the level of individual patients (e.g., monitoring HRQOL over time), a reliability coefficient of at least 0.90 is preferred. Although some scales may function well as single-item measures, in general, a health concept is better measured by a set of questions than by a single question. Multi-item measures are thus more reliable.

Validity refers to how well the scale or instrument measures the attribute it is intended to measure. Content validity, sometimes referred to as face validity, involves qualitative assessment of the scope, completeness, and relevance of a proposed scale *(39)*. Criterion validity is a more quantitative approach to assessing the performance of scales and instruments. It requires the correlation of a scale's score with other measurable health outcomes (predictive validity) and with results from established tests (concurrent validity). Generally accepted standards also dictate that validity statistics should exceed 0.70 *(38)*. Construct validity, perhaps the most valuable assessment of a survey instrument, is a measure of how meaningful the scale or survey instrument performs in a multitude of settings and populations over a number of years. Construct validity comprises two other forms of validity: convergent and divergent. Convergent validity implies that several different methods for obtaining the same information about a given trait or concept produce similar results. Divergent validity means that the scale does not correlate too closely with similar but distinct concepts or traits. Because instruments are not simply valid or invalid, the task of validating them is always ongoing.

Responsiveness of a HRQOL instrument refers to how sensitive the scales are to change over time. That is, a survey may be reliable and valid when used at a single point in time, but in some circumstances it must also be able to detect meaningful improvements or decrements in quality of life during longitudinal studies. The instrument must "react" in a time frame that is relevant for patients over time. Because HRQOL may change over time, longitudinal measurement of these outcomes is important *(40,41)*. Different domains may become more or less prominent over time as the course of disease and recovery evolves. Although their perception of cure waxes and wanes with time since treatment or the latest prostate specific antigen level, patients may feel more or less affected by their HRQOL impairments. In addition, patients may experience what is known as a response shift as they learn to adapt to the chronicity of HRQOL alterations *(42)*.

3.4. Interpreting HRQOL Scores in the Context of Clinically Meaningful Differences

Most, though not all, contemporary HRQOL instrument domains are scored with a range of 0–100, with higher values representing better outcomes. To make useful inferences regarding absolute scores or change scores over time, it is important to determine what meaning different numerical values have *(43)*. When no such thresholds have been established, one can roughly approximate the smallest difference that is important to the patient as one-third to one-half of a standard deviation *(44,45)*. A more quantitative approach involves calculating an effect size, or Guyatt statistic, typically expressed as the ratio of the raw change in score among those who change to the standard deviation of the change among those who did not change *(46,47)*.

3.5. General Vs Disease-Specific HRQOL Assessment

HRQOL instruments may be general or disease–specific. General HRQOL domains address the components of overall well–being, whereas disease–specific domains focus on the impact of particular organic dysfunctions that affect HRQOL *(6,48)*. General HRQOL instruments typically address general health perceptions, sense of overall well–being, and function in the physical, emotional, and social domains. Disease–specific HRQOL instruments focus on more directly relevant domains, such as anxiety about cancer recurrence; urinary, sexual, and bowel impairment; and any bother caused by these dysfunctions. Disease-specific and general HRQOL domains often impact each other, leading to important interactions that must be considered in the interpretation of

HRQOL data *(49)*. Further research is needed in many disciplines to explore how much of the variation in overall HRQOL is explained by variation in the disease-specific domains.

In some conditions, such as cirrhosis with ascites, advanced renal failure, and stroke, general HRQOL may be so profoundly affected that disease-specific HRQOL assessment is unnecessary. In many indolent conditions, however, the treatments may alter bodily functions that are not be fully appreciated by assessing only the broader domains of general HRQOL. Conversely, in patients with advanced cancer, HRQOL may be affected predominantly by pain, fatigue, and other constitutional symptoms that are well captured by general HRQOL instruments.

3.6. Instrument Length and Translations

Investigators must be parsimonious when selecting HRQOL instruments. Although longer instruments may provide richer datasets, researchers must recognize that fatigue may limit the ability of patients to provide useful information. This phenomenon, known as response burden, must be considered when assessing HRQOL in clinical or research settings.

Cultural issues must be taken into account when administering HRQOL instruments. Although an instrument may have been linguistically translated into a new language, it may not have the same meaning in that culture *(50)*. This is particularly relevant when studying quality of life, social attitudes, and health behaviors in different countries or cultures. Different nations and cultures may have very different concepts of health, well-being, illness, and disease. Therefore, a well-developed concept in one group of people may not even exist in another. Even with an instrument that is well-validated in English, various English-speaking populations across the world may not approach the concept with the same ideas *(51)*. Specific methodologies have been developed for cross-validating HRQOL instruments in other languages *(52, 53)*. In addition, one must distinguish between measures that have been developed for use in one culture or language and then translated into another from those which have been developed from the outset in with a cross-cultural perspective. Failing to be attentive to multicultural issues may result in significant bias when collecting and interpreting data. New instrument development should always be undertaken with an eye toward eventual international translation and cultural adaptation.

In addition to varying cultural perspectives on disease and health, international differences in health systems may also have a substantial impact on the way patients view their quality of life. For example, in countries where patients are required to pay all or most of the treatment costs, spending a lot of money for marginally better survival rates may have a larger effect on quality of life than the disease or its treatments. Furthermore, in cultures where the patient's relatives are compelled to absorb the costs of care, the quality of life of the entire family unit should also be considered.

4. VALIDATED HRQOL INSTRUMENTS

When studying quality of life for clinical or research purposes, it is preferable to use published instruments that have been previously validated in the relevant population. In general, one should avoid extracting single items or scales from different instruments to construct a new one unless they have been independently psychometrically validated. The development and validation of a new HRQOL instrument is an arduous task. Hence, investigators should first examine existing instruments to determine if they adequately capture the domains of interest before developing a new instrument. Although there is often little empirical basis to choose one instrument over another *(54)*, a variety of validated instruments is available for use in research and clinical settings. Most of the available instruments can be self-administered.

4.1. General HRQOL Instruments

General quality of life instruments have been extensively studied and validated in many types of patients, sick and well. Examples include the RAND Medical Outcomes Study 36-Item Health Survey, also known as the SF-36 *(55–57)*; the Quality of Well-Being scale (QWB) *(58–63)*; the Sickness Impact Profile *(64, 65)*; and the Nottingham Health Profile *(66–68)*. Each assesses various components of HRQOL, including physical and emotional functioning, social functioning, and symptoms. Each has been thoroughly validated and tested.

The SF-36 is one of the most commonly used instruments and is regarded by some as a "gold standard" measure of general HRQOL. It is a 36-item, self-administered instrument that takes less than 10 min to complete and quantifies HRQOL in multi-item scales that address eight different health concepts—physical function, role limitation because of physical problems, bodily pain, general health perceptions, social function, emotional well-being, role limitation because of emotional problems, and energy/fatigue. The SF-36 may also be scored in two summary domains—physical and mental. Recently, a shorter 12-item version, the SF-12, has been developed for use in studies requiring greater efficiency. It provides a somewhat narrower view of overall health status and is scored only in the two summary domains *(69–71)*.

The QWB summarizes three aspects of health status—mobility, physical activity, and social activity—in terms of quality–adjusted life-years, quantifying HRQOL as a single number that may range from death to complete well–being. The original QWB contains only 18 items, but it requires a trained interviewer. A newer self-administered version of the QWB is now available and has been shown to produce scores which are equivalent to the interviewer-administered version and stable over time *(60)*.

The Sickness Impact Profile measures health status by assessing the impact of sickness on changing daily activities and behavior. It is self–administered but contains 136 items and can take 30 min or longer to complete. Test-retest reliability is consistently high (0.88–0.92) in validation populations.

The Nottingham Health Profile covers six types of experience that may be affected by illness: pain, physical mobility, sleep, emotional reactions, energy, and social isolation by using a series of weighted yes or no items. It contains 38 self–administered items and can be completed fairly quickly.

Mental health is often measured with the Profile of Mood States *(72–75)*, a 65-item, self-administered instrument that measures dimensions of affect or mood in six domains including anxiety, depression, anger, vigor, fatigue, and confusion. A validated short form is also available *(76)*. A list of important domains that are measured using general HRQOL instruments is shown in Table 2.

4.2. Cancer-Specific HRQOL Instruments

Because of the well–documented impact of malignancies and their treatment on HRQOL, cancer–specific quality of life also has been investigated extensively. Numerous

Table 2
Domains Included in General Health-Related Quality of Life Instruments

SF-36	Independent categories
Physical function	Sleep and rest
Role limitations from physical problems	Eating
Bodily pain	Work
Energy/fatigue	Home management
Mental health	Recreation and pastimes
Role limitations from emotional problems	Nottingham Health Profile
General health perceptions	Pain
SF-12	Physical mobility
Physical component summary	Sleep
Mental component summary	Emotional reactions
Quality of Well-Being Scale	Energy
Mobility	Social isolation
Physical activity	Profile of Mood States
Social activity	Tension-anxiety
Sickness Impact Profile	Depression-dejection
Physical	Anger-hostility
Ambulation	Vigor-activity
Mobility	Fatigue-inertia
Body care and movement	Confusion-bewilderment
Psychosocial	
Social interaction	
Alertness behavior	
Emotional behavior	
Communication	

instruments have been developed and tested that measure the special impact of cancer (regardless of primary site) on patients' routine activities. These instruments are particularly relevant to surgeons because surgery is often a cornerstone of treatment for many malignancies. Examples of cancer-specific instruments include the European Organization for the Research and Treatment of Cancer Quality of Life Questionnaire (EORTC QLQ-C30) *(77)*, the Functional Assessment of Cancer Therapy (FACT) *(78)*, and the Cancer Rehabilitation Evaluation System (CARES) and its short form (CARES-SF) *(79–81)*. Each has been validated and tested in patients with various types of cancer. Readers are directed to the Quality of Life Instruments Database (http://www.qolid.org) for guidance when selecting an instrument for quality of life measurement in studies of prostate or other cancers.

The EORTC QLQ-C30 was designed to measure cancer-specific HRQOL in patients with a variety of malignancies. Its 30 items address domains that are common to all cancer patients. The questionnaire includes five functional scales (physical, role, emotional, cognitive functioning, and social functioning), a global health scale, three symptom scales (fatigue, nausea/vomiting, and pain), and six single items concerning dyspnea, insomnia, appetite loss, constipation, diarrhea, and financial difficulties because of disease. The EORTC QLQ-C30 does not include domains specific to prostate cancer, but it

Table 3
Domains Included in Cancer-Specific Health-Related Quality of Life Instruments

European Organization for the Research and Treatment Quality of Life Questionnaire QLQ-C30
Physical
Role
Emotional
Cognitive
Social functioning
Functional Assessment of Cancer Therapy – General
Physical
Social/family
Relationship with doctor
Emotional
Functional
Cancer Rehabilitation Evaluation System – Short Form
Physical, psychosocial
Medical interaction
Marital interaction
Sexual function
Rotterdam Symptom Checklist
Psychosocial distress
Physical distress

has performed well in this population *(82)*. Disease-specific modules for cancers of the breast *(83)*, lung *(84)*, prostate *(85)*, and head and neck *(86)* have been developed according to methodologically rigorous techniques. Other disease-specific modules are under development.

The FACT is usually applied as a two–part instrument that includes a general item set pertaining to all cancer patients (FACT–G) and one of several item sets containing special questions for patients with specific tumors. Each item is a statement that a patient may agree or disagree with across a five–point range. The FACT–G domains include well–being in five main areas: physical, social/family, relationship with doctor, emotional, and functional. The FACT–G includes 28 items and is easily self–administered. Disease-specific modules are available for colorectal *(87)*, breast *(88)*, prostate *(89)*, ovary *(90)*, and other cancers and for issues specific to bone marrow transplantation *(91)*, anemia, and fatigue *(92, 93)*.

The CARES Short Form (CARES–SF) is a 59–item, self–administered instrument that measures cancer–related quality of life with five multi-item scales: physical, psychosocial, medical interaction, marital interaction, and sexual function. A large and valuable database of patients with many different tumors, including urologic tumors, has been collected by the instrument's authors *(80)*. These data are helpful when comparing the experience of prostate cancer patients with that of patients with other types of cancer.

The Rotterdam Symptom Checklist contains 27 items that are scored in two domains (psychosocial and physical distress), as well as several miscellaneous items relevant to cancer patients. Its two dimensions are reliable across populations *(94)*.

The UCLA Prostate Cancer Index has been popularized as a reliable, valid instrument to measure disease-targeted HRQOL in men treated for early-stage prostate cancer *(95)*,

Table 4
Recommendations

For clinicians

- When reading the literature, integrate findings of health-related quality of life (HRQOL) studies into the overall assessment of treatment outcomes

For clinical investigators

- Use validated instruments to measure HRQOL
- Select instruments based on study focus—some instruments may be better than others, depending on the clinical question to be addressed
- Obtain multidisciplinary expertise on HRQOL measurement early (i.e., during the design phase of studies)
- When possible, assess HRQOL longitudinally, beginning at baseline
- When possible, use controls in studies of therapeutic effectiveness including HRQOL measurements

For methodologists

- Develop, translate, and validate a core set of prostate cancer–specific HRQOL questions to facilitate comparisons of study outcomes from different countries and cultures
- Conduct studies to determine the clinical meaning of absolute HRQOL scores and HRQOL score changes
- Conduct studies to calibrate HRQOL instrument scores against each other in various populations
- Establish the optimal frequency and timing of HRQOL measurement for longitudinal studies
- Develop situation-specific patient education methods to inform and facilitate medical decision making

It is a self-administered, 20-item questionnaire that quantifies disease-specific HRQOL in the six domains of urinary, sexual, and bowel function and bother. Cross-cultural translations of the Prostate Cancer Index are available in Spanish *(96)*, French *(97)*, Japanese *(98)*, and Dutch *(99)*. Table 3 provides a list of domains commonly included in cancer-specific instruments.

5. PRACTICAL GUIDELINES

Investigators considering measuring HRWOL in a clinical study involving patients with acute or chronic diseases should obtain early consultation (in the design phase) from an expert in this area. The choice of an instrument (or instruments) for the study will depend on the particular population being studied and the clinical questions being asked. Using previously validated instruments, to the extent they are applicable and appropriate, obviates the need for an arduous process of instrument development and validation. A general and a disease-specific module in combination will be suitable for most studies. However, if a particular domain (e.g., pain) is the focus of the study, specific, expanded questionnaires should be sought focusing on the area of interest. Respondent burden needs to be considered, particularly for longitudinal studies in which subjects will complete the same instruments multiple times. Pretesting instruments that will be used in clinical studies is always advisable.

5.1. Clinical Applications of HRQOL Research

The increased popularity of HRQOL measurement in clinical trials has led to improvements in the quality of patient care. When physicians are attuned to the quality of life concerns of their patients, care is more comprehensive at the bedside and in the clinic. As HRQOL studies are extended to the screening environment, we may learn that quality of life is affected by anxiety in the prediagnosis phase. This factor must be considered in assessments of the value of screening programs.

Beyond the descriptive analysis, HRQOL outcomes must be compared in patients undergoing different modes of therapy. General and disease-specific HRQOL must be measured to facilitate comparison with patients treated for other common chronic conditions. Quality of life outcomes must also be controlled for variations in comorbidity or in sociodemographic variables such as age, race, education, income, insurance status, geographic region, and access to health care. In this context, HRQOL may be linked with many factors other the traditional medical ones. Research initiatives must rely on established, reliable, valid HRQOL instruments administered by objective third parties. Quality of life can have many different definitions and interpretations, but its measurement must adhere to the strict application of psychometric science. A list of recommendations regarding HRQOL research is provided in Table 4.

With better information on quality of life, in addition to clinical outcomes, we will develop a rich database that encompasses the entire spectrum of clinical disease outcomes. We will then improve our ability to evaluate new treatment modalities, educate our patients, and counsel them individually on what to do and expect from medical care.

REFERENCES

1. Fayers PM, Jones DR. Measuring and analysing quality of life in cancer clinical trials: a review. Stat Med 1983;2(4):429–446.
2. Feeny DH, Torrance GW. Incorporating utility-based quality-of-life assessment measures in clinical trials. Two examples. Med Care 1989;27(3 Suppl.):S190–S204.
3. Sadura A, Pater J, Osoba D, et al. Quality-of-life assessment: patient compliance with questionnaire completion. J Natl Cancer Inst 1992;84(13):1023–1026.
4. Fossa SD. Quality of life after palliative radiotherapy in patients with hormone- resistant prostate cancer: single institution experience. Br J Urol 1994;74(3):345–351.
5. Moinpour CM, Lovato LC. Ensuring the quality of quality of life data: the Southwest Oncology Group experience. Stat Med 1998;17(5–7):641–651.
6. Patrick DL, Erickson P. Assessing health-related quality of life for clinical decision-making. In. Walker SR, Rosser RM, eds. Quality of life assessment: key issues in the 1990s. Dordrecht: Kluwer Academic Publishers, 1993:11–64.
7. Osoba D. Measuring the effect of cancer on quality of life. Boca Raton: CRC Press, 1991.
8. WHO. Constitution of the World Health Organization, basic documents. Geneva: WHO, 1948.
9. Fitzpatrick JM, Kirby RS, Krane RJ, et al. Sexual dysfunction associated with the management of prostate cancer. Eur Urol 1998;33(6):513–522.
10. Chang VT, Thaler HT, Polyak TA, et al. Quality of life and survival: the role of multidimensional symptom assessment. Cancer 1998;83(1):173–179.
11. Cassileth BR, Soloway MS, Vogelzang NJ, et al. Patients' choice of treatment in stage D prostate cancer. Urology 1989;33(5 Suppl.):57–62.
12. Tulsky DS. An introduction to test theory. Oncology (Huntingt) 1990;4(5):43–48.
13. Testa MA, Simonson DC. Assessment of quality-of-life outcomes. N Engl J Med 1996;334(13): 835–840.
14. Guyatt GH, Naylor CD, Juniper E, et al. Users' guides to the medical literature. XII. How to use articles about health-related quality of life. Evidence-Based Medicine Working Group. JAMA 1997;277(15): 1232–1237.

15. McSweeny AJ, Creer TL. Health-related quality-of-life assessment in medical care. Dis Mon 1995;41(1):1–71.
16. Aaronson NK. Methodologic issues in assessing the quality of life of cancer patients. Cancer 1991;67 (3 Suppl.):844–850.
17. Deyo RA, Diehr P, Patrick DL. Reproducibility and responsiveness of health status measures. Statistics and strategies for evaluation. Control Clin Trials 1991;12(4 Suppl.):142S–158S.
18. Adler NE, Epel ES, Castellazzo G, Ickovics JR. Relationship of subjective and objective social status with psychological and physiological functioning: preliminary data in healthy white women. Health Psychol 2000;19(6):586–592.
19. Nelson EC, Landgraf JM, Hays RD, Wasson JH, Kirk JW. The functional status of patients. How can it be measured in physicians' offices? Med Care 1990;28(12):1111–1126.
20. Finlay WM, Lyons E. Methodological issues in interviewing and using self-report questionnaires with people with mental retardation. Psychol Assess 2001;13(3):319–335.
21. Fossa SD, Aaronson NK, Newling D, et al. Quality of life and treatment of hormone resistant metastatic prostatic cancer. The EORTC Genito-Urinary Group. Eur J Cancer 1990;26(11–12):1133–1136.
22. Litwin MS, Lubeck DP, Henning JM, Carroll PR. Differences in urologist and patient assessments of health related quality of life in men with prostate cancer: results of the CaPSURE database. J Urol 1998;159(6):1988–1992.
23. Slevin ML, Plant H, Lynch D, Drinkwater J, Gregory WM. Who should measure quality of life, the doctor or the patient? Br J Cancer 1988;57(1):109–112.
24. Fossa SD, Moynihan C, Serbouti S. Patients' and doctors' perception of long-term morbidity in patients with testicular cancer clinical stage I. A descriptive pilot study [see comments]. Support Care Cancer 1996;4(2):118–128.
25. Lampic C, von Essen L, Peterson VW, Larsson G, Sjoden PO. Anxiety and depression in hospitalized patients with cancer: agreement in patient-staff dyads. Cancer Nurs 1996;19(6):419–428.
26. Sneeuw KC, Aaronson NK, Sprangers MA, et al. Value of caregiver ratings in evaluating the quality of life of patients with cancer. J Clin Oncol 1997;15(3):1206–1217.
27. Sprangers MA, Aaronson NK. The role of health care providers and significant others in evaluating the quality of life of patients with chronic disease: a review. J Clin Epidemiol 1992;45(7):743–760.
28. Kornblith AB, Herr HW, Ofman US, Scher HI, Holland JC. Quality of life of patients with prostate cancer and their spouses: the value of a data base in clinical care. Cancer 1994;73:2791–2802.
29. Testa MA. Parallel perspectives on quality of life during antihypertensive therapy: impact of responder, survey environment, and questionnaire structure. J Cardiovasc Pharmacol 1993;21(Suppl. 2):S18–S25.
30. Weeks JC, Nelson H, Gelber S, Sargent D, Schroeder G. Short-term quality-of-life outcomes following laparoscopic-assisted colectomy vs open colectomy for colon cancer: a randomized trial. JAMA 2002;287(3):321–328.
31. Talcott JA, Rieker P, Clark JA, et al. Patient-reported symptoms after primary therapy for early prostate cancer: results of a prospective cohort study. J Clin Oncol 1998;16(1):275–283.
32. Herrmann D. Reporting current, past, and changed health status. What we know about distortion. Med Care 1995;33(4 Suppl.):AS89–AS94.
33. Aseltine RH, Jr., Carlson KJ, Fowler FJ, Jr., Barry MJ. Comparing prospective and retrospective measures of treatment outcomes. Med Care 1995;33(4 Suppl.):AS67–AS76.
34. Litwin MS. How to measure survey reliability and validity. Thousand Oaks, CA: Sage Publications, 2002.
35. Sprangers MA, Cull A, Groenvold M, et al. The European Organization for Research and Treatment of Cancer approach to developing questionnaire modules: an update and overview. EORTC Quality of Life Study Group. Qual Life Res 1998;7(4):291–300.
36. Collins D. Pretesting survey instruments: an overview of cognitive methods. Qual Life Res 2003;12(3):229–238.
37. Cronbach LJ. Coefficient alpha and the internal structure of tests. Psychometrika 1951;16:297–334.
38. Nunnally JC. Psychometric theory. 2nd ed. New York: McGraw-Hill, 1978.
39. Messick S. The once and future issues of validity: assessing the meaning and consequences of measurement. In: Wainer H, Braun HI, eds. Test validity. Hillside, NJ: Lawrence Erlbaum Associates, 1988.
40. Zwinderman AH. The measurement of change of quality of life in clinical trials. Stat Med 1990;9(8):931–942.
41. Olschewski M, Schumacher M. Statistical analysis of quality of life data in cancer clinical trials. Stat Med 1990;9(7):749–763.
42. Sprangers MA. Response-shift bias: a challenge to the assessment of patients' quality of life in cancer clinical trials. Cancer Treat Rev 1996;22(Suppl. A):55–62.

43. Samsa G, Edelman D, Rothman ML, et al. Determining clinically important differences in health status measures: a general approach with illustration to the Health Utilities Index Mark II. Pharmacoeconomics 1999;15(2):141–155.
44. Norman GR, Sloan JA, Wyrwich KW. Interpretation of changes in health-related quality of life: the remarkable universality of half a standard deviation. Med Care 2003;41(5):582–592.
45. Sloan JA, Dueck A. Issues for statisticians in conducting analyses and translating results for quality of life end points in clinical trials. J Biopharm Stat 2004;14(1):73–96.
46. Guyatt G, Walter S, Norman G. Measuring change over time: assessing the usefulness of evaluative instruments. J Chronic Dis 1987;40(2):171–178.
47. Guyatt GH, Osoba D, Wu AW, Wyrwich KW, Norman GR. Methods to explain the clinical significance of health status measures. Mayo Clin Proc 2002;77(4):371–383.
48. Patrick DL, Deyo RA. Generic and disease-specific measures in assessing health status and quality of life. Med Care 1989;27(3 Suppl.):S217–S232.
49. Fossa SD, Woehre H, Kurth KH, et al. Influence of urological morbidity on quality of life in patients with prostate cancer. Eur Urol 1997;31(Suppl. 3):3–8.
50. Kagawa-Singer M, Kassim-Lakha S. A strategy to reduce cross-cultural miscommunication and increase the likelihood of improving health outcomes. Acad Med 2003;78(6):577–587.
51. Sagnier PP, Girman CJ, Garraway M, et al. International comparison of the community prevalence of symptoms of prostatism in four countries. Eur Urol 1996;29(1):15–20.
52. Boyle P. Cultural and linguistic validation of questionnaires for use in international studies: the nine-item BPH-specific quality-of-life scale. Eur Urol 1997;32(Suppl. 2):50–52.
53. Vela Navarrete R, Martin Moreno JM, Calahorra FJ, et al. [Cultural and linguistic validation, in Spanish, of the International Prostatic Symptoms Scale (I-PSS)]. Actas Urol Esp 1994;18(8):841–847.
54. Gill TM, Feinstein AR. A critical appraisal of the quality of quality of life measurements. JAMA 1994;272(8):619–626.
55. Ware JE, Kosinski M, Keller SK. SF-36 physical and mental health summary scales: a user's manual. Boston: The Health Institute, New England Medical Center, 1994.
56. Ware JE, Jr., Sherbourne CD. The MOS 36-item short-form health survey (SF-36). I. Conceptual framework and item selection. Med Care 1992;30(6):473–483.
57. Gandek B, Ware JE, Jr., Aaronson NK, et al. Tests of data quality, scaling assumptions, and reliability of the SF- 36 in eleven countries: results from the IQOLA Project. International Quality of Life Assessment. J Clin Epidemiol 1998;51(11):1149–1158.
58. Kaplan RM, Bush JW, Berry CC. Health status: types of validity and the index of well-being. Health Serv Res 1976;11(4):478–507.
59. Kaplan RM, Ganiats TG, Sieber WJ, Anderson JP. The Quality of Well-Being Scale: critical similarities and differences with SF-36. Int J Qual Health Care 1998;10(6):509–520.
60. Kaplan RM, Sieber WJ, Ganiats TG. The quality of well-being scale: comparison of the interviewer-administered version with a self-administered questionnaire. Psychol Health 1997;12:783–791.
61. Kaplan RM, Bush JW. Health-related quality of life measurement for evaluation research and policy analysis. Health Psychol 1982;1:61–80.
62. Kaplan RM, Anderson JP. A general health policy model: update and applications. Health Serv Res 1988;23(2):203–235.
63. Anderson JP, Kaplan RM, Berry CC, Bush JW, Rumbaut RG. Interday reliability of function assessment for a health status measure. The Quality of Well-Being scale. Med Care 1989;27(11):1076–1083.
64. Bergner M, Bobbitt RA, Carter WB, Gilson BS. The Sickness Impact Profile: development and final revision of a health status measure. Med Care 1981;19(8):787–805.
65. Bergner M, Bobbitt RA, Pollard WE, Martin DP, Gilson BS. The sickness impact profile: validation of a health status measure. Med Care 1976;14(1):57–67.
66. Hunt SM, McEwen J, McKenna SP. Measuring health status: a new tool for clinicians and epidemiologists. J R Coll Gen Pract 1985;35(273):185–188.
67. McDowell IW, Martini CJ, Waugh W. A method for self-assessment of disability before and after hip replacement operations. BMJ 1978;2(6141):857–859.
68. Martini CJ, McDowell I. Health status: patient and physician judgments. Health Serv Res 1976;11(4): 508–515.
69. Ware J, Jr., Kosinski M, Keller SD. A 12-Item Short-Form Health Survey: construction of scales and preliminary tests of reliability and validity. Med Care 1996;34(3):220–233.
70. Ware JE, Kosinski M, Keller SD. SF-12: How to score the SF-12 physical and mental health summary scales. Boston, MA: The Health Institute, New England Medical Center, 1995.

71. Gandek B, Ware JE, Aaronson NK, et al. Cross-validation of item selection and scoring for the SF-12 Health Survey in nine countries: results from the IQOLA Project. International Quality of Life Assessment. J Clin Epidemiol 1998;51(11):1171–1178.
72. Norcross JC, Guadagnoli E, Prochaska JO. Factor structure of the Profile of Mood States (POMS): two partial replications. J Clin Psychol 1984;40(5):1270–1277.
73. Jacobson AF, Weiss BL, Steinbook RM, Brauzer B, Goldstein BJ. The measurement of psychological states by use of factors derived from a combination of items from mood and symptom checklists. J Clin Psychol 1978;34(3):677–685.
74. Cella DF, Jacobsen PB, Orav EJ, et al. A brief POMS measure of distress for cancer patients. J Chronic Dis 1987;40(10):939–942.
75. Albrecht RR, Ewing SJ. Standardizing the administration of the Profile of Mood States (POMS): development of alternative word lists. J Pers Assess 1989;53(1):31–39.
76. Baker F, Denniston M, Zabora J, Polland A, Dudley WN. A POMS short form for cancer patients: psychometric and structural evaluation. Psychooncology 2002;11(4):273–281.
77. Aaronson NK, Ahmedzai S, Bergman B, et al. The European Organization for Research and Treatment of Cancer QLQ-C30: a quality-of-life instrument for use in international clinical trials in oncology. J Natl Cancer Inst 1993;85(5):365–376.
78. Cella DF, Tulsky DS, Gray G, et al. The Functional Assessment of Cancer Therapy scale: development and validation of the general measure. J Clin Oncol 1993;11(3):570–579.
79. Schag CA, Ganz PA, Heinrich RL. CAncer Rehabilitation Evaluation System—short form (CARES-SF). A cancer specific rehabilitation and quality of life instrument. Cancer 1991;68(6):1406–1413.
80. Schag CA, Ganz PA, Wing DS, Sim MS, Lee JJ. Quality of life in adult survivors of lung, colon and prostate cancer. Qual Life Res 1994;3(2):127–141.
81. Schag CA, Heinrich RL. Development of a comprehensive quality of life measurement tool: CARES. Oncology (Huntingt) 1990;4(5):135–138.
82. Curran D, Fossa S, Aaronson N, Kiebert G, Keuppens E, Hall R. Baseline quality of life of patients with advanced prostate cancer. European Organization for Research and Treatment of Cancer (EORTC), Genito-Urinary Tract Cancer Cooperative Group (GUT-CCG). Eur J Cancer 1997;33(11):1809–1814.
83. Sprangers MA, Groenvold M, Arraras JI, et al. The European Organization for Research and Treatment of Cancer breast cancer-specific quality-of-life questionnaire module: first results from a three-country field study. J Clin Oncol 1996;14(10):2756–2768.
84. Bergman B, Aaronson NK, Ahmedzai S, Kaasa S, Sullivan M. The EORTC QLQ-LC13: a modular supplement to the EORTC Core Quality of Life Questionnaire (QLQ-C30) for use in lung cancer clinical trials. EORTC Study Group on Quality of Life. Eur J Cancer 1994;5:635–642.
85. Aaronson NK, van Andel G. An international field study of the reliability and validity of the EORTC QLQ-C30 version 3 and a disease-specific questionnaire module (QLQ-PR25) for assessing the quality of life of patients with prostate cancer. Brussels: EORTC Data Center, 2001. (EORTC protocol 15011-30011).
86. Bjordal K, Kaasa S, Mastekaasa A. Quality of life in patients treated for head and neck cancer: a follow-up study 7 to 11 years after radiotherapy. Int J Radiat Oncol Biol Phys 1994;28(4):847–856.
87. Ward WL, Hahn EA, Mo F, Hernandez L, Tulsky DS, Cella D. Reliability and validity of the Functional Assessment of Cancer Therapy-Colorectal (FACT-C) quality of life instrument. Qual Life Res 1999;8(3):181–195.
88. Brady MJ, Cella DF, Mo F, et al. Reliability and validity of the Functional Assessment of Cancer Therapy-Breast quality-of-life instrument. J Clin Oncol 1997;15(3):974–986.
89. Esper P, Mo F, Chodak G, Sinner M, Cella D, Pienta KJ. Measuring quality of life in men with prostate cancer using the functional assessment of cancer therapy-prostate instrument. Urology 1997;50(6):920–928.
90. Basen-Engquist K, Bodurka-Bevers D, Fitzgerald MA, et al. Reliability and validity of the functional assessment of cancer therapy-ovarian. J Clin Oncol 2001;19(6):1809–1817.
91. McQuellon RP, Russell GB, Cella DF, et al. Quality of life measurement in bone marrow transplantation: development of the Functional Assessment of Cancer Therapy-Bone Marrow Transplant (FACT-BMT) scale. Bone Marrow Transplant 1997;19(4):357–368.
92. Yellen SB, Cella DF, Webster K, Blendowski C, Kaplan E. Measuring fatigue and other anemia-related symptoms with the Functional Assessment of Cancer Therapy (FACT) measurement system. J Pain Symptom Manage 1997;13(2):63–74.
93. Cella D. The Functional Assessment of Cancer Therapy-Anemia (FACT-An) Scale: a new tool for the assessment of outcomes in cancer anemia and fatigue. Semin Hematol 1997;34(3 Suppl. 2):13–19.

94. de Haes JC, van Knippenberg FC, Neijt JP. Measuring psychological and physical distress in cancer patients: structure and application of the Rotterdam Symptom Checklist. Br J Cancer 1990;62(6):1034–1038.
95. Litwin MS, Hays RD, Fink A, et al. The UCLA Prostate Cancer Index: development, reliability, and validity of a health-related quality of life measure. Med Care 1998;36(7):1002–1012.
96. Krongrad A, Perczek RE, Burke MA, et al. Reliability of Spanish translations of select urological quality of life instruments. J Urol 1997;158(2):493–496.
97. Karakiewicz PI, Kattan MW, Tanguay S, et al. Cross-cultural validation of the UCLA prostate cancer index. Urology 2003;61(2):302–307.
98. Kakehi Y, Kamoto T, Ogawa O, et al. Development of Japanese version of the UCLA Prostate Cancer Index: a pilot validation study. Int J Clin Oncol 2002;7(5):306–311.
99. Korfage IJ, Essink-Bot ML, Madalinska JB, et al. Measuring disease specific quality of life in localized prostate cancer: the Dutch experience. Qual Life Res 2003;12(4):459–464.

14 Measuring Patient Satisfaction

Arvin Koruthu George
and Martin G. Sanda, MD

CONTENTS

The past three decades have seen an evolution in the evaluation of cancer care. Supply has made a determined effort to catch up with the demand for a more comprehensive assessment of factors that ascertain quality of care, striving to provide better service to a consumer-driven society. This is mainly the result of patients becoming more knowledgeable and savvy to the type of care and treatment options they may receive, health care services striving to attain higher standards of care to compete for consumers, and increased interest in health services research *(1)*. The quality of care has been defined as "the degree to which health services for individual or populations increase the likelihood of desired health outcomes and are consistent with current professional knowledge" *(2,3)*.

Cancer care quality can be characterized from the patient perspective, or from the provider perspective (Figure 1). In domains where they are nonoverlapping, these two perspectives (that of the patient and the care providers) can reflect distinct and complementary views of care quality (light shaded area of Figure 1): the patient and provider may have different perspectives regarding the severity of cancer treatment side effects, wherein these separate perspectives may reflect distinct components of the quality of care received. For example, the provider may understand how a patient's lack of bleeding requiring transfusion reflects quality, whereas the patient may not recognize the benefit of averted transfusion, but instead may recognize the full impact of erectile dysfunction in ways not immediately evident in the provider perspective.

From: *Clinical Research for Surgeons*
Edited by: D. F. Penson and J. T. Wei © Humana Press Inc., Totowa, NJ

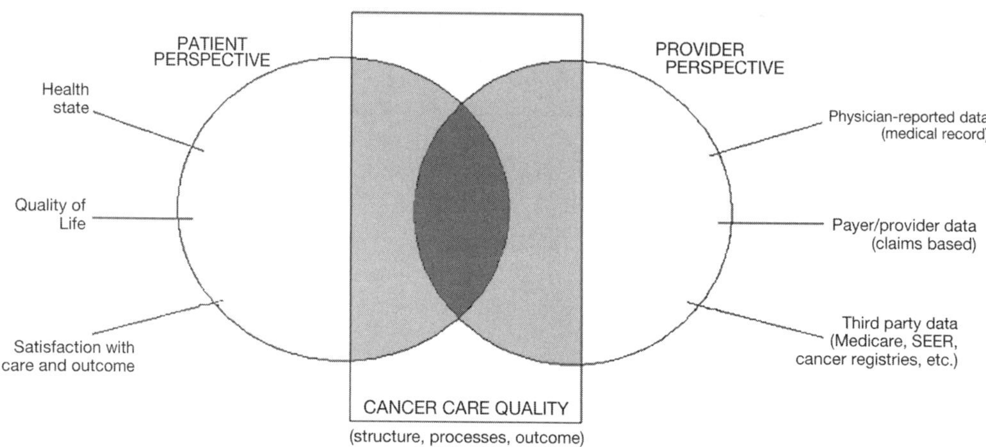

Figure 1: Quality of health care can be divided into the domains of patient perspective and provide perspective. Patient perspective is composed of the patient's health state at the time of evaluation, health-related quality of life, and satisfaction with care and outcome. The provider perspective can be assessed via physician-reported, payer/provider-reported, or third party–reported outcomes. Each domain/subdomain interacts with the others and hence has its own value in judging quality care and predicting patient satisfaction with cancer care services.

On the other hand, the most robust components of care quality may be represented by domains wherein the perspective of the patient and provider converge or overlap (dark shaded area of Figure 1). For example, the patient and provider may share similar perspective regarding whether a cancer is in remission or has recurred after treatment. Other components of cancer care quality may remain elusive and not readily determined by the patient or provider (unshaded region of the Cancer Care Quality rectangle in Figure 1). This "unmeasurable" component of cancer care quality can be exemplified by *bone fide* uncertainties in cancer care, such as lack of knowledge about whether a novel treatment being evaluated in a clinical trial represents a superior or inferior treatment alternative.

The patient perspective is pivotal to measuring satisfaction with care received and is in turn determined by the patient's quality of life and current health state. The provider perspective or, more broadly, the perspective of the health care system, is determined by the assortment of individuals and factors that avail or impart health care services to the patient. Patient-reported outcomes and provider-reported outcomes are the measurements used to gauge satisfaction with care and can be in the form of a questionnaire, medical records, or even data collected from cancer registries. Acquiring, measuring, and comparing patient and provider reported outcomes are invaluable steps in evaluating the quality of care. The contributing factors of both perspectives intersect with each other, creating a complete picture of the quality of cancer care (Figure 1).

1. THE BROADER CONTEXT OF SATISFACTION ASSESSMENT: PATIENT AND PROVIDER PERSPECTIVE IN EVALUATING QUALITY OF CARE

Patient and provider perspectives can be integrated to contribute complementary insight regarding structure, process, and outcome in the classic model of health care evaluation developed by Donabedian. In this model, "structure" refers to the institution delivering care or provider, "process" refers to the activities associated with providing care, and "outcome" refers to the change in the patient's health status from medical intervention *(4, 5)*. Each domain has its own value in judging the quality of care, and predicting patient satisfaction with cancer care services.

Treatment involves interaction within the various arms of health care as is the nature of oncology care, with patients interacting with health care professionals of varying specialties such as primary care physicians, surgeons, radiologists, chemotherapy staff, nurses, administrators, and other care givers meaning that care-provider perspectives cannot be clustered together. In actuality, each perspective contributes to the care providers' evaluation of the patients' quality of care. In contrast, though the patient perspective is also multifaceted in character, the patient perspective comprises quality of life, health state, and satisfaction—facets that are patient centered. Hence, satisfaction is a component of quality of care that interrelates with quality of life and health state, with each domain contributing to the other (Figure 1).

1.1. Provider Perspective

The health care system perspective's contribution to quality of care is composed of those who directly and indirectly interact with the patient in the care setting. Cancer patients receive care from health professionals in multiple specialities, and each provider perspective can be used in satisfaction assessment. Physician-reported outcomes have the advantage of being a first-hand evaluation of the patient's care, coming from an individual that is knowledgeable about the issues that affect both the patient and the health care system. The physician's view also gives an in-depth report of each patient's care from an integral part of the health care service. The disadvantage of considering physician-reported outcomes is that they are unintentionally biased views, which may differ from patient opinion *(6)*. Additionally, physicians can only report on a limited number of cases, meaning that a lesser proportion of the community is represented. Payer-reported outcomes are those derived from managed care datasets and payer-protected databases, which may provide a bounty of information on care services, but are often inaccessible for evaluation purposes. Third-party reported outcomes also play a lesser role in health care quality evaluation. Data reported by external care services such as Medicare and external reporting such as the National Cancer Institute's Surveillance, Epidemiology, and End Results program and other cancer registries, though relevant, are more closely related to the domain of provider perspective than patient perspective in which satisfaction lies. Such datasets can be used to assess the quality of care by providing survival outcome data *(7)*. Unfortunately, gaps in data make it difficult to like and evaluate information and only monitoring survival rates falls short of comprehending the vast spectrum of factors that contribute to quality care assessment.

1.2. Patient Perspective/Patient-Reported Outcomes

Numerous ways of appraising satisfaction with care have been employed, but the benchmark to satisfaction evaluation has been the use of patient-reported outcomes. Self-reported outcomes have been previously used with satisfaction of adequate care provision. Patient-reported outcomes make available a unique and essentially pertinent outlook on satisfaction with care, because when patient perspective is not taken into account, the evaluation of services is biased toward the provider's perspective. Health care is dependent on physician-patient communication. The quality of doctor-patient interactions must be evaluated with respect to patient preference and as a result the patient's own evaluation of such components is indispensable *(1)*. Traditional methods used as indicators of quality of oncology care, such as survival outcomes and processes of care or adherence of treatment guidelines, are inadequate to measure the multidimensional nature of satisfaction with care provision *(2)*. Patients often will have different views in terms of expectation of care and outcome, satisfaction, and quality of care, highlighting the critical need for assessment of satisfaction based on patient opinion. Patient-reported measures also provide insight in the relative value of various risks and benefits of care that should be assigned. Patient-reported outcomes in terms of health care assessment demonstrate more concrete treatment factors such as side effects, pain symptoms, discomfort, and functional status. This same approach must be used in assessing satisfaction with care so that focus is maintained on the patient's experience with illness and experience with care given. Patient reported outcomes are scientific measures that evaluate change in health outcomes *(2, 8)*. This means that self-reported measures must be customized to operate within the specific framework it is being used (i.e., satisfaction) and by using reliable and validated measures *(9)*. Ideally, assessment of satisfaction with cancer care should be tailored to the disease-specific measures influencing care, because receiving outcomes from individual patients is an inherently subjective process. But broader measures impart a more comprehensive view of the impact of satisfaction with treatment.

Patient-reported outcomes can be thought of in three components listed in Figure 1. The first and most pivotal domain of the patient perspective is the health state. For example, the cancer-related health state can be considered as symptoms and well-being or lack thereof from cancer or its treatment. Knowing and understanding the patient's health state allows one to understand the position from which the patient is coming, because it will affect health-related quality of life and overall satisfaction with care.

One aspect of patient-reported outcomes influenced by health state but also affected by other factors is the patient-reported health-related quality of life. An abundance of research has been done in relation to quality of life in cancer patients, with some research addressing its interconnected relationship with satisfaction. As one would expect, a better quality of life or a superior health state is accompanied by greater satisfaction with care. For example, higher global satisfaction scores for quality of life on the European Organization for Research and Treatment of Cancer (EORTC)-QLQ30, a 30-item, self-assessment of various areas of functioning, symptoms, and global quality of life, predicted higher satisfaction with all aspects of care. Additional research has demonstrated that physical function and psychologic distress are also significantly associated with global satisfaction with treatment outcome *(10, 11)*.

Satisfaction is the final subdomain of the patient perspective and is jointly influenced by health-related quality of life and health state. Patient-reported satisfaction with care services is the ultimate validator of quality care, and its quantification is thus a crucial step in assessing the health care system *(12)*. However, satisfaction with cancer care is a subjective topic because of a combination of each patient's differing personality, expectations, and care outcome. Patient-reported satisfaction provides a system of checks and balances against the biased provider perspective.

2. COMPONENTS AND DETERMINANTS OF SATISFACTION

Patient satisfaction can be influenced by a variety of external and internal influences, and in turn satisfaction with care itself can be thought of as comprising two distinct components: satisfaction with the care process and satisfaction with care outcome. The components of the care process may include waiting time, provision of information, access to care, adequacy of care environment, and speed of treatment, each of which can be further broken down to demonstrate the multifaceted nature of satisfaction. The other major element of the satisfaction domain is satisfaction with care outcome. Individuals whose outcomes are either below their own perception of expectation or who experience adverse treatment effects may be less satisfied with the care they have been provided.

Patient satisfaction is directly correlated with outcome of care, and it has been revealed that this relationship is stronger for absolute outcomes than for relative ones *(13, 14)*. The factors that influence satisfaction with care in relation to cancer include cancer control, health-related quality of life, side effects of treatment, recovery, financial outcomes, and quality of death, demonstrating the multidimensionality of satisfaction with cancer care. Physical function, ability to perform daily activities, psychologic well-being, and other influences are often different and of greater concern for the cancer patient. Side effects of treatment contain both symptoms (pain, fatigue, shortness of breath) and toxicities, which themselves may require specific treatment strategies and would inevitably compound and complicate the patient's satisfaction with outcome of care *(13)*. Recovery, financial outcomes, and quality of death have particular significance in the cancer patient, because their needs and outcomes are often much different. The psychologic well-being of cancer patients needs to be taken into consideration because of the ramifications of disease and its impact not only on the patient, but also the family and even community.

Satisfaction has an essential role in that patients' opinion represents an end point in quality of care evaluation. Satisfaction contributes to patients' quality of life, providing additional evidence to its value. The role of satisfaction within the context of cancer care means farther reaching consequences exist that are unique or more relevant to cancer patients and that can impact the quality of care. Thus the role of satisfaction can provide beneficial insight in terms of assessing the unmet needs of cancer patients.

3. STRUCTURE AND COMPONENTS OF INSTRUMENTS/QUESTIONNAIRES MEASURING PATIENT SATISFACTION WITH HEALTH CARE QUALITY

Satisfaction has been described as the intervening variable between service provision and its ultimate outcome *(15)*. It is a construct that can be evaluated using either qualitative or quantitative techniques. Qualitative measures allow the patient to respond to open-ended questions, which are then used to determine the unmet needs of the patient. Examples of qualitative measurements may include management observation, employment feedback programs, use of work teams and quality circles and focus groups—each with specific advantages and disadvantages *(9)*. Quantitative evaluation has been most prominent and several patient satisfaction surveys have been produced over the past two decades to evaluate the satisfaction with health care, and to a lesser extent satisfaction with cancer care. Quantitative surveys of satisfaction provides patients with the opportunity to give a personal account of their care and give a far more accurate measure of the patients' service experience. Examples of quantitative management techniques for measuring patient satisfaction with care includes comment cards, mail surveys, onsite personal interviews, and telephone interviews. The typical quantitative questionnaire will provide the patient with multiple items and give responses that may be rated on a scale. The scale may vary from questionnaire to questionnaire, but each question is assigned a score according to the response provided and factor analysis is completed to give a quantitative value of a patient's satisfaction with various aspects of health care. Notably, however, even the most biometrically robust quantitative assessment of patient satisfaction can be skewed by factors such as a patient's personality, expectations, age, education, and sociopsychologic phenomena *(16)*. Among quantitative instruments, satisfaction with care can be expressed as either a single measure of global satisfaction, wherein items regarding conceptually distinct care components are combined, or as multidimensional subscales that quantify conceptually discrete components of care satisfaction. Global satisfaction scales have had limited value in terms of quality of care assessment because of the various domains of satisfaction, but essential to the evaluation of overall satisfaction. Multidimensional patient satisfaction assessment allows improved response variability and contrast of satisfaction ratings between different aspects of care.

The majority of studies has either developed new instruments to evaluate satisfaction or has modified an existing instrument, demonstrating the need for a standardized and specific survey of satisfaction with care *(16)*. Selection of items for evaluation of satisfaction with cancer care must be determined using diverse approaches. Wiggers et al have demonstrated that item selection can be completed with a three-pronged approach. Requesting a panel of cancer care providers to give its views of what is relevant to cancer care, requesting the same of patients in terms of what is most important to them, and finally an extensive literature review indicating the aspects of care considered important to cancer patients must be integrated and refined for item selection. These items are those that directly or indirectly influence patient satisfaction with care. Technical competence, communication skills, interpersonal skills, accessibility of care, continuity of care, hospital and clinic care nonmedical care, finances, family care, psychologic well-being, information provision, care organization, and quality of life are items that have been previously outlined *(17, 18)*.

The method of administration of such surveys also varies and the applicability of its measurements is determined by rate of response, time, and various other methodologic concerns. It has been shown that computer administration of questionnaires is simple, time-effective, and acceptable means of improving patient-provider communication *(19)*. These methodologic concerns may also be specific to the disease, because different situations would require assessment with different parameters *(20)*. Questionnaires may be administered via personal interview, telephone interview, and, most commonly, mail survey. The route of administration must be chosen according to the boundaries of the study, so that it may be tailored with the best specifications for the project, because different methods are useful for different purposes. A study comparing postal vs interviewer administration of a questionnaire measuring satisfaction with care services showed that though response rate did not differ significantly between these groups, postal questionnaires had significantly more missing data *(21)*. The concerns listed previously carry much weight when assessing the quality of data that are taken in, because they represent factors that may unintentionally distort data.

4. INSTRUMENTS/QUESTIONNAIRES USED TO MEASURE GENERAL SATISFACTION WITH CARE

Measurement of satisfaction of health care quality has often been broad, administering routinely a core of items assessing common factors, and thus allowing cross-setting comparability *(15)*. Table 1 provides a summary of a few of the general satisfaction instruments that are available and their potential strengths and weaknesses *(6, 15, 22–31)*. The University of California San Francisco client satisfaction scale, the Client Satisfaction Questionnaire brought forth by Attkisson and Greenfield is a direct measure of the individual's personal experience with a specific service (such a health care) and has shown success in multiple areas of service evaluation. Though comprehensive, the Client Satisfaction Questionnaire may be too general in measuring specific health care services, and its global assessment of satisfaction may not always be sufficient (Table 1). The Consumer Assessment of Health Care Plans Study 2.0 adult core survey is an example of a general satisfaction questionnaire that is visit-specific. The Consumer Assessment of Health Care Plans Study strongly evaluates global satisfaction and provides details on background factors that may influence responses. The strength of this instrument is its ability to evaluate satisfaction with care process, but seems to be incomplete with an evaluation of satisfaction with care outcome a pivotal component of satisfaction with care *(27)*.

5. INSTRUMENTS/QUESTIONNAIRES USED TO MEASURE SATISFACTION WITH CANCER CARE

Measuring satisfaction of cancer care carries the burden of additional issues that may be of greater concern or even exclusive to this disease-specific patient population. Cancer patients often have lengthier periods of treatment, which translates to greater interaction and the importance of interpersonal skills of health care professionals. Additionally, the longer term treatments may have a greater impact on quality of life, follow-up care, recovery from treatment, and rehabilitation (both physical and psychologic) *(32, 33)*. Cancer patients are faced with the uncertainty of a disease with shifting course and prognosis of illness. These patients are placed in a position of greater dependency on the health care system; thus, continuity of care becomes a standing issue. The nature of this disease is such that a greater emotional need is placed on the patient, family, and care provider. Most current assessments of satisfaction with cancer care are supplemented surveys of satisfaction with general health care and therefore lack the all-inclusive view that is necessary.

Because of previous evaluations of satisfaction with cancer care, more specific surveys have been produced to reevaluate areas that have scored low on satisfaction scales or that have not been taken into consideration, yet have a major impact on care. Satisfaction with cancer genetic services, information provision, family satisfaction, pain management, and spousal satisfaction have opened subcategories that have been assessed *(34–37)*. These subcategories have been explored because of evidence of dissatisfaction in those areas. Several studies have shown that patient satisfaction with information provision has been lacking; thus, attempts to remedy this problem and reassess the affects of adapted care policy has taken place *(28, 38)*. Several available multi-item instruments for measuring satisfaction with cancer care and their potential strengths and limitations are summarized in Table 2 *(18, 24, 34–37, 39–53)*. The FAMCARE scale has been developed to measure family satisfaction with advanced cancer care because the patient's family is not isolated from the effects of the effects of the patient's illness *(34)*. Such scales provide an indirect evaluation of quality of care and may show disparity. It has been shown that the patient is often more satisfied with level of care than family members. These niches cut out by need of assessment though indirect approaches to quality of care comprise a relevant part in assessment. The Comprehensive Assessment of Satisfaction with Care (CASC) survey has been successfully used in assessing satisfaction with cancer

Table 1
Instruments for Measuring Satisfaction With General Health Care

Instrument	*Author/ Citation*	*No. of Items*	*Components of Satisfaction Measured*	*Strengths/Weaknesses*
CSQ-8	Attkisson, Greenfield	8	Global satisfaction with care	Less focused but provides good overall evaluation of satisfaction
CSQ-18	Attkisson, Greenfield	18	Multiple dimensions of satisfaction with care	More detailed than CSQ-18, but still broad
SSS-30	Attkisson, Greenfield	30	Multiple dimensions of satisfaction with care	Direct measure of service satisfaction but not specific to health care
PSS	Linder-Pelz, Ware	25	Satisfaction with affective care, communication, and technical care	Potential for response bias
Unnamed	Osterweiss-Howell, Ware	21	Multiple dimensions of satisfaction with care	Use of "satisfaction" within response criteria skews response distribution
VSQ	Ware	51	Satisfaction with visit-specific health care services	
QSP	Rahmquist	21	Multiple dimensions of satisfaction with care	Used to analyze impact of background factors on satisfaction with care
CAHPS 2.0	Hargraves	43	Satisfaction with health care process	Includes global satisfaction ratings but focus on use of services rather than satisfaction
IHF	Woodward	14	Satisfaction with health care process	Employs use of open-ended questions
PACE	Atherly		Satisfaction with capitated care	Also includes family satisfaction
PPE-15	Jenkinson et al.	15	Satisfaction with information, communication, and pain management	Concise and extensively used but not a comprehensive evaluation

CSQ, Client Satisfaction Questionnaire; CAHPS, Consumer Assessment of Health Care Plans Study.

care in patients with endocrine gastrointestinal tumors as well as general cancer in-patients *(33, 54)*. The Comprehensive Assessment of Satisfaction with Care survey takes into consideration additional items in the process of care, specific to cancer, that are excluded from other instruments such as quality of life, attention to psychosocial problems, and continuity of care, though it lacks assessment of cost of care and treatment efficacy satisfaction *(43)*. The PSQ-III has been described as "the best developed and

Table 2
Instruments for Measuring Satisfaction With Cancer Care

Instrument	*Author*	*No. of Items*	*Components of Satisfaction Measured*	*Subjects*	*Strengths/Weaknesses*
FAMCARE	Kristjanson	20	Family satisfaction with advanced cancer care	Family members of cancer patient	Provides different perspective, but indirectly evaluates satisfaction with care
SEQUS®	Gourdji et al.	31	Satisfaction with care process	Cancer outpatients	Online format increases ease of use and allows for additional comments
Hall	Lubeck et al., Hall	12	Satisfaction with care process	Prostate cancer	Brief survey containing subscales and has balance between phrasing but can't distinguish providers
MISS	Wolf, Ware et al., Holloway	17	Satisfaction with cancer geneticservices	Genetic counseling recipients	Satisfaction with cancer genetics testing
CASC	Bredart et al.	60	Cancer care in cross-cultural settings	Cancer patients	Items created by cancer patient interviews and by oncology specialists, but does not evaluate satisfaction with cost of care or treatment efficacy
College of Health (UK)	Grunfeld et al.	15	Satisfaction with care process and primary care vs. specialist care	Breast cancer patients	Adapted for both the hospital and general practice setting
QLQ-SAT32	Kavadas et al., Bredart et al.	32	Satisfaction with care process and outcome	Esophageal and gastric cancer patients	Relates the health related quality of life and satisfaction with cancer care
Unnamed	Schoen et al.	23	Satisfaction with care process and outcome (pain and distress)	Sigmoidal endoscopy patients	Relates care process and outcome
PMH/ PSQ-MD	Loblaw et al., Bitar et al.	24	Satisfaction with physician in a cancer care setting	Cancer outpatients	Four-prong item development process and takes into consideration extra burdens of cancer patient
PSQ III	Ware, Hagedoorn et al.	36	Satisfaction with care process	Cancer patients	Extensively tested measure but excludes satisfaction with treatment outcome
ISQ	Thomas et al.	13	Background factors, satisfaction with information provision	Cancer patients	Considers how information provided impacts cancer care and evaluates demographics

CASC, Comprehensive Assessment of Satisfaction with Care.

most extensively measure available" and has been determined an appropriate measure of cancer patients' satisfaction *(52)*; however, this instrument has shown significant response bias requiring amended versions.

An ideal survey that assesses satisfaction with cancer care would be both a multidimensional and global assessment of satisfaction. It would also have to measure how important each aspect of care is to the patient so that priority would be given to what affects the consumer most. Such a survey would have to compensate for methodologic concerns such timing of survey, method of administration, and sources of data error including age, education level of patient, and sex *(44)*. It is also essential that this instrument would consider satisfaction with the care process and care outcome, which is presently lacking in all instruments. Such factors have been shown to create disparity between categories. The most critical challenge is creating a standardized satisfaction questionnaire that will allow comparability of data despite differences in care for different diseases *(55)*. A reliable and validated scale of such nature has yet to be formed.

6. CURRENT STATUS AND POTENTIAL OF INSTRUMENTS MEASURING CANCER CARE SATISFACTION

An abundance of instruments exist in measuring satisfaction with cancer care. A greater focus on the patient has yielded instruments that are tailored to include issues fundamentally part of the cancer experience. Current instruments have taken into account the impact of the disease on family, the psychologic concerns of the patient, and experiences such as continuity of care uniquely associated with the cancer-affected individual. These assessments have led to greater knowledge of adequacy in cancer care and have served to identify avenues for improving care quality. Different studies have assessed specific areas of cancer care such as satisfaction with information, cancer screening and diagnosis, satisfaction with care process, and satisfaction with treatment outcome. But none has evaluated all aspects of the satisfaction continuum, covering the perspective of both the patient and the care provider. Characterization of how component domains contribute to the sphere of satisfaction and what each contribution means in terms of the end result of improving care has also been elusive.

Improved and broadly applied assessment of satisfaction with cancer care could yield significant advances n cancer care quality. Data can be used to focus attention on what is lacking in cancer care in terms of what is most important to the patient. Provider-patient interactions can be personalized to patient preference and benefit the overall care experience. Care providers can use this information to allocate resources to those areas that are in need and thus be more competitive in provision of care. Greater knowledge of what is needed will inevitably lead to greater efficiency and productivity, improved patient compliance, and superior continuity of care *(9)*. Such changes will inevitably benefit the consumer and the health care provider and have led to tangible changes in health care provision *(56)*. Attention must be taken in forming a reliable and validated survey, which would optimize such effects.

ACKNOWLEDGMENT

This work is supported by in part by NIH 5R01CA095662–05 (Sanda PI).

REFERENCES

1. Cleary PD, McNeil BJ. Patient satisfaction as an indicator of quality care. Inquiry 1988;25(1):25–36.
2. Dennison CR. The role of patient-reported outcomes in evaluating the quality of oncology care. Am J Manag Care 2002;8(18 Suppl):S580–S586.
3. Lohr KN, Schroeder SA. A strategy for quality assurance in Medicare. N Engl J Med 1990;322:707–712.
4. Donabedian A. Explorations in quality assessment and monitoring. Vol. 1. The definition of quality and approaches to its assessment. Ann Arbor, MI: Health Administration Press, 1980.
5. Sitzia J, Wood N. Patient satisfaction: a review of issues and concepts. Soc Sci Med 1997;45(12): 1829–1843.
6. Larsen DL, Attkisson CC, Hargreaves WA, Nguyen TD. Assessment of client/patient satisfaction: development of a general scale. Eval Prog Plann 1979;2(3):197–207.
7. Brook RH, McGlynn EA, Cleary PD. Quality of health care. Part 2: measuring quality of care. N Engl J Med 1996;335(13):966–970.
8. Santanello N. 'PRO' methodological issues. Patient reported outcomes 'PRO' symposium: conceptual and methodological issues. Paper presented at: International Society for Pharmacoeconomic and Outcomes Research (ISPOR) 7th International Meeting; 2002; Arlington, VA.
9. Ford RC, Bach SA, Fottler MD. Methods of measuring patient satisfaction in health care organizations. Health Care Manage Rev 1997;22(2):74–89.
10. Homma Y, Kawabe K, Hayashi K. Urologic morbidity and its influence on global satisfaction. Int J Urol 1998;5(6):556–561.
11. Hoffman RM, Hunt WC, Gilliland FD, Stephenson RA, Potosky AL. Patient satisfaction with treatment decisions for clinically localized prostate carcinoma. Results from the Prostate Cancer Outcomes Study. Cancer 2003;97(7):1653–1662.
12. Schwartz CE, Sprangers MA. An introduction to quality of life assessment in oncology: the value of measuring patient-reported outcomes. Am J Manag Care 2002;18(Suppl):S550–S559.
13. Ganz PA. What outcomes matter to patients: a physician-researcher point of view. Med Care 2002;40(6 Suppl):III1 1–9.
14. Kane RL, Maciejewski M, Finch M. The relationship of patient satisfaction with care and clinical outcomes. Med Care 1997;35(7):714–730.
15. Greenfield TK, Attkisson CC. Steps toward a multifactorial satisfaction scale for primary care and mental health services. Eval Prog Plann 1989;12:271–278.
16. Sitzia J. How valid and reliable are patient satisfaction data? An analysis of 195 studies. Int J Qual Healthcare 1999;11(4):319–328.
17. Wiggers JH, Donovan KO, Redman S, Sanson-Fisher RW. Cancer patient satisfaction with care. Cancer 1990;66(3):610–616.
18. Bredart A, Razavi D, Robertson C, et al. A comprehensive assessment of satisfaction with care: preliminary psychometric analysis in an oncology institute in Italy. Ann Oncol 1999;10(7)839–846.
19. Taenzer P, Bultz BD, Carlson LE, et al. Impact of computerized quality of life screening on physician behaviour and patient satisfaction in lung cancer outpatients. Psychooncology 2000;9(3):203–213.
20. Mandelblatt JS, Ganz PA, Kahn KL. Proposed agenda for the measurement of quality-of-care outcomes in oncology practice. J Clin Oncol 1999;17(8):2614–2622.
21. Addington-Hall J, Walker L, Jones C, Karlsen S, McCarthy M. A randomised controlled trial of postal versus interviewer administration of a questionnaire measuring satisfaction with, and use of, services received in the year before death. J Epid Comm Health 1998;52(12):802–807.
22. Attkisson CC, Greenfield TK. The UCSF client satisfaction scales: I. The Client Satisfaction Questionnaire-8. In: Maruish, ed. The use of psychological testing for treatment planning and outcomes assessment. 2nd ed. London: Lawrence Erlbaum Associates, 1999; 1333–1346.
23. Linder-Pelz S, Struening EL. The multidimensionality of patient satisfaction with a clinic visit. J Comm Health 1985;10:42.
24. Ware JE Jr, Hays RD. Methods for measuring patient satisfaction with specific medical encounters. Med Care 1988;26(4):393–402.
25. Osterweis M, Howell JR. Administering satisfaction questionnaires at diverse ambulatory care sites. J Amb Care Man 1979;67.
26. Rahmqvist M. Patient satisfaction in relation to age, health status and other background factors: a model for comparisons of care units. Int J Qual Health Care 2001;13(5):385–390.
27. Hargraves JL, Hays RD, Cleary PD. Psychometric properties of the Consumer Assessment of Health Plans Study (CAHPS) 2.0 adult core survey. Health Serv Res 2003;38(6 Pt 1):1509–1527.

28. Woodward CA, Ostbye T, Craighead J, Gold G, Wenghofer EF. Patient satisfaction as an indicator of quality care in independent health facilities: developing and assessing a tool to enhance public accountability. Am J Med Qual 2000;15(3):94–105.
29. Atherly A, Kane RL, Smith MA. Older adults' satisfaction with integrated capitated health and long-term care. Gerontologist 2004;44(3):348–357.
30. Jenkinson C, Coulter A, Reeves R, Bruster S, Richards N. Properties of the Picker Patient Experience questionnaire in a randomized controlled trial of long versus short form survey instruments. J Pub Health Med 2003;25(3):197–201.
31. Jenkinson C, Coulter A, Bruster S. The Picker Patient Experience Questionnaire: development and validation using data from in-patient surveys in five countries. Int J Qual Health Care 2002;14:353–358.
32. Kahn KL, Malin JL, Adams J, Ganz PA. Developing a reliable, valid, and feasible plan for quality-of-care measurement for cancer: How should we measure? Medical Care;40(6) Suppl. III-73–III-85.
33. Bredart A, Razavi D, Robertson C, et al. Assessment of quality of care in an oncology institute using information on patients' satisfaction. Oncology 2001;61(2):120–128.
34. Kristjanson LJ. Validity and reliability testing of the FAMCARE Scale: measuring family satisfaction with advanced cancer care. Soc Sci Med 1993;6(5):693–701.
35. Holloway S, Porteous M, Cetnarskyj R, et al. Patient satisfaction with two different models of cancer genetic services in southeast Scotland. Br J Cancer 2004;90(3):582–589.
36. Hwang SS, Chang VT, Kasimis B. Dynamic cancer pain management outcomes: the relationship between pain severity, pain relief, functional interference, satisfaction and global quality of life over time. J Pain Symptom Manage 2002;23(3):190–200.
37. Thomas R, Kaminski E, Stanton E, Williams M. Measuring information strategies in oncology—developing an information satisfaction questionnaire. Eur J Cancer Care 2004;13(1):65–70.
38. Molenaar S, Sprangers MA, Rutgers EJ, et al. Decision support for patients with early-stage breast cancer: effects of an interactive breast cancer CDROM on treatment decision, satisfaction, and quality of life. J Clin Oncol 2001;19(6):1676–1687.
39. Gourdji I, McVey L, Loiselle C. Patients' satisfaction and importance ratings of quality in an outpatient oncology center. J Nurs Care Qual. 2003;18(1):43–55.
40. Lubeck DP, Litwin MS, Henning JM, et al. An instrument to measure patient satisfaction with healthcare in an observational database: results of a validation study using data from CaPSURE. Am J Manag Care 2000;6(1):70–76.
41. Hall JA, Feldstein M, Fretwell MD, et al. Older patients' health status and satisfaction with medical care in an HMO population. Med Care 1990;28:261–269.
42. Wolf MH, Putnam SM, James SA, Stiles WB. The medical interview satisfaction scale: development of a scale to measure patient perceptions of physician behaviour. J Behav Med 1978;1:391–401.
43. Bredart A et al. A comprehensive assessment of satisfaction with care: preliminary psychometric analysis in French, Polish, Swedish, and Italian oncology patients. Patient Educ Couns 2001;43(3):243–252.
44. Bredart A, Razavi D, Robertson C, et al. Timing of patient satisfaction assessment: effect on questionnaire acceptability, completeness of data, reliability and variability of scores. Patient Educ Couns 2002;46(2):131–136.
45. Bredart A, Razavi D, Delvaux N, et al. A comprehensive assessment of satisfaction with care for cancer patients. Support Care Cancer 1998;6(6):518–523.
46. Grunfeld E, Fitzpatrick R, Mant D, et al. Comparison of breast cancer patient satisfaction with follow-up in primary care versus specialist care: results from a randomized controlled trial. Br J Gen Pract 1999;49(446):705–710.
47. Kavadas V, Barham CP, Finch-Jones MD, et al. Assessment of satisfaction with care after inpatient treatment for oesophageal and gastric cancer. Br J Surg 2004;91(6):719–723.
48. Bredart A, Mignot V, Rousseau A, et al. Validation of the EORTC QLQ-SAT32 cancer in-patient satisfaction questionnaire by self-versus interview-assessment comparison. Patient Educ Couns 2004;54(2):207–212.
49. Schoen RE, Weissfeld JL, Bowen NJ, Switzer G, Baum A. Patient satisfaction with screening flexible sigmoidoscopy. Arch Intern Med 2000;160(12):1790–1796.
50. Loblaw DA, Bezjak A, Bunston T. Development and testing of a visit-specific patient satisfaction questionnaire: the Princess Margaret Hospital Satisfaction With Doctor Questionnaire. J Clin Oncol 1999;17(6):1931–1938.
51. Bitar R, Bezjack A, Mah K, et al. Does tumor status influence cancer patients' satisfaction with doctor-patient interaction? Support Care Cancer 2004;12(1):34–40.

52. Hagedoorn M, Uijl SG, Van Sonderen E, et al. Structure and reliability of Ware's Patient Satisfaction Questionnaire III: patients' satisfaction with oncological care in the Netherlands. Med Care 2003;41(2):254–263.
53. Pun Wong DK, Chow SF. A qualitative study of patient satisfaction with follow-up cancer care: the case of Hong Kong. Patient Educ Couns 2002;47(1):13–21.
54. Von Essen L, Larsson G, Oberg K, Sjoden PO. 'Satisfaction with care': associations with health-related quality of life and psychological function among Swedish patients with endocrine gastrointestinal tumours. Eur J Cancer Care 2002;11(2):91–99.
55. Ware JE Jr, Philips J, Yody BB, Adamczyk J. Assessment tools: functional health status and patient satisfaction. Am J Med Qual 1996;11(1):S50–S53.
56. Davis SW, Quinn S, Fox L, MacElwee N, Engstrom PF. Satisfaction among cancer outpatients. Progress in clinical and biological research 1988;278:227–232.

IV

Special Topics in Surgical Clinical Research

15 Quality of Care

Jessica B. O'Connell, MD
and Clifford Y. Ko, MD, MS, MSHS

Contents

1. WHAT IS QUALITY OF CARE?

It is sometimes difficult for surgeons (or any clinician) to define quality of care—particularly as a policy measure. In practice, we tend to know it when we see it (or when we do not), but there are still controversies as to what is and what is not quality care. For example, should high-quality care be defined as state-of-the-art care? Reasonable care? Or given the trend toward levels of evidence, should high-quality care be only evidence-based care?

The American Medical Association defined high-quality care as that "which consistently contributes to the improvement or maintenance of quality and/or duration of life" *(1)*. Donabedian, who was considered an authority on quality of care, defined high-quality care as "that kind of care which is expected to maximize an inclusive measure of patient welfare, after one has taken account of the balance of expected gains and losses that attend the process of care in all its parts" *(2)*. Finally, a recent and familiar definition holds that quality consists of the "degree to which health services for individuals and populations increase the likelihood of desired health outcomes and care consistent with current professional knowledge" *(3)*.

With these descriptions in mind, this chapter will address some of the key issues relevant to quality of care. More specifically, this chapter will include the following: a conceptual framework for measuring quality of care; a discussion regarding the concept of use of services (e.g., appropriateness, underuse, overuse) with particular regard to surgery; and finally, we will end with a discussion of some of the ongoing quality evaluations in surgery.

From: *Clinical Research for Surgeons*
Edited by: D. F. Penson and J. T. Wei © Humana Press Inc., Totowa, NJ

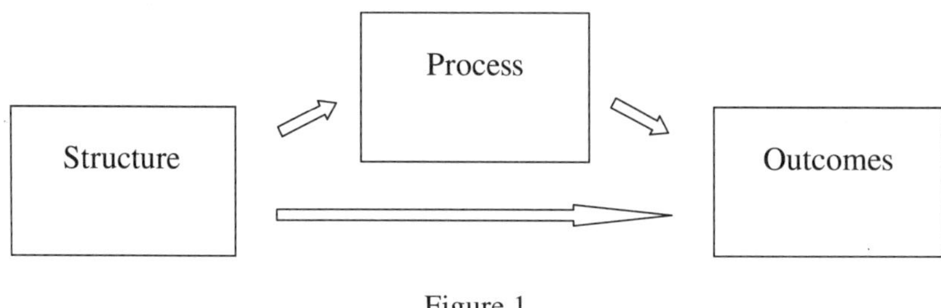

Figure 1

2. FRAMEWORK FOR MEASURING QUALITY OF CARE

2.1. A Conceptual Framework for Measuring Quality of Care: Structure, Process, and Outcomes

Avedis Donabedian (1919–2000) is considered to be the father of quality of care in the health services research field. He proposed a conceptual model for evaluating "quality of care," which consisted of three distinct dimensions: *structure, process,* and *outcomes.* Within his basic paradigm, structure and process are the critical aspects affecting health care outcomes. These relationships provide a framework for measuring and deriving quality evidence supporting specific treatments and care over others *(2)*. As depicted in Fig. 1, structure items can affect process, which in turn can affect outcomes; however, structure items in-and-of themselves can also affect outcomes as well.

2.1.1. Structure

Structure is defined as a combination of "raw materials" required for medical care. These materials may be as basic as equipment, such as the type of ventilation systems or operating room beds. Structural factors may be more complex, with "administrative structure" variables such as clinic staffing, nurse-to-patient ratio, administrative policies and arrangements, or with "clinical structure" including patients' severity of illness and comorbidities, patient case-mix, and risks of specific outcomes before surgery *(4)*. Many of the structural components are easily measurable, and some have lately been used as a proxy for measuring quality. Two structural variables that have received much attention in recent literature include hospital procedure volume and surgeon procedure volume.

2.1.2. Process

Process of care is defined as what is done to and for the patient. For the surgeon, quality-related processes of care may involve factors in the preoperative, intraoperative, or postoperative settings *(4)*. These items can include things such as performance of specific intraoperative surgical techniques (e.g., nerve-sparing prostatectomy), the appropriate use of deep venous thrombosis prophylaxis, or the appropriate use of prophylactic antibiotics. Processes can be broken down into smaller components. For example, for appropriate use of prophylactic antibiotics, this process can be subdivided into starting the prophylactic antibiotic within 1 hr before incision time, use of the appropriate type of antibiotic, and discontinuing the prophylactic antibiotic within 24 hr.

Process items may also include less tangible, but important items such as patients seeking out medical care and providers performing the correct tests to make a correct diagnosis or recommendation. Items such these are often more difficult to measure and evaluate.

Overall, process items tend to have a stronger association with outcomes, but for purposes of evaluation, are often more difficult (and expensive) to measure because of the detail that is required to evaluate the processes. One advantage of process measures, however, is that, unlike many structural variables, process measures generally do not require risk adjustment.

2.1.3. Outcomes

The crux of what health services researchers, the surgical community, and patients themselves are interested in is the "outcomes." Outcomes can be described as the effects of care on the health status of patients and populations and include a variety of clinical and physiologic measures, patient-reported items, and economic variables *(4)*. More specifically, outcomes may include broad items such as overall mortality, morbidity, length of stay and overall cost—none of which are new concepts to the surgeon. Of course, outcomes can be more focused and include such things as 30-d mortality or anastomotic leak rates. The area of patient-reported outcomes is becoming particularly important in light of the community and individual's perception of medical and surgical quality. Such patient-derived outcome measures can include patient satisfaction, functional health status, and quality of life measures.

A leading example of using outcomes to promote surgical quality is the National Surgical Quality Improvement Program (NSQIP), which is "the first national, validated, outcome-based, risk-adjusted, and peer-controlled program for the measurement and enhancement of the quality of surgical care" *(5)*. One of the claims to fame of NSQIP is that it provides "risk-adjusted outcomes" by collecting detailed "clinical and demographic data on the operations performed. Another advantage of NSQIP is that it measures outcomes, both mortality and complications, for a 30-d period instead of stopping at the time of discharge. As will be discussed later in this chapter, NSQIP has demonstrated its usefulness for both measuring and improving surgical quality in the VA [Veterans Administration] system. It is now being piloted in selected non-VA university hospitals across the United States" *(6–9)*.

2.2. Quality Evaluation Studies

Given Donabedian's model, it is noteworthy that most quality of care evaluators discuss and evaluate quality (and safety) in terms of structure, process, and outcomes. Although there have been relatively few surgical studies that have addressed quality of care in this manner, such studies are becoming increasingly seen in the literature. This next section will highlight some of these studies by discussing three aspects of the Donabedian model: structure-outcome associations, process-outcome associations, and structure–process associations.

2.2.1. Structure–Outcome Associations

Structure is a relatively easy way to look at quality of care in terms of attaining good outcomes. We in the health care community already have several systems in place to monitor "structure." For example, it is relatively easy to obtain data regarding how many beds a hospital has, the nurse-to-patient ratio on each floor, and how many operations are performed at each hospital each year.

For surgery, there have been several studies reporting the structural variables of *hospital volume* and *surgeon volume* and their relationship to better outcomes (i.e., postoperative mortality, complications, and length of stay). This had been shown for a variety of surgical interventions including colorectal surgery, hepatic resection, radical prostatectomy, total hip replacement, thyroidectomy, carotid endarterectomy (CEA), and even palatoplasty *(10–19)*. Many of these studies make up the body of literature that Leapfrog's evidence-based hospital referral criteria is based on.

To briefly introduce this topic, several studies have shown hospital volume to be predictive of good outcomes. A good example is a study on hepatic resection. This study used the Nationwide Inpatient Sample (1996–1997) to look at in-hospital mortality and length of stay. They found that high-volume hospitals had a mortality rate of 3.9% vs 7.6% at low-volume hospitals ($p < 0.001$). Multivariate analysis further found that "high-volume hospitals had a 40% lower risk of in-hospital mortality compared with low-volume hospitals (odds ratio (OR), 0.60; 95% confidence interval (CI), 0.39–0.92; $p = 0.02$)" *(13)*. Other predictors of mortality in the multivariate analysis included age older than 65 years, hepatic lobectomy (vs wedge resection), primary hepatic malignancy (vs metastases), and the severity of underlying liver disease. Their main conclusion was that "hospital procedural volume is an important predictor of mortality after hepatic resection [and that] patients who require resection of primary and secondary liver tumors should be offered referral to a high-volume center" *(13)*. It should also be noted, however, that their study also points to the fact that there are other aspects unrelated to volume itself that are significant to patient outcomes (e.g., specific operation [hepatic lobectomy versus wedge resection]), the severity of the patients' underlying liver disease, and other patient comorbidities. Some of these factors identified in this study as well as others fall under the *process* heading within Donabedian's framework.

Other than hospital volume, the volume of the actual surgeon has been associated with outcomes. Hu et al. found that "for radical prostatectomy, high-volume surgeons had half the complication risk OR = 0.53; 95% CI, 0.32–0.89) and shorter lengths of stay (4.1 versus 5.2 days, $p = 0.03$) compared with low-volume surgeons." They concluded that "surgeon volume is inversely related to in-hospital complications and length of stay in men undergoing radical prostatectomy, [and that] hospital volume is not significantly associated with outcomes after adjusting for physician volume" *(20)*.

Still other studies have examined both surgeon and hospital volume as predictors of outcome. An interesting examination of colorectal cancer outcomes was performed using Maryland state discharge data from 1992 to 1996. The authors divided 9739 total patient cases into three groups based on annual surgeon case volume per year: low (≤5), medium (5–10), and high (>10) and hospital volume per year: low (<40), medium (40–70), and high (≥70). Using regression analyses and adjusting for variations in type of resections performed, cancer stage, patient comorbidities, urgency of admission, and patient demographic variables, they found that the majority of surgeons (81%) and hospitals (58%) were in the low-volume group. The low-volume surgeons operated on 36% of patients at an average rate of 1.8 cases per yr. They also found that higher surgeon volume was associated with improved outcomes, and that medium-volume surgeons were able to achieve results equivalent to high-volume surgeons when they operated in high- or medium-volume hospitals. Furthermore, low-volume surgeons' outcomes improved with increasing hospital volume but never equaled those of the high-volume surgeons *(12)*.

In keeping with colorectal cancer, Schrag et al. used the Surveillance, Epidemiology, and End Results Medicare-linked database and evaluated 24,166 colon cancer resection patients age 65 years and older from 1991 to 1996. Their outcome measures included 30-d and 2-yr mortality, overall survival, and the frequency of operations requiring an ostomy. Multiple confounding factors were controlled for such as: age, sex, race, comorbid illness, cancer stage, socioeconomic status, emergent hospitalization, and the presence of obstruction or perforation. They found that both surgeon volume and hospital volume were independently important predictors of surgical outcomes ($p < 0.01$), however unlike the Hu et al. study mentioned previously, Schrag et al. reported that the effect of surgeon volume was attenuated after adjusting for hospital volume ($p < 0.03$). They concluded that "both hospital and surgeon-specific procedure volume predict outcomes following colon cancer resection; but hospital volume may exert a stronger effect. Therefore, efforts to optimize the quality of colon cancer surgery should focus on multidisciplinary aspects of hospital care rather than solely on intraoperative technique" *(11)*. This study points to the fact that there are many factors involved in good patient outcomes, including the patients themselves. Furthermore, this study highlights the variability of the volume-outcome association, which can be dependent on both the disease being studied and the specific outcome measure used.

The reader should know that not all studies have demonstrated a volume-outcome relationship. Some of the best literature in this regard comes from the NSQIP data analyses. Khuri's 1999 evaluation of the volume-outcome relationship using NSQIP data found that there was no association between high surgical volume and better outcomes (risk-adjusted 30-d death, and 30-d stroke rates in CEA) in eight commonly performed operations (n = 68,631) of intermediate complexity *(21)*. These operations included: nonruptured abdominal aortic aneurysmectomy, vascular infrainguinal reconstruction, CEA, lung lobectomy/pneumonectomy, open and laparoscopic cholecystectomy, partial colectomy, and total hip arthroplasty. They concluded that "volume of surgery in these operations should not be used as a surrogate for quality of surgical care" *(21)*.

The volume-outcome debate will likely continue, because there are good data on both sides of the argument. Although inherent limitations are associated with each of the data sources that are used to perform these volume-outcome analyses, it is important for the researcher to understand the strengths and weaknesses of these analyses overall. In terms of strengths, volume-outcome studies are relatively easy to perform. Data are readily available, often with quite large sample sizes, and several studies to date have shown an association between higher volume and better outcomes. However, there are also limitations. For example, the data are imperfect. Most of these databases are administrative and limited in that regard. One of the best nonadministrative data sources is the NSQIP, which has been demonstrated to have applicability even outside the Veterans Administration system. Finally, many people feel that the health policy implications of the volume-outcome relationship are to "regionalize" surgical care, sending patients to high-volume surgical specialty centers or high-volume surgeons. Whether this is a feasible option in the United States is unknown. Additionally, deciding which operations to "regionalize" is also extremely difficult.

Along similar lines to volume is another structure-outcome relationship: that of specialty (e.g., specialty surgeon, specialty hospital). Whether a specialist, as opposed to a generalist, performs better and obtains better outcomes has also been the subject of significant controversy in the literature and the press. We will examine this topic by examining some of the literature in this regard. In the field of vascular surgery, Hannan et al. found that patients who were operated on by a vascular surgeon had lower odds of having an adverse outcome compared with those operated on by a general surgeon. Their study was based on a voluntary registry of 3644 patients undergoing CEA between April

1, 1997, and March 31, 1999, in New York hospitals. They found that "patients undergoing surgery performed by vascular surgeons had lower odds of experiencing an adverse outcome (OR = 0.36, $p = 0.009$)." Importantly, they further noted that "processes of care and surgical specialty were highly correlated with one another [and] are significant interrelated determinants of adverse outcome for CEA" *(22)*.

On the other hand, there are also studies available that show that generalists do not necessarily provide substandard care as compared to specialists. Iglesias et al. found for 4587 appendectomies (3624 performed by specialist surgeons and 963 by general practitioner surgeons) that rates of comorbidity, diagnostic accuracy, and mean lengths of stay were similar for patients of general practitioner and specialist surgeons. They did, however, note that patients operated on by specialists were older and more likely to have perforations and to require second intra-abdominal or pelvic procedures *(23)*.

It may be likely that for more complex procedures and more complex disease processes that specialty care is preferred. For example, one study compared survival outcomes for patients with breast cancer cared for by specialist and nonspecialist surgeons in a retrospective study of 3786 female patients younger than age 75 years treated between 1980 and 1988 in a geographically defined population in urban west of Scotland *(24)*. They found 5-yr survival rates to be 9% higher and the 10-yr survival 8% higher for patients cared for by specialist surgeons. Furthermore, they found a reduction in risk of dying of 16% after adjustment for age, tumor size, socioeconomic status, and nodal involvement *(24)*. More specific research is needed to elucidate what processes of care the specialist surgeons are using that lead to better outcomes.

2.2.2. Process–Outcome Links

Although structural variables (e.g., volume, specialty training) are relatively easy to study and associate to outcomes, most researchers and clinicians agree that it is the processes of care that make the difference. However, studying processes of care is difficult for many reasons, including the detail required to appropriately study it, the paucity of evidence available to identify the important processes, and the expense. Regarding the surgical discipline, however, there are starting to be increasingly better performed studies looking at surgical process. For example, there is fairly good evidence showing that preoperative antibiotics should be given within a certain defined time before incision to decrease postoperative wound infections. Classen et al. evaluated the rate of surgical wound infections in 2847 patients undergoing elective "clean" or "clean-contaminated" surgical procedures and found the lowest wound infection rate to occur in the group who received prophylactic antibiotics within the 2 hr before incision *(25)*.

Regarding specific surgical specialties, some of the best process-outcomes research available is in the field of cardiovascular surgery. In the mid-1980s the New York State Department of Health developed a program to track the quality of care in cardiac surgery. In 1989, it began publishing annual data on risk-adjusted mortality after coronary artery bypass graft surgery (CABG) in the New York State Cardiac Surgery Reporting System. With the outcomes information from the first report, the department began implementing several structure and process-based changes. Some of the process-based changes included changing the specific processes within the clinical management protocols for stabilizing patients in the early postmyocardial infarction period before surgery and having weekly conferences between cardiologists and cardiac surgeons to increase the communication and discussion of plans for patient care. These changes helped to improve patient outcomes immensely with a 41% decline in death rate over the first 4 yr of the project *(26)*.

In vascular surgery, there are studies to support using one or more specific processes of care in the operating room, including eversion endarterectomy, the use of protamine, or the use of shunts. These processes overall were found to be associated with lower odds of an adverse outcome (OR = 0.42, p = 0.006) *(22)*. Another recent study evaluated Medicare patients who underwent 10,561 CEA procedures from 1995 to 1996 in 10 different states. The outcome measures included 30-d stroke or mortality post-CEA and they found that the processes of using preoperative antiplatelet agents (OR = 0.70, $p < 0.05$), intraoperative heparin (OR = 0.49, $p < 0.05$), and patch angioplasty (OR = 0.73, $p < 0.05$) were all associated with a lower hazard of complication *(27)*.

Although these are good examples of process variables being associated with outcome, in surgery, there are numerous processes that are probably important to obtaining good outcomes, but have not been studied with randomized controlled studies (RCTs). Such processes might include ambulation of the appropriate patient after surgery, intraoperatively palpating the liver during a cancer resection, or even appropriately informing the patient of options and risks of surgery. The issue that needs to be addressed for the quality of care researcher is that such processes measures are probably very important to providing high-quality surgical care but they have not been studied, nor will they likely ever, by RCTs. The dilemma remains that although the surgical community is focusing on evidence-based care, many of the processes that are likely important to good care are not proven with Level I evidence. Surgical investigators need to identify ways to validate these non-RCT studied, yet vital process measures.

2.2.3. Structure–Process Links

One of the arguments for using structural variables as a marker for better quality is that if the "correct" structural factor is chosen, then it is more likely that the correct process is performed. This notion is well exemplified in the previously mentioned study looking at the surgeon specialty–outcomes relationship/processes of care for CEA. In the Hannan et al. study that reviewed 3644 patients from a voluntary CEA registry, they found process outcome links in that the use of >1 specific processes of care (eversion endarterectomy, protamine, or shunts) was found to be associated with lower odds of an adverse outcome relative to patients undergoing CEA without the processes (OR = 0.42, p = 0.006). Similarly, they found structure–outcome links in that patients undergoing surgery performed by vascular surgeons had lower odds of experiencing an adverse outcome *(22)*. Results such as these suggest that there is something intrinsic to specialty surgeons and the performance of specific processes that lead to better outcomes.

A similar study but one based on "hospital" instead of "surgeon" showed that the structure of a specialty "cancer center" leads to receiving better treatment, or to the process of chemotherapy, as compared to receiving care at a "noncancer hospital." More specifically, their "results showed that patients admitted first to a noncancer hospital were less than half as likely to go on to receive chemotherapy as those first admitted to a cancer unit or centre (OR = 0.28). This result was not explained by distance between hospital of first admission and nearest cancer center or by increasing age or severity of illness" *(28)*.

Although these two studies provide some evidence of a structure–process link, very little research on this relationship has been performed for surgical outcomes. There are still many unknowns, such as how do administrative and organization structures affect or link to processes of care? How can the environment, personnel, equipment be optimized to promote the most appropriate and effective delivery of care? These are some of the many important issues that still need to be addressed in this arena *(4)*.

3. "CORRECT USE OF SERVICES"—APPROPRIATENESS

The concept of "appropriateness" stems from the issue of processes of care. Appropriateness is essentially the "correct use" of health care services. Along these same lines are the concepts of inappropriateness (or misuse of specific processes of care), underuse (e.g., specific processes of care not used enough) and overuse (e.g., specific processes of care being used too much). Regarding the latter, "if one could extrapolate from the available literature, then perhaps one fourth of hospital days, one fourth of procedures, and two fifths of medications could be done without" *(29)*. If this is true, then our national's annual health care bill could be cut by potentially \$100 billion without harming the public *(30)*.

The question naturally is, how do we know what care is "appropriate?" Who determines appropriateness, and, more importantly, how are they determined? Ideally, we as physicians would like Level I evidence to be the basis for our quality process measures, but what happens if there isn't any Level I evidence available?

Much of this discussion on the topic of quality of care has been re-ignited by the Institute of Medicine report, *To Err is Human*. Misuse or errors in health care delivery was shown to lead to many adverse effects including delayed diagnoses, higher costs, and unnecessary injuries and deaths *(31)*. Along these same lines, a study of New York State hospitals found that 1 in 25 patients was injured by inappropriate care, and that 13.6% of those injured actually died secondary to the error. Negligence was cited for 27% of the injuries and 51% of the deaths. These results lead to the estimate that 180,000 deaths per yr are caused by preventable errors *(32)*.

Appropriateness in surgery is important—particularly with regard to indications for performing a surgical procedure. It is not so easy to set specific criteria—although worthy attempts have been made. For example, in the late 1980s, consensus guidelines were published with regard to explicit criteria for the appropriateness of cholecystectomy. Since then, because many new diagnostic and treatment techniques were developed, an updated study in 2002 was performed on the appropriateness of indications for cholecystectomy. Two separate panels of six experts in gastroenterology and six in surgery were used, with the study creating an algorithm tool for assessing the appropriateness of cholecystectomy. In brief, a total of 210 scenarios were evaluated in the report; interestingly, 51% were deemed appropriate, 26% uncertain, and 23% inappropriate *(33)*.

From the 1980s to present, several appropriateness studies have been published involving the "procedural" fields. For example, some of these studies found that 17% of coronary angiograms, 32% of CEA, 17% of upper endoscopies, and 2.4% of CABG were performed for inappropriate indications *(34–37)*. Other studies have demonstrated that the rates of inappropriate care vary as 4% of percutaneous transluminal coronary angioplasty, 1.6% of CABG, and 10.6% of CEA being deemed inappropriate *(38,39)*. Outside the field of cardiovascular medicine, there have been several studies in gynecology, including a recent study from 2000, which found that up to 16% of hysterectomies were performed for inappropriate reasons *(40)*. Partially because of the difficulty of defining appropriateness, there is marked variation in these results. The important message to take from these studies, however, is that some level of inappropriate/misuse was identified.

3.1. Underuse and Overuse of Services

After appropriate processes of care are identified, one can measure *underuse* as well as *overuse* of that specific process of care. In this regard, underuse of services can be defined as not using an appropriate medical treatment or intervention when indicated. It may lead to many potential adversities in medical care including potential complications, need for more health care services, higher costs, and premature deaths. Recent studies evaluating the underuse of medical care found that more than 50% of heart attack patients did not receive appropriate postmyocardial infarction treatment including beta-blocker medications *(41)*. Furthermore, another study found 34% of cardiac patients do not receive percutaneous transluminal coronary angioplasty when deemed appropriate by a panel of experts *(42)*. A recent study on breast disease, focused on understanding and preventing the underuse of effective breast cancer therapies, found that up to 16% of women with early-stage breast cancer did not receive adjuvant therapy *(43)*. Although there are likely several different reasons for the underuse of proven therapy, these are the critical issues that require study. Whether the provider was not aware of the need for therapy, whether there were communication problems among providers, or whether the patient refused treatment are some of the possible areas for further study. These issues need to be further elucidated in order to improve care. Much work of this type is needed in surgery overall.

Overuse of medical and surgical services can lead to unnecessary health costs and potential complications that are detrimental to patient health. One area in which this topic is well researched is with regard to prophylactic antibiotics in surgery. Virtually everyone knows that overuse of antibiotics has been shown to lead to bacterial resistance, morbidity (e.g., *Clostridium difficile* colitis), and excess and unnecessary costs. Yet studies seem to continually show antibiotic overuse (i.e., one study found that 71% of antibiotics prescribed were overused, particularly with regard to redundant coverage for gram-positive organisms that were found in 56% of cases evaluated) *(44)*. Procedurally, whereas we specified that one study found 16% of hysterectomies were performed inappropriately, another found up to 70% of hysterectomies were performed unnecessarily *(40,45)*.

3.2. Variations of Services

Having had a discussion regarding underuse and overuse, it is not surprising that there are studies that show marked variations in the use of services. Such variations have been shown to occur with regard to differences in numerous demographic factors such as age, gender, race/ethnicity, insurance status, and even geographic location. For example, one study of several surgical procedures found wide variations in surgical rates across geographic areas. Birkmeyer et al. studied patients ages 65 to 99 enrolled in Medicare in 1995 (excluding those enrolled in risk-bearing health maintenance organizations), and calculated rates of 11 common inpatient procedures for each of 306 US hospital referral regions. They found that the rates of hip fracture repair, resection for colorectal cancer, and cholecystectomy varied only 1.9- to 2.9-fold across hospital referral regions; CABG, transurethral prostatectomy, mastectomy, and total hip replacement had intermediate variation profiles, varying 3.5- to 4.7-fold across regions; and lower extremity revascularization, CEA, back surgery, and radical prostatectomy had the highest variation profiles, varying 6.5- to 10.1-fold across hospital referral regions. They concluded that "although the use of many surgical procedures varies widely across geographic areas, rates of "discretionary" procedures are most variable. To avoid potential overuse or underuse, efforts to increase consensus in clinical decision making should focus on these high variation procedures" *(46)*. Another study found that even controlling for rates of use, there were still regional variations in the use of coronary angiography, CEA, and upper gastrointestinal tract endoscopy *(34)*. It is not just in this country that variation occurs: a study from Israel found similar results when evaluating rates of cholecystectomy in four different regional hospitals *(47)*.

The subject of racial variations in care is one of ongoing interest, with many articles citing medical and surgical differences in treatment of ethnic minorities. One study of 10,073 African-American and 123,127 Caucasian women diagnosed with Stage I, IIA, or IIB breast carcinoma in the SEER database (1988–1998) found that "African-American women were significantly less likely to receive follow-up radiation therapy in every 10-year age group except in the older than 85 age group" *(48)*. Another study of 6437 Medicare beneficiaries from Arizona, Illinois, New Mexico, or Texas who underwent a primary total hip replacement (THR), as compared with 12,874 controls, found that the "odds of THR decreased as the probability of Hispanic ethnicity increased, from an OR of 1.00 among beneficiaries with non-Hispanic surnames, to an OR of 0.36 among those with heavily Hispanic surnames (95% CI, 0.31–0.43)." They concluded that "Hispanic persons with Medicare receive THR at lower rates than do non-Hispanic persons . . . [and that] . . . because Medicare covers THR . . . under utilization of THR by Hispanic persons cannot be attributed to lack of health insurance alone *(49)*. And yet another study using SEER found that in comparing black and white patients age 65 years or older with a diagnosis of resectable non–small-cell lung cancer (stage I or II, 1985–1993, $n = 10{,}984$), the rate of surgery was 12.7% lower for black patients than for white patients (64.0% vs 76.7%, $p < 0.001$), and that the 5-yr survival rate was also lower for blacks (26.4% vs 34.1%, $p < 0.001$). They did find, however, that "among the patients undergoing surgery, survival was similar for the two racial groups, as it was among those who did not undergo surgery." They concluded that "the lower survival rate among black patients with early-stage, non–small-cell lung cancer, as compared with white patients, is largely explained by the lower rate of surgical treatment among blacks" *(50)*.

In summary, the issue of appropriateness is an important one for the study of quality of care. Similar to studying processes of care, the study of appropriateness is difficult because of the paucity of level I evidence. Also, even when RCT evidence is available, patient care is often more complex because of contributing factors such as level of comorbid disease and other patient-related issues. Still, more work is needed in this area, and such work needs to be performed in an unbiased manner by the clinicians that perform these operations.

4. EVALUATING QUALITY IN SURGERY WITH "REPORT CARDS"

Particularly since the publication of the Institute of Medicine's report, *To Err is Human*, "report cards" on the quality of health care have become increasingly popular. Report cards have addressed numerous issues and have a variety of layouts regarding how quality is reported, as well as recommendations for obtaining quality care. In this section, we will briefly discuss three groups that have evaluated quality.

Especially for surgeons, one well-known evaluator of quality has been the Leapfrog Group. The Leapfrog Consortium is a "program aimed at mobilizing employer purchasing power to alert America's health industry that big leaps in patient safety and customer value will be recognized and awarded" *(51)*. Specifically, it is a collaborative of Fortune 500 companies and other large public and private health care purchasers that provide health benefits to more than 34 million Americans in all 50 states. The Leapfrog Group encourages its members to choose hospitals that adhere to three hospital safety measures:

(1) computer physician order entry, in which physicians enter all orders into a computer system; (2) intensive care unit physician staffing, in which all intensive care units are staffed full-time by a physician who is credentialed in critical care, and (3) evidence-based hospital referral. The evidence-based hospital referral criteria are most important to surgeons in that it is based on the volume-outcome relationship and recommends receiving the following four operations at hospitals that perform more than a certain threshold number of cases per year: (1) CABG (≥450/yr); (2) esophagectomy (≥13/yr); (3) abdominal aortic aneurysm repair (≥50/yr); and (4) pancreatic resection (≥11/yr) *(51)*. Leapfrog represents one of the first economically based attempts at changing and improving health care.

Because of the relative ease to obtain structural data, the lay press often uses such variables to evaluate and grade quality. One of the more well-known examples in this regard is the annual *US News and World Report* rating of medical centers. The hospitals are ranked based on a set of criteria consisting of mainly structure and some outcome information to create the "U.S. News Index" that "combines mortality, number of discharges, and other measures to summarize quality of care" *(52)*. The specific components of this index include a variety of data points including: *discharges*: number of Medicare patients discharged in 1999, 2000, and 2001 after receiving specified care; *RNs to beds*: ratio of full-time, on-staff registered nurses to hospital beds; *technology services*: number of key specialty-specific technologies offered; *National Cancer Institute cancer center*: designated a "clinical" or "comprehensive" cancer center by the National Cancer Institute; *hospice, palliative care*: presence of a hospice program or palliative-care program; *trauma center*: presence of a certified trauma care center; *discharge planning*: of three services (patient education, case management, patient representatives), the number offered; *service mix*: of nine patient and community services (such as hospice or home healthcare), the number offered; *geriatric services*: of seven (such as adult day care and an arthritis center), the number offered; *gynecology services*: of four (obstetric care, reproductive healthcare, birthing rooms, women's health center), the number offered; and *medical/surgical beds*: intensive-care surgical beds (only in kidney disease); amongst other *(52)*. Again many of the quality indicators used by this report are structure-based variables. As stated earlier, structure data are easy to access and report but stand as a relatively crude measure for quality.

Finally, the previously mentioned New York State Cardiac Surgery Reporting System is another excellent example of how a "report card" program can have an excellent effect on improving patient outcomes when structural and process-based changes are implemented based on the findings of the report. This New York State Cardiac Surgery Reporting System was developed in the mid-1980s by the New York State Department of Health in order to track the quality of care in cardiac surgery. They began publishing annual data on risk-adjusted mortality after CABG in 1989 and with the outcomes results from the first report, they implemented several structure and process-based changes, which led to a 41% decline in the post-CABG death rate over the first 4 yr of the project *(26)*.

As a result of these studies, current quality indicators are largely empiric. These include a systematic literature review to draft a list of potential indicators based on evidence that they influence the quality of care. This list of candidate indicators is then reviewed by patient and expert focus groups to demonstrate and enhance face and content validities. Experts are also asked to rate the indicators in terms of likelihood of being able to measure them in a clinical setting. In this regard, the Delphi methodology has been successfully applied. After an indicator set has been proposed, the next step is demonstration of feasibility of assessment and validity in a clinical setting. This may comprise field testing these indicators at a small number of institutions and settings to show that they can be readily measured either using administrative data or chart review. The amount of work to measure these indicators are assessed at this stage and will in part determine if use of these indicators on a wider scale will be practical. It is also helpful to demonstrate associations of these indicators with actual outcomes (and thereby demonstrate construct validity) and to identify covariates (e.g., age, disease severity) that may influence these indicators. Las, these indicators are applied in a larger network or in a national evaluation to identify current level of quality among a broad spectrum of institutions, practice environments, and geographic regions. The result of these data may then be used to develop benchmarks and also for continued demonstration of validity by simultaneously assessing outcomes and correlation better outcomes with greater indicator compliance. Ultimately, quality of care indicators will be adopted by third-party payers such as Medicare in "Pay for Performance" programs that will reward physicians and institutions for measuring quality or providing better quality of care. Similarly, withholding revenue or other punitive actions may be taken for nonparticipants or those who consistently demonstrate poor quality.

5. CONCLUSIONS

Improving quality of care, patient safety, and eliminating medical errors are becoming increasingly essential in today's health care system—especially in surgery. Although volumes can be written regarding surgical quality of care, we have organized this chapter to introduce some of the important concepts on quality of care—specifically, the Donabedian model of structure, process, and outcomes, as well as the concept of appropriateness. As the reader can see, there are many areas where further work is needed and it is important that surgeons perform this work.

REFERENCES

1. American Medical Association, Council of Medical Service. Quality of care. JAMA 1986;256:1032–1034.
2. Donabedian A. The definition of quality and approaches to its assessment. Ann Arbor, MI: Health Administration Press, 1980.
3. Lohr KN, Donaldson MS, Harris-Wehling J. Medicare: a strategy for quality assurance, V: quality of care in a changing health care environment. QRB Qual Rev Bull 1992;18:120–126.
4. Rubin H. Framework for evidence-based sugery. In: Gordon T, Cameron, JL, ed. Evidence-based surgery. Lewiston, NY: B.C. Decker, Inc, 2000; 47–59.
5. National Surgical Quality Improvement Program (NSQIP). (Accessed August 13, 2003, at http://www.solutions.starmountain.com/nsqip/PDF_Files/Appendix_G-Informational_Brochure-draft_2.pdf).
6. Fink AS, Campbell DA Jr, Mentzer RM Jr, et al. The National Surgical Quality Improvement Program in non-veterans administration hospitals: initial demonstration of feasibility. Ann Surg 2002;236: 344–353.
7. Liu JH, Etzioni DA, O'Connell JB, Maggard MA, Ko CY. Using volume criteria: do California hospitals measure up? J Surg Res. In press.
8. Khuri SF, Daley J, Henderson W, et al. The Department of Veterans Affairs' NSQIP: the first national, validated, outcome-based, risk-adjusted, and peer-controlled program for the measurement and enhancement of the quality of surgical care. National VA Surgical Quality Improvement Program. Ann Surg 1998;228:491–507.
9. Daley J, Khuri SF, Henderson W, et al. Risk adjustment of the postoperative morbidity rate for the comparative assessment of the quality of surgical care: results of the National Veterans Affairs Surgical Risk Study. J Am Coll Surg 1997;185:328–340.

10. Ko CY, Chang JT, Chaudhry S, Kominski G. Are high-volume surgeons and hospitals the most important predictors of in-hospital outcome for colon cancer resection? Surgery 2002;132:268–273.
11. Schrag D, Panageas KS, Riedel E, et al. Surgeon volume compared to hospital volume as a predictor of outcome following primary colon cancer resection. J Surg Oncol 2003;83:68–78.
12. Harmon JW, Tang DG, Gordon TA, et al. Hospital volume can serve as a surrogate for surgeon volume for achieving excellent outcomes in colorectal resection. Ann Surg 1999;230:404–411.
13. Dimick JB, Cowan JA Jr, Knol JA, Upchurch GR Jr. Hepatic resection in the United States: indications, outcomes, and hospital procedural volumes from a nationally representative database. Arch Surg 2003;138:185–191.
14. Cowan JA Jr, Dimick JB, Thompson BG, Stanley JC, Upchurch GR Jr. Surgeon volume as an indicator of outcomes after carotid endarterectomy: an effect independent of specialty practice and hospital volume. J Am Coll Surg 2002;195:814–821.
15. Yao SL, Lu-Yao G. Population-based study of relationships between hospital volume of prostatectomies, patient outcomes, and length of hospital stay. J Natl Cancer Inst 1999;91:1950–1956.
16. Ellison LM, Heaney JA, Birkmeyer JD. The effect of hospital volume on mortality and resource use after radical prostatectomy. J Urol 2000;163:867–869.
17. Katz JN, Losina E, Barrett J, et al. Association between hospital and surgeon procedure volume and outcomes of total hip replacement in the United States Medicare population. J Bone Joint Surg Am 2001;83-A:1622–1629.
18. Sosa JA, Bowman HM, Tielsch JM, et al. The importance of surgeon experience for clinical and economic outcomes from thyroidectomy. Ann Surg 1998;228:320–330.
19. Witt PD, Wahlen JC, Marsh JL, Grames LM, Pilgram TK. The effect of surgeon experience on velopharyngeal functional outcome following palatoplasty: is there a learning curve? Plast Reconstr Surg 1998;102:1375–1384.
20. Hu JC, Gold KF, Pashos CL, Mehta SS, Litwin MS. Role of surgeon volume in radical prostatectomy outcomes. J Clin Oncol 2003;21:401–405.
21. Khuri SF, Daley J, Henderson W, et al. Relation of surgical volume to outcome in eight common operations: results from the VA National Surgical Quality Improvement Program. Ann Surg 1999;230:414–432.
22. Hannan EL, Popp AJ, Feustel P, et al. Association of surgical specialty and processes of care with patient outcomes for carotid endarterectomy. Stroke 2001;32:2890–2897.
23. Iglesias S, Saunders LD, Tracy N, Thangisalam N, Jones L. Appendectomies in rural hospitals. Safe whether performed by specialist or GP surgeons. Can Fam Physician 2003;49:328–333.
24. Gillis CR, Hole DJ. Survival outcome of care by specialist surgeons in breast cancer: a study of 3786 patients in the west of Scotland. BMJ 1996;312:145–148.
25. Classen DC, Evans RS, Pestotnik SL, et al. The timing of prophylactic administration of antibiotics and the risk of surgical-wound infection. N Engl J Med 1992;326:281–286.
26. Chassin MR. Achieving and sustaining improved quality: lessons from New York State and cardiac surgery. Health Aff (Millwood) 2002;21:40–51.
27. Kresowik TF, Bratzler D, Karp HR, et al. Multistate utilization, processes, and outcomes of carotid endarterectomy. J Vasc Surg 2001;33:227–234.
28. Pitchforth E, Russell E, Van der Pol M. Access to specialist cancer care: is it equitable? Br J Cancer 2002;87:1221–1226.
29. Brook RH. Practice guidelines and practicing medicine. Are they compatible? JAMA 1989;262: 3027–3030.
30. Phelps CE. The methodologic foundations of studies of the appropriateness of medical care. N Engl J Med 1993;329:1241–1245.
31. Kohn LT, Corrigan J, Richardson WC, Donaldson MS. To err is human: building a safer health system. Washington DC: National Academy Press, 2000.
32. Richardson WC, Berwick DM, Bisgard JC. The Institute of Medicine report on medical errors. N Engl J Med 2000;343:663–4; author reply 665.
33. Quintana JM, Cabriada J, de Tejada IL, et al. Development of explicit criteria for cholecystectomy. Qual Saf Health Care 2002;11:320–326.
34. Chassin MR, Kosecoff J, Park RE, et al. Does inappropriate use explain geographic variations in the use of health care services? A study of three procedures. JAMA 1987;258:2533–2537.
35. Winslow CM, Solomon DH, Chassin MR, et al. The appropriateness of carotid endarterectomy. N Engl J Med 1988;318:721–727.

36. Kahn KL, Kosecoff J, Chassin MR, Solomon DH, Brook RH. The use and misuse of upper gastrointestinal endoscopy. Ann Intern Med 1988;109:664–670.
37. Leape LL, Hilborne LH, Park RE, et al. The appropriateness of use of coronary artery bypass graft surgery in New York State. JAMA 1993;269:753–760.
38. Leape LL, Hilborne LH, Schwartz JS, et al. The appropriateness of coronary artery bypass graft surgery in academic medical centers. Working Group of the Appropriateness Project of the Academic Medical Center Consortium. Ann Intern Med 1996;125:8–18.
39. Hilborne LH, Leape LL, Bernstein SJ, et al. The appropriateness of use of percutaneous transluminal coronary angioplasty in New York State. JAMA 1993;269:761–765.
40. Broder MS, Kanouse DE, Mittman BS, Bernstein SJ. The appropriateness of recommendations for hysterectomy. Obstet Gynecol 2000;95:199–205.
41. Ellerbeck EF, Jencks SF, Radford MJ, et al. Quality of care for Medicare patients with acute myocardial infarction. A four-state pilot study from the Cooperative Cardiovascular Project. JAMA 1995;273: 1509–1514.
42. Hemingway H, Crook AM, Feder G, et al. Underuse of coronary revascularization procedures in patients considered appropriate candidates for revascularization. N Engl J Med 2001;344:645–654.
43. Bickell NA, McEvoy MD. Physicians' reasons for failing to deliver effective breast cancer care: a framework for underuse. Med Care 2003;41:442–446.
44. Glowacki RC, Schwartz DN, Itokazu GS, et al. Antibiotic combinations with redundant antimicrobial spectra: clinical epidemiology and pilot intervention of computer-assisted surveillance. Clin Infect Dis 2003;37:59–64.
45. Bernstein SJ, McGlynn EA, Siu AL, et al. The appropriateness of hysterectomy. A comparison of care in seven health plans. Health Maintenance Organization Quality of Care Consortium. JAMA 1993;269:2398–2402.
46. Birkmeyer JD, Sharp SM, Finlayson SR, Fisher ES, Wennberg JE. Variation profiles of common surgical procedures. Surgery 1998;124:917–923.
47. Pilpel D, Fraser GM, Kosecoff J, Weitzman S, Brook RH. Regional differences in appropriateness of cholecystectomy in a prepaid health insurance system. Public Health Rev 1992;20:61–74.
48. Joslyn SA. Racial differences in treatment and survival from early-stage breast carcinoma. Cancer 2002;95:1759–1766.
49. Escalante A, Barrett J, del Rincon I, et al. Disparity in total hip replacement affecting Hispanic Medicare beneficiaries. Med Care 2002;40:451–460.
50. Bach PB, Cramer LD, Warren JL, Begg CB. Racial differences in the treatment of early-stage lung cancer. N Engl J Med 1999;341:1198–1205.
51. The Leapfrog Group. (Accessed August 13, 2003, at http://www.leapfroggroup.org/).
52. America's Best Hospitals. (Accessed August 13, 2003, at http://www.usnews.com/usnews/nycu/health/hosptl/tophosp.htm).

16 Cost-Effectiveness Analyses

Lynn Stothers, MD, MHSc, FRCSC

Contents

Surgical research has traditionally focused on comparing health outcome measures of a new technique with accepted practice. As health care resources become scarcer and options for newer, more expensive diagnostic tests and surgical interventions increase, incorporating cost-effectiveness analysis (CEA) into surgical research studies becomes increasingly important. CEA provides the information necessary to allow resource allocation decisions to be based on the best balance between health outcomes and cost. Thus research data collection and analysis must include not just direct health outcome, but also financial costs and both positive and negative changes in life expectancy and in quality of life *(1)*.

Acquiring skills in, and a thorough understanding of, various types of economic analysis are important for today's surgical researchers. Although the costs associated with surgery—personnel, facilities and equipment—are high, surgical interventions have gained credibility in their ability to enhance quality of life, extend life years, and reduce disability time for workers. CEA incorporated into research can provide the necessary impetus for clinicians to change their patterns of practice, for department heads and service providers to change program funding, and for governments to change health care policy. At a basic level, CEA can even help patients choose between surgical interventions.

From: *Clinical Research for Surgeons*
Edited by: D. F. Penson and J. T. Wei © Humana Press Inc., Totowa, NJ

Other types of economic analysis can also be used by researchers to justify expenditure of research dollars and to help in selecting the most cost-effective research design or research program. In addition, the information from CEA will help practicing surgeons decide which of the growing number of techniques and equipment options warrant their time to learn and maintain skills and which options will complement their practice *(2)*.

1. BASIC CONCEPTS AND TERMINOLOGY

A number of basic types of economic analysis exists, and the researcher should choose the most appropriate for their needs. In its simplest form, economic analysis measures and reports costs related to a particular treatment or treatment pathway. More complex forms of analysis, including CEA and cost-utility analysis (CUA), report costs related to a particular health outcome. The outcomes form the denominator in the reported ratio. For example, CEA analysis may focus on cost per cases of disease prevented or years of life gained, whereas CUA analysis focuses on cost per quality adjusted life years gained (QALY).

Economic analyses involve the techniques of *decision analysis*, a tool originally developed in gaming theory. Decision analysis is a method of determining the best decision by examining how different scenarios affect an overall decision. It assigns a numeric value to each of many possible outcomes, and thus allows for quantitative analysis of decision making when there are many possible pathways *(3)*.

1.1. Cost Analysis

The primary purpose of all forms of cost analysis is to compare the costs and values of different outcomes. Initially, cost analysis was limited to financial costs, but analysts and investigators now realize that other factors, particularly the intervention's value in prolonging life and improving quality of life, must be included for the interpretation to be relevant. Recognition by investigators of the importance of cost analysis has led to an increasing number of reports on cost analysis in the medical literature.

Cost analysis, initially developed in the 1960s and 1970s as an economics tool, has evolved in the health field from basic "cost of illness" calculations, through "cost-benefit analysis," to its current level of complexity—cost-effectiveness and cost-utility analyses.

Cost of illness analysis typically quantifies the burden of medical expenses (direct costs) and the resulting value of lost productivity (indirect costs) attributable to a specific condition such as an illness or injury *(4)*.

CEA measures the costs and consequences of two or more diagnostic or treatment pathways related to a single common effect or health outcome and summarizes the results in ratios that demonstrate the cost of achieving a unit of health effect for diverse types of patients and for variations of the intervention *(5)*.

Health outcomes that might form the denominator of the cost-effectiveness ratio include lives saved, cases found, or disability days prevented. For example, both live and cadaveric kidney donor transplant procedures are options for prolonging the life of a patient with renal failure. CEA could be used to calculate the cost per life-year gained of live vs cadaveric renal transplantation. Although costs are typically reported in clinical journals as "cost per life year gained," they could equally be calculated as "life years gained per dollar amount spent." The latter approach may be helpful for health administrators or those working with a fixed budget.

Although CEA is often used to compare different surgical options for a given disease, it can also be used to compare various surgical treatments that have a common outcome.

For example, the cost per life-years saved could be compared for kidney transplant vs open heart surgery. CUA is a form of CEA in which particular attention is paid to the quality of health outcome related to treatment. In CUA, health effects are expressed in terms of QALYs. A QALY is a measure of health outcome that assigns to a given period of time a weighting that corresponds to the health-related quality of life during that period and then aggregates these weights across time periods. Results of CUA are expressed as a cost per QALY gained *(6)*.

The QALY is important because it considers both quantity and quality of life. CUA should be considered the analysis of choice when the health outcome of interest is improvement in quality of life. For example, CUA would be useful in studying surgical interventions such as treatments for urinary incontinence or arthritis that aim to improve not only physical function but also social function and psychologic well-being.

Cost-benefit analysis estimates the net social benefit of an intervention by comparing the benefit of the intervention with the cost, with all benefits and costs measured in dollars *(6)*. Health outcomes are converted into monetary values using "willingness to pay" (the value an individual would pay for reduction in illness severity) or "risk of death" or "human capital" methods (an individual's value to society based on productivity or future wages) *(7,8)*. This chapter focuses on CEA/CUA, the most advanced forms of cost analysis and, in general, the most relevant for surgical research.

2. HOW IS CEA PERFORMED?

In simple terms, a CEA is performed by conducting a simulated clinical trial and comparing the costs related to a common outcome at the end of the treatment pathway. The data used can be obtained from the literature or from an actual clinical trial conducted by the investigator. The trial "cohort" of patients, designed by the researcher to reflect the actual makeup of the population being studied, moves through a predefined, finite sequence of clinical alternatives (e.g., various surgical or medical approaches to treatment) at predetermined intervals. How the patients move through the chain of treatment options is based on the best available information on the probabilities of success with each intervention *(9)*. An example of the application of CEA in a clinical trial is Chang et al's study of total hip arthroplasty *(10)*.

To be useful on a long-term basis, CEA research on similar topics need to be comparable and to use comparable terminology and well-designed methodology. The need for standardization of CEA led the US Public Health Service to convene the Panel on Cost-Effectiveness in Health and Medicine, which has made recommendations on how cost analyses should be conducted and reported. Their recommendations are published in a series of three articles reported in the *Journal of the American Medical Association (5,11,12)*. The Panel recommends that researchers define a "base case," a model that incorporates all the information that the investigator thinks best represents the interventions and choices being compared. The Panel also recommends the use of a "reference case," a model that considers the comparison from a societal perspective and uses the standardized methods and assumptions defined by the Panel. The "base case" and the "reference case" may be the same model if the investigation being undertaken is from a societal perspective. The Panel also recommends that all investigators conduct and report the "reference case" analysis in addition to the "base case" analysis, if they differ, to contribute to the knowledge base. The use of standards for the costs and health effects that should be included and the ways in which they should be valued provides analysts and users a way of comparing the result of different studies, even if the area of research differs (e.g., surgical intervention vs public health promotion) *(5)*.

3. GENERAL STEPS FOR PERFORMING A CEA

3.1. Define the Cost-Effectiveness Research Question

In CEA, a research question should be clearly defined that compares the consequences of various diagnostic or treatment options in terms of costs. The following is an example of a CEA research question: "Is Achilles tendon repair cost effective compared with conservative treatment, from a societal perspective, as measured by quality of life years?" *(3)*. To formulate a research question, the following parameters need to be defined as follows.

1. The perspective from which the study will be done, which determines the costs to be included
2. The time frame
3. Effectiveness measures
4. Relevant treatment options
5. Relevant outcomes

3.1.1. Perspective

Decide whose perspective will serve as the basis for the analysis—society as a whole, the funding source, the patient/family, or the physician. Who will be using the CEA ratios and for what purpose? The perspective is reflected in the research question and goals of the analysis. The perspective determines which costs and health effects should be included in the cost-effectiveness ratio and how the costs and effects should be valued *(7,13)*. This step is important because, since differing costs are included depending on the perspective, studies based on different perspectives are not comparable *(13)*. If the analysis being conducted is a "reference case," according to the US Panel on Cost Effectiveness, then the perspective will be societal.

Typical costs and outcomes to consider *(7,13–15)*, defined by the chosen perspective, include the following.

- *The societal perspective*, in which all costs and outcomes that affect everyone in society, regardless of who pays, are included (total net cost of medical and other payments for resource use, time away from work and out-of-pocket expenditures).
- *The funding source perspective*, in which the true costs to the funding source of providing a service are included; items borne by the patient/family such as time away from work, out-of-pocket expenses are excluded.
- *The patient/family perspective*, in which only those costs and outcomes relevant to the patient/family, such as copayment of health care costs, time away from work, and out-of-pocket expenses, are included.
- *The program perspective*, in which all direct costs to the program are considered.
- *The clinician perspective*, in which all aspects of surgical resource utilization, such as operating room time, office time, direct costs to surgical budget, and costs of learning and maintaining skills in a new technique, are included.

3.1.2. Time Frame

Define the period of time for which costs and benefits will be determined. The time frame should be long enough to capture future health outcome and the economic impact of an intervention, which may be short term or last the duration of the patient's life. However, because many other factors affect long-term outcome, long-term costs should be limited to those that can be directly attributed to the intervention *(7,16)*.

3.1.3. Effectiveness Measures

Select the specific measures of interest, whether these are *final outcomes* such as life years gained or lost or QALYs, or so-called *intermediate outcomes,* such as patients appropriately treated. A search of the literature can assist with compiling a relevant list of the dimensions of success to be considered. For example, in the case of outcomes for surgical treatment of urinary incontinence, the investigator could consider "the number of dry patients," or the number of patients who no longer need protection against incontinence. When reviewing the literature, consider how the outcomes are measured. Life years calculated from mortality data are usually comparable, but outcomes such as "the number of dry patients" can vary by whether the outcome was measured using objective testing measurements or by survey.

3.1.4. Relevant Treatment Options

Decide what particular treatment options will be considered. Options may include other surgical procedures, medical interventions, complementary medical approaches such as diet and exercise or herbal interventions, or "doing nothing." It is worthwhile to differentiate "doing nothing" from treatment schemes such as "watchful waiting." As an example, "watchful waiting" when applied to a patient with low-risk prostate cancer would still include the costs of physician visits and other tests needed during a patient's follow up.

3.1.5. Relevant Outcomes

Decide which outcomes to include by reviewing reports on efficacy/effectiveness, side effects, and complications *(13)*.

3.2. Define the Possible Pathways (Schematic Model)

After compiling the data, develop a schematic model that will describe the sequence in which interventions occur, how the course of a health condition is affected, complications, and health outcomes. The most commonly used formats for structuring the schematic model are decision trees and spreadsheets.

3.2.1. Structuring a Decision Tree

To make a decision tree, begin with the question, conventionally on the left, working to the right *(3)* (Figure 1). Draw lines leading from the question for each possible pathway. Each decision point or node is represented by a square; each chance point (when outcome is uncertain) is represented by a circle. The probability of an outcome is written below the line for the outcome, and the utility of the outcome is written beside the triangle. A sample decision tree from Kocher et al. *(3)* is shown.

A Markov model may be used if a simple decision tree is inadequate. A Markov model is a form of decision tree that allows cycling through the process or pathways more than once. There are many approaches to Markov modeling, most of which involve some form of Monte Carlo simulation (*see* the next paragraph). The Markov process is complex, particularly if the medical condition being modeled is complex and occurs over a long period, or if changes in health status occur frequently. As a result, Markov processes require sophisticated, computerized calculations, and will likely require the involvement

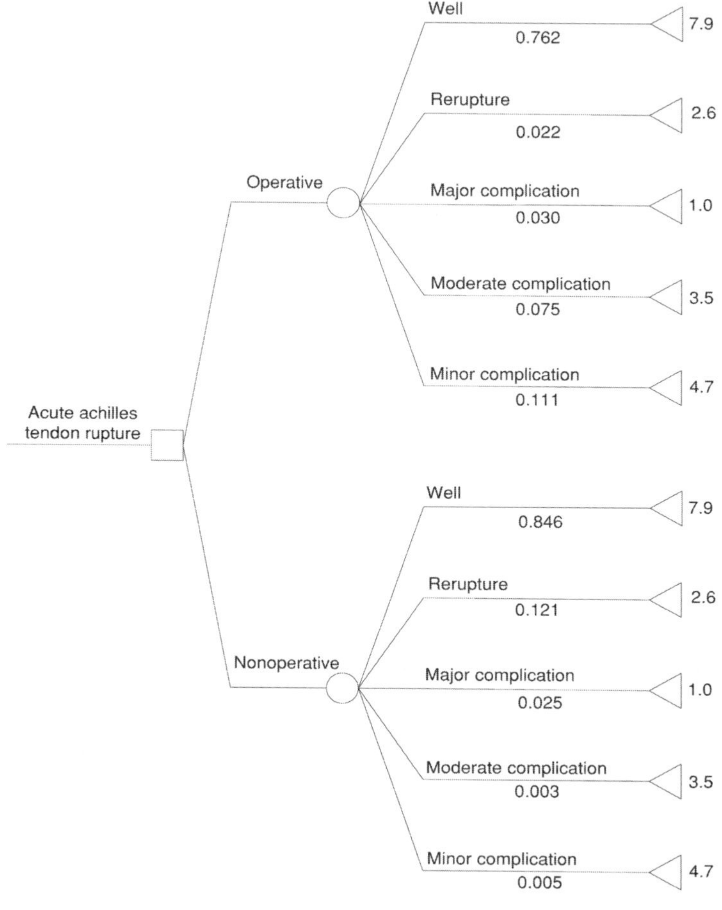

Figure 1

of an expert in the field for the design of the study *(17)*. An example of a Markov model in surgical CEA research can be found in Romangnuolo et al *(18)*.

Monte Carlo simulation is a form of statistical analysis in which the probability of different outcomes is calculated repeatedly, using different scenarios for each calculation. Although the mathematical calculations are complex, computer software is available to perform the calculations as a single operation, providing information about the full range of possible outcomes, and the likelihood of each *(19)*.

3.3. Estimate the Probabilities of the Various Outcomes to Be Included

Estimate the probability that each given outcome will occur, based on the best available evidence—a systematic review of the literature, consultation with experts, or independent research. The goal is to find the most accurate estimate of the probability for each event in the schematic model. The best estimate, or "baseline estimate," is used to perform the "base case" analysis. In order of strength of evidence, probabilities may be obtained from:

- Large randomized, controlled trials (RCTs) with clear-cut results
- Small RCTs with uncertain results

- Nonrandomized, contemporaneous controls
- Nonrandomized, historical controls
- No controls, case series only
- Expert opinion *(7,13)*

3.4. For CUA, Determine Outcome Utilities

For studies that emphasize changes in both quality and quantity of life as a result of the treatment, utility values need to be determined. Utility values are determined in one of three ways: (1) values taken from the literature, (2) judgment of the investigator who proposes a range of reasonable utilities, or (3) direct measurement on a sample of patients. In the latter, the patient can take part in a standardized interview regarding time tradeoff, or in a standard reference gamble, in which the patient is asked how much time with the disease they would trade for time without disease, or what chance of death they would take to be free of disease. In this way, the patient is given a choice of alternatives in which the patient is presented with a set of scenarios and is asked to choose between various pairs of alternatives *(3)*.

An example of a utilities values table in the literature is provided from Summerfield et al *(20)* (Table 1).

Table 1
Utilities of Health States Estimated by Patients and Volunteers[a]

Utilities of Health States Estimated by Patients and Volunteers[a]					
Health State	Informants	Sample Size	Mean Utility (95% CI)[b]	Mean Utility (95% CI)[c]	Loss of Utility (95% CI)[d]
Profoundly hearing impaired, no benefit from acoustic hearing aids (traditional candidate)	Patients	87	0.562 (0.527-0.596)	0.843 (0.805-0.880)	0.281 (0.255-0.308)
Profoundly hearing impaired, marginal benefit from acoustic hearing aids (marginal hearing aid user)	Patients	115	0.725 (0.693-0.757)	0.870 (0.839-0.900)	0.145 (0.123-0.167)
Traditional candidate benefiting from a unilateral cochlear implant	Patients	87	0.750 (0.705-0.794)	0.813 (0.769-0.857)	0.063 (0.048-0.078)
Marginal hearing aid user benefiting from a unilateral cochlear implant	Patients	115	0.802 (0.767-0.838)	0.851 (0.815-0.887)	0.049 (0.039-0.059)
Profoundly hearing impaired, no benefit from acoustic hearing aids (traditional candidate)	Volunteers	70	0.765 (0.730-0.800)	1[e]	0.235 (0.200-0.270)
Severely profoundly hearing impaired, marginal benefit from acoustic hearing aids (marginal hearing aid user)	Volunteers	70	0.836 (0.807-0.865)	1[e]	0.164 (0.135-0.193)
Benefiting from a unilateral cochlear implant	Volunteers	70	0.934 (0.915-0.954)	1[e]	0.066 (0.046-0.085)
Benefiting from bilateral cochlear implants	Volunteers	70	0.965 (0.952-0.978)	1[e]	0.035 (0.022-0.048)

[a] CI indicates confidence interval
[b] Mean utility measured with Mark II Health Utilities Index (patients and with time trade-off technique [volunteers])
[c] Mean utility recalculate with Mark II Health Utilities Index after placing patients at the highest levels of the hearing and speech dimensions
[d] Mean loss of utility due to impaired hearing and speech
[e] Assuming a utility of unity in the absence of impairments to hearing and speech

An alternative is to use a visual analog scale, usually given to a cohort of patients with the condition under investigation. Here is an example of a visual analog scale from Kocher et al. *(3)*:

How would you value the following possible scenarios after treatment for Achilles tendon rupture?

0 = the worst possible medical outcome for me; 10 = best possible medical outcome for me.

Place an X on the line at the appropriate location.

____

0 10

"Doing well" was defined as: No complications. No rerupture. Return to work at 10.0 weeks. 73% return to same level of athletics. At least 80% strength recovery.

The first two techniques can be difficult to apply when a change in quality of life rather than mortality is the outcome under consideration.

Utility values will vary depending on the individual because of personal preferences for such parameters as cost savings, ability to work, improved function, pain relief, and minimization of complications *(3)*.

3.4.1. How to Calculate QALYs

QALYs put a numerical value on quality and quantity of life. The first step is to determine the possible states of health that the intervention could achieve. Next, a weight, ranging from 1 to 0, is assigned to each possible health state, corresponding to the health-related quality of life, in which a weight of 1 corresponds to optimal health and 0 corresponds to a health state judged equivalent to death. This is often determined by direct questioning of real patients. The duration that a patient is likely to spend in each state of health as a result of the intervention in question is then estimated based on the literature. Finally, the value of the health state is multiplied by the amount of time the patient will be in that state and the totals are summed to obtain QALY expected from this intervention *(6)*.

The following is an example of how to calculate QALYs *(21)*.

Treatment of metastatic cancer with three different drugs results in three different life expectancies and three different quality of life scores:

Treatment	*Length of Time in Health State*	*Utility Value for Health State*	*QALY (Equivalent to Number of Years in Perfect Health)*
Drug 1	0.8 yr	0.67	0.8*0.67 = 0.54
Drug 2	2.4 yr	0.53	2.4*0.53 = 1.3
Drug 3	5.2 yr	0.84	5.2*0.84 = 4.4

The mathematics involved in calculating QALYs are straightforward. However, determining how much weight should be placed on patient preferences for various health outcomes and calculating the probability that a given outcome will occur is much more difficult. Attention to detail at this stage is essential if the results of the CEA are to be useful *(5)*.

3.5. Collect All Relevant Cost Data, as Determined By the Research Question

Collect cost data as comprehensively as possible. The costs of a particular health care intervention involve three basic divisions of cost—direct costs, indirect costs, and intangible costs. The costs themselves are typically divided into direct, indirect and intangible costs.

Direct costs of urinary incontinence are borne by both the health sector and by individual patients and their families. Direct costs related to operating costs for the health sector include both inpatient and outpatient services, particularly in the areas of supplies, equipment, and health professionals. Some direct health sector costs are *variable*, such as the cost of supplies and health professionals' time, whereas others are *fixed*, such as the overhead incurred in running a hospital or clinic. Not all patients will incur direct costs to the health sector. For example, in the case of urinary incontinence, it has been estimated that 2% of individuals living in the community and 5% of those living in institutions seek treatment *(22)*. Direct costs carried by the patient include medication and supplies used to manage a given medical condition. For example, in the case of urinary incontinence, padding and devices are used as protection against incontinence. Some devices used to manage a given medical condition are gender specific. Some men use gender-specific protective undergarments, often more costly than female garments, or condom drainage or an external device such as a penile clamp.

Indirect costs include lost earnings for both the patient and family or friends who provide care for the affected person. Age and working status are particularly important related to indirect costs. For example, because the prevalence of urinary incontinence increases dramatically with age, the working status of the 60+ age group is of particular importance. Sources for this information include hospital discharge surveys, insurance claims (Medicare), survey data (National Medical Expenditure Survey), patient records, and diaries *(13)*. A study by Stylopolous is an example of a large-scale CUA that uses national database data *(23)*.

Intangible costs include the monetary value of pain, suffering, and anxiety from the disease in question. Intangible costs are difficult to determine in most cases, and are generally the least well measured in the literature.

Costs vary to some extent by year, region, practice patterns in the area, and country. Medical expenses are often calculated using existing medical claims databases (e.g., Medicare) and typically rely on totaling costs, charges, or payments for those claims in which either a primary or secondary diagnosis involves the condition of interest. In incidence-based studies, an attempt is made to eliminate those claims that may have resulted from late effects of a condition, such as treatment for past injuries.

Productivity losses (work and leisure time lost from illness or premature mortality) are usually quantified by average annual wages, with adjustments for household productivity. Prevalence-based analysis quantifies lost productivity for the base year, whereas incidence-based analysis looks at the present value of all future lost productivity because of a relevant change in a patient's condition occurring in a specified time period *(4)*.

Consideration should be given to future benefits and costs for interventions expected to prolong life. For example, significant costs would likely accrue over the long term for patients who would not be expected to survive without the intervention being studied.

3.5.1. Discounting

Although there is controversy about whether and how discounting is to be applied *(24)*, most authorities on CEA recommend its use *(1,25)*. Discounting, a systematic method for calculating the present value of money that will be spent and health states that will occur in the future *(7)*, is performed because both costs and benefits of health care interventions can take place over a prolonged period, but those that accrue much later are less reliable and less likely to be the direct result of the intervention. Also, costs and benefits now are considered to be more valuable than those in the future because of "positive time preference"—people prefer to have things now than at some nebulous point in the future *(6)*.

Economic costs are weighted by a discount rate according to the year in which they accrue. Most countries specify the applicable discount rate, but future costs and utilities are usually "discounted" to present value at a rate of 3–5% per year *(7)*. The discounting of outcomes is more complex and controversial. Health does not have a true monetary value, and research indicates that many people do not place a high value on future health, as evidenced by behaviors such as smoking and substance abuse where the negative health outcomes are known. This can be very useful in discounting where an intervention is initially expensive, but there are life-long health benefits (e.g., a vaccination program) or future health benefits (e.g., an antismoking campaign). Adopting a zero discount rate increases the cost effectiveness of such programs *(6)*.

3.6. Calculate the Cost-effectiveness Ratio and the Incremental Cost-effectiveness Ratio

With the information in place, the costs related to the particular research question and perspective are totaled and related to a given denominator for effectiveness, creating a cost effectiveness ratio (dollars per life year saved for treatment A vs dollars per life year saved for treatment B).

For easier comparison between treatment strategies, an incremental analysis can be completed by comparing the various interventions to the base case strategy or intervention. Incremental cost-effectiveness represents the additional cost and effectiveness when one health care option is compared with another. Each option is then compared with the next most effective option. Incremental costs and effectiveness are the differences between the two options in costs and effectiveness or the extra cost per unit of outcome *(1,7,8,26)*. For example, the incremental cost-effectiveness ratio of strategy A compared with strategy B indicates how much money will be spent for each additional unit of health gain achieved by choosing strategy A over B *(27)*.

This ratio can be simply expressed as:

$$\text{CE ratio} = \frac{\text{cost}_{\text{new strategy}} - \text{cost}_{\text{current practice}}}{\text{effect}_{\text{new strategy}} - \text{effect}_{\text{current practice}}}$$

CEA provides guidance concerning what is both practical and possible in a given surgical situation by identifying *preferred strategies* based on the cost-effectiveness ratio. There is no agreement about what constitutes a preferred strategy, although society's cost-effectiveness ratio threshold (i.e., for any given intervention, how much is an improvement of one QALY worth to society?) is commonly estimated to be US$20,000–$100,000 per QALY *(13)*. One organization argues that any intervention with a cost-effectiveness ratio of <$20,000/QALY should be considered highly desirable; an intervention with a cost-effectiveness ratio of $20,000 to $100,000/QALY should be considered potentially acceptable, and an intervention with a cost-effectiveness ratio of >$100,000/QALY should be considered economically unacceptable, but there is no agreement concerning this "rule of thumb" *(6)*.

3.7. Perform Sensitivity Analysis

A sensitivity analysis considers which estimates in the analysis are most subject to debate because the estimates were not based on hard data, were subject to variation given their method of measurement, or were based on the investigator's value judgments. It involves a series of mathematical calculations that isolate factors or variables to indicate the degree of influence each factor has on the outcome of the entire analysis. Although the data used in performing CEA inevitably involve some uncertainties, sensitivity analysis can demonstrate whether changes in the variables would change the decision and may increase the level of confidence in decisions or suggest future directions for research to increase certainty *(7,13)*. Areas of uncertainty that arise include lack of RCT data and comparison of costs where not all costs are known *(6)*.

To perform sensitivity analysis, the values for probabilities, utilities, and costs are varied within plausible ranges, and the cost-effectiveness ratio is recalculated *(6)*. To be "plausible," the ranges used are usually within the 95% confidence interval around the mean, or are based on a literature review, or consultation with experts concerning clinical feasibility. In univariate or one-way sensitivity analysis, one variable at a time is changed. In multivariate sensitivity analysis, several variables are changed simultaneously. The analysis demonstrates whether the CEA results are robust within the plausible range of assumptions, or whether the analysis depends on specific assumptions.

Consideration should be given to performing *threshold analysis*, which is a type of analysis that determines the point at which the costs of two strategies are equal (i.e., the incremental cost-effectiveness is zero).

4. INTERPRETATION OF CEA FINDINGS AND OUTPUT

To be useful for all stakeholders on a long-term basis, CEA reports on similar topics need to be comparable. The Panel on Cost-Effectiveness in Health and Medicine provides a clear description of what should be included in a report on cost effectiveness and is shown Table 2 *(12)*.

When evaluating someone else's CEA *(13)*, look for a well-defined research question and consider what competing alternative therapies were compared and whether all competing alternatives were included. How did the authors define "effectiveness" in their analysis, and is the definition of effectiveness well established in the literature? Check whether a list of all relevant consequences and costs were included and whether a clear description of the valuation of these was provided. Look for sensitivity analysis, which should be considered mandatory. Check whether the authors justified the boundaries they used to vary different parameters for the discounting process.

5. TOOLS FOR PERFORMING CEA

CEA can be done using most commonly available statistical packages, such as SPSS or SASS. A free spreadsheet tool and encyclopedia from Solution Matrix Ltd. are available online at http://www.solutionmatrix.com/business-case-tools.html#Free.

Table 2
Checklist for Reporting the Reference Case Cost-Effectiveness Analysis

Framework	Background of the problem General framing and design of the analysis Target population for intervention Other program descriptors (e.g., care setting, model of delivery, timing of intervention) Description of alternative programs Boundaries of the analysis Time horizon Statement of the perspective of the analysis
Data and Methods	Description of event pathway Identification of outcomes of interest in analysis Description of model used Modeling assumptions Diagram of event pathway (schematic model) Software used Complete description of estimates of effectiveness, resource use, unit costs, health states and quality-of-life weights and their sources Methods for obtaining estimates of effectiveness, costs and preferences Critique of data quality Statement of year of costs Statement of method used to adjust costs for inflation Statement of type of currency Source and methods for obtaining expert judgment Statement of discount rates
Results	Results of model validation Reference case results (discounted at 3% and undiscounted); total costs and effectiveness, and incremental cost-effectiveness ratios Results of sensitivity analyses Other estimates of uncertainty, if available Graphical representation of cost-effectiveness results Aggregate cost and effectiveness information Disaggregated results, as relevant Secondary analyses using 5% discount rate Other secondary analyses, as relevant
Discussion	Summary of reference case results Summary of sensitivity of results to assumptions and uncertainties in the analysis Discussion of analysis assumptions having important ethical implications Limitations of the study Relevance of study results for specific policy questions or decision Results of related cost-effectiveness analyses Distributive implications of an intervention

Statistics Canada developed a program called Population Health Model *(28)* for the evaluation of cancer control interventions and policy decision making. Models of the costs of diagnosis and treatment of lung and breast cancer were developed and incorporated into the Population Health Model, and the program was then used to evaluate the economic impact of treatment. Using Monte Carlo microsimulation methods, the program generates and then ages over time a sample of synthetic individuals to whom demographic and labor force characteristics, health risk factors, and individual health histories typical of the population are assigned. This allows for the implementation of a competing risk framework, by which the event with the shortest time to transition is deemed to happen. The limitation of the model is that it is only as good as the data entered. Its strength is that it can be used as a policy analysis tool to answer "what-if" questions that go beyond cost issues to incorporate outcome measures.

6. LIMITATIONS OF CEA

Depending on the nature of the intervention being studied, the number of alternative interventions, and the time frame, CEA can be very complex and laborious, and help from an expert in the field may be required. It may be difficult to develop a decision tree that truly represents a patient's choice, course of illness, or health outcomes. In this regard, research into patient choices for treatment, and the reporting of "treatment pathways" for various illnesses is helpful. The process of simplifying the course of illness itself may eliminate important outcomes. There may be inadequate information available on which to make accurate assumptions about the probabilities of events, costs, and utilities *(7)*. Small errors in calculations may lead to incorrect conclusions *(3)*.

CEA is an evolving process, and experts do not agree about many issues, such as discounting *(18)*, the arbitrary assignment of values in defining QALYs, and the inclusion of future lost earnings in cost calculations *(4,29)*.

Despite the limitations and controversies surrounding CEA, it remains a useful tool, and when performed with care, it is far more accurate than the intuitive decision-making process.

REFERENCES

1. Weinstein MC, Stason WB. Foundations of cost-effectiveness analysis for health and medical practices. N Engl J Med 1977;296:716–721.
2. Swanstrom LL. Laparoscopic hernia repairs. The importance of cost as an outcome measurement at the century's end. Surg Clin North Am 2000;80(4):1341–1351.
3. Kocher MS, Henley MB. It is money that matters: decision analysis and cost effectiveness analysis. Clin Orth Rel Res 2003;413:106–116.
4. Finkelstein E, Corso P. Cost-of-illness analyses for policy making: a cautionary tale of use and misuse. Expert Rev Pharmacoecon Outcomes Res 2003;3(4):367–369.
5. Russell LB, Gold MR, Siegel JE, Daniels N, Weinstein MC. The role of cost-effectiveness analysis in health and medicine. JAMA 1996;276(14):1172–1177.
6. Gallin JI. Principles and practices of clinical research. New York: Academic Press, 2002.
7. Subak LL, Caughey AB. Measuring cost-effectiveness of surgical procedures. Clin Obstet Gynecol 2000;43(3):551–560.
8. Gold MR, Siegel JE, Weinstein MC, Russell LB. Cost effectiveness in health and medicine. New York: Oxford University Press, 1996.
9. Stein SC, Burnett MG. Decision analysis to estimate cost effectiveness in neurosurgery. Neurosurg Focus 2002;12(4):1–5.
10. Chang RW, Pellisier JM, Hazen GB. A cost-effectiveness analysis of total hip arthroplasty for osteoarthritis of the hip. JAMA 1996;20;275(11):858–865.

11. Weinstein MC, Siegel JE, Gold MR, Kamlet MS, Russell LB. Recommendations of the Panel on Cost-effectiveness in Health and Medicine. JAMA 1996;276(15):1253–1258.
12. Siegel JE, Weinstein MC, Russell LB, Gold MR. Recommendations for reporting cost-effectiveness analyses. Panel on Cost-Effectiveness in Health and Medicine. JAMA 1996;276(16):1339–1341.
13. Bradley CJ. Cost-effectiveness analysis in health care. Department of Medicine, Michigan State University. (Accessed December 22, 2003, at www.healthteam.msu.edu/ccc/epidemiology1/CostEffectivenessAnalysis1.pdf).
14. Eisenberg JM. Clinical economics: a guide to the economic analysis of clinical practices. JAMA 1989;262:2879–2886.
15. Drummond MF. Allocating resources. Int J Technol Assess Health Care 1990;6:77–92.
16. Pettiti DB. Meta-analysis, decision analysis, and cost-effectiveness analysis: methods for quantitative synthesis in medicine. New York: Oxford University Press, 1994.
17. Cher DJ, Lenert, LA. Rapid approximation of confidence intervals for Markov process decision models—applications in decision support systems. J Am Med Inform Assoc 1997;4(4):301–312.
18. Romangnuolo J. Meier MA. Medical or surgical therapy for erosive reflux esophagitis: cost-utility analysis using a Markov model. Ann Surg 2002;236(2):191–202.
19. Smith RL. Risk assessment. Technical guidance manual. EPA903-F-94-001. Philadelphia: United States Environmental Protection Agency, Hazardous Waste Management Division, Office of Superfund Programs, 1994.
20. Summerfield AQ. A cost utility scenario analysis of bilateral cochlear implantation. Arch Otolaryngol Head Neck Surg 2002;128(11);1255.
21. Implementing QALYs. (Accessed December 28, 2003, at http://www.jr2.ox.ac.uk/bandolier/painres/download/whatis/ImplementQALYs.pdf).
22. Wagner TH, Hu TW. Economic costs of urinary incontinence. Int Urogynecol J Pelvic Floor Dysfunct 1998;9(3):127–128.
23. Stylopoulos N, Gazele GS, Rattner DW. A cost-utility analysis of treatment options for inguinal hernia. Surg Endosc 2003;17(2);180–189.
24. Archarya AK. Rethinking discounting of health benefits in cost effectiveness analysis. (Accessed December 28, 2003, at http://www.sussex.ac.uk/Units/economics/dp/arnab1.pdf).
25. Eisenberg JM. Clinical economics: a guide to the economic analysis of clinical practices. JAMA 1989;262:2879–2886.
26. Drummond MF, O'Brien B, Torrance GW, Stoddart GL. Methods for the economic evaluation of health care programs. New York: Oxford University Press, 1997.
27. Giffin RB. Cost-benefit analysis: a primer for community health workers. Tucson: AZ: University of Arizona Rural Health Office, 1999.
28. Will BP, Berthelot J-M, Nobrega KM, Flanagan W, Evans WK. Canada's Population Health Model (POHEM): a tool for performing economic evaluations of cancer control interventions. Eur J Cancer 2001;37:1797–1804.
29. Garber AM. Advances in cost-effectiveness analysis of health interventions. NBER Working Paper Series. Cambridge, MA: National Bureaus of Economic Research, 1990.

17 Qualitative Research Techniques

Donna L. Berry, PhD, RN, AOCN, FAAN
Sally L. Maliski, PhD, RN,
and William J. Ellis, MD

CONTENTS

1. WHAT IS QUALITATIVE RESEARCH?

There are aspects of the human experience that cannot be enumerated or represented by a summary score. Clinicians in the surgical disciplines intuitively know this, yet often are not certain how to evaluate the perspectives and circumstances of their patients' experiences. Qualitative research is systematic inquiry that focuses on exploring and understanding the experiences of individuals and groups. Both the perceptions of patients and health care consumers plus those of providers have been studied widely using qualitative research methods.

The philosophical foundations of qualitative research include the work of twentieth century authors such as Heidegger *(1)*, Merleau-Ponty *(2)*, and Habermas *(3)* and have been applied by anthropologists, sociologists, and, more recently, health care scientists to rich programs of research *(4)*. Over the last three decades, an increasing number of nurse researchers have embraced the use of qualitative research methods to study complex processes, meanings, and human experiences relevant to health and illness conditions *(5)*. Grounded theory, phenomenology, and ethnography are qualitative research strategies often implemented by those researchers who are compelled to gather rich descriptions of their interest areas *(4)*.

From: *Clinical Research for Surgeons*
Edited by: D. F. Penson and J. T. Wei © Humana Press Inc., Totowa, NJ

Clinicians who accept a world view that has assumptions of holism, particularly with regard to individuals in the context of health care, tend to use qualitative research as one approach to scientific inquiry. This holistic world view is embodied in a number of fundamental beliefs: reality is different for each person, based on his or her perceptions and interpretations; meaning is always embedded in a given context *(5)* and; the various aspects of an individual (e.g., physiology, emotion) cannot be separated when studying human health responses *(6)*.

The methods and procedures of qualitative research are driven by assumptions relevant to the primacy of interpretation and always involve data directly gathered from the research participant's perspective or the participant's natural context and environment. Notably, these methods can be applied to multiple research designs and have been combined with quantitative approaches by many clinical researchers in a technique referred to as *triangulation (7)*. Clinicians in surgical specialties should be aware of the opportunities to study both quantitative variables (e.g., disease-free survival) and the qualitative experiences of patients (e.g., treatment decision making).

2. WHY USE QUALITATIVE RESEARCH METHODS?

Qualitative research methods are appropriate in any scientific inquiry that seeks to understand the "unique nature of human thoughts, behaviors, negotiations and institutions" *(8)*. Because clinical research involves these aspects of being human, most clinical studies could incorporate some aspect of a qualitative method. Scientists are taught that the research question determines the appropriate method. Therefore there are particular research questions to which qualitative methods not only could be applied, but also *should* be applied. Strang *(9)* claimed that qualitative research methods are complementary to those of quantitative research, providing opportunities to emphasize meanings and experiences of participants. He suggested a range of applications for qualitative research methods including induction of new hypotheses, exploring complex phenomena, and developing conceptual constructs for future quantitative questionnaires and validation of quantitative results. Studies may employ solely qualitative methods or combine qualitative and quantitative methods in a variety of designs. Miller and Crabtree *(6)* argued for multimethod clinical trials in which qualitative data can uncover and illuminate hidden theoretical assumptions and suggest new conceptualizations of the condition and human reactions.

When a clinical researcher wants to study human response, behavior, or experience, but not enough is known about the response to validly measure or assess it with a quantitative instrument, then qualitative methods can be applied to understand the concept or phenomenon of interest. In other words, use qualitative methods when you "don't know what you don't know," particularly when you, as the researcher, have never experienced the phenomenon yourself. For example, clinicians might expect that going back to the workplace after treatment for cancer would be problematic, possibly including anxiety, fatigue, even fear of discrimination. Yet, after study of this experience with qualitative methods in a sample of individuals with genitourinary cancers, the return-to-work process was found to be positive. The participants reported a sense of well being associated with various aspects of their experience of returning to work *(10,11)*. In another qualitative study of men postprostatectomy for prostate cancer, investigators found that incontinence, within the context of successful removal of the prostate cancer, was interpreted as part of the healing process and not a negative symptom *(12)*.

Several widely used quality of life questionnaires and symptom assessments have been developed empirically based on qualitative methods that provided rich description of the experience in a selected patient population. The European Organization for Research and Treatment of Cancer uses a systematic approach to instrument development for its QLQ series of quality of life questionnaires that includes analysis of patient input into content and item generation *(13)*. The initial work that resulted in the University of California Los Angeles Prostate Cancer Index *(14)* began with focus groups of men with prostate cancer and their spouses. The Symptom Distress Scale *(15)* and the Cancer Related Fatigue Distress Scale *(16)* are examples of quantitative symptom assessments that began with qualitative data production to establish questionnaire content grounded in patients' experiences and first hand reports of symptom sensations.

Other uses of qualitative methods increasingly are seen in health care literature. The study of health care delivery is a most appropriate area to implement qualitative methods because of the complexity of experiences confronted within the health care system *(17)*. Exploring perceptions and experiences of clinicians is not uncommon in qualitative research as investigators attempt to acknowledge the multiple facets of a health care encounter *(18)*. We will present original data from a qualitative study of physicians who counsel men with localized prostate cancer regarding treatment options (Section 4).

Analysis of textual data created without the intervention of a researcher is often termed a "narrative" or content analysis. In the particular case of a recorded and transcribed conversation between a patient and his or her clinician, analytic techniques of conversation analysis *(19)* can be applied. A key point to acknowledge with applications of such methods is that if the investigator is simply making a list of words or concepts that have been pre-determined as "of interest," then the approach is more investigator centered and may lack the qualitative emphasis on participants' interpretations, implicit meanings, and contexts *(20)*.

In summary, qualitative research methods are valuable, theory-based, rigorous approaches to clinical research. Understanding the life context, perspectives, and experiences of our patients, other health care consumers, and ourselves is a necessary adjunct to empiric evaluation of objective outcomes whenever the integrated human response is of interest.

3. QUALITATIVE METHODS OF INQUIRY

3.1. Ethnography: Understanding Culture and Context

Ethnography is the study of a cultural group. This can include a social group or system. Ethnography has its roots in anthropology. Early anthropologic studies often included extensive time in the field observing and living with the group being studied.

3.1.1. Types of Issues Addressed

Ethnography is useful to understand a group's learned patterns of behavior, beliefs, customs, attitudes, and ways of life. Groups of interest may be an ethnic group sharing meanings of behavior and language or may be an organizational group such as surgeons working in a particular institutional setting. There will be shared behaviors, beliefs, and attitudes that constitute a form of culture *(21)*.

3.1.2. Sampling

The sample consists of various forms of information regarding the group to be studied. This information may come from an individual experience being part of the group, observations of the group from an outsider's perspective, verbal or written stories, written and unwritten rules and procedures, publications, and literature. The investigator samples as wide a range of information sources as feasible to most completely describe the group. Spradley *(22,23)* provides more detail on sampling and data collection in ethnography.

3.1.3. Data Collection

Data collection is accomplished in ethnography through extensive fieldwork. During this process, the investigator immerses himself or herself within the group. The investigator may become a member of the group and function as a participant-observer or may gain access to a group through gatekeepers and then locate individuals who can provide valuable insights as key informants. Because data collection involves long-term, extensive contact with a group, investigators may become part of the group and are no longer observers, but participants. Interviewing, observing, reviewing documents, and assembling artifacts are all forms of data collection that may be used in an ethnographic study. However, interviews and observation are the most widely used.

3.1.4. Analysis

Analysis of ethnography generally occurs in three phases; description, analysis, and interpretation. Description is a straightforward presentation of the setting and events. In developing this description, the investigator may focus on "a day in the life" of the group or an individual, a key event, or developing a story with plot and characters. From the description, the investigator moves into the analysis phase that is a sorting process. This can be accomplished through highlighting specific material using tables, charts, diagrams, and figures. Patterns in the data are identified that may represent themes, allowing for comparisons with other cultural groups or theoretical frameworks. The analysis phase typically includes a critique of the research process and suggestions for redesigning the study.

Interpretation is the process of transforming the data. During this phase, the investigator speculates and presents his or her reflection of the meaning or influence that the themes have relative *to* the group's behavior, structure, and interactions. Interpretations are structured by inferences from the data or theory. Ethnography is the product of an ethnographic study. Typically, this takes the form of a book, although results of ethnographic studies can be found in journal articles.

3.1.5. Rigor

Although verification is not an appropriate standard to apply to ethnography, quality can be assured in a number of ways. Triangulation of data sources is used to compare information coming from different sources and from different phases of the study. Respondent validation is accomplished by ascertaining from the respondents whether the accounts accurately reflect their experience. Personal reflection by the investigator on experiences, in the form of a written or spoken record, is necessary to prevent personal values and ideologies from influencing the work and to maintain a research perspective while embedded within a group's culture.

3.2. Grounded Theory: Understanding Process From the Ground Up

Grounded theory is a qualitative research method that inductively builds theory from the data. The data, not theory, are the starting point. Grounded theory had its beginnings in sociology in the late 1960s with the publication of Glaser and Strauss's book, *The*

Discovery of Grounded Theory: Strategies for Qualitative Research (3). The researchers claimed that much of the methodologic work in sociology focused on the verification of existing theory and little considered the generation of new theory. Thus Glaser and Strauss developed and presented the rudiments of grounded theory while studying awareness of dying. The method has been elaborated and clarified over the years, and a process for conducting grounded theory research has been delineated *(24–26)*. Because there are sets of procedures to follow in the conduct of grounded theory, it is often considered the best method for new qualitative researchers to use. Grounded theory is used not only in the social sciences, but in health services research, nursing studies, and education.

3.2.1. Types of Research Issues Addressed

Grounded theory is most appropriate to explore processes and develop situation-specific theory. The method is well-suited to developing concepts that lead to theoretical explanations about a specific phenomenon and how people respond to it within a circumscribed context about which little is known. Some examples of research questions that would be answered by grounded theory method are: What theory explains the process of how men with early stage prostate cancer decide among treatment options? Or, what are the major processes in the transition to survivorship after pneumonectomy for lung cancer?

3.2.2. Sampling Procedures

After the phenomenon or condition of interest is identified, the investigator will choose participants who have experience with the phenomenon and are able to articulate it. As the study progresses, theoretical sampling will be used to select participants who may give additional or diverse perspectives based on the categories identified in the data. Sampling will continue until no new categories are revealed and each category is completely described. At this point, theoretical saturation has been achieved. Although it is difficult to state an exact sample size, typically 20–30 interview participants are adequate to reach saturation.

3.2.3. Data Collection

Data collection is accomplished primarily by individual interviews. Additional data from other sources may be used such as the medical record, observations of the setting and behaviors, interviewer and participant journaling, and focus groups. When interviewing, it is important to let the participants tell their story while keeping the conversation related to the topic of interest.

3.2.4. Analysis

Grounded theory data analysis is conducted concurrently with ongoing data collection identifying the concepts and categories that begin to frame the process being described. There is a series of steps that begins with reviewing the transcribed data line-by-line and ends with articulation of a theoretical explanation that describes the concepts and their relationship to one another within the study sample. A constant comparative process is used to continually scrutinize new data relevant to the themes that have already been identified until all themes have been fully described and no new themes are appearing. The result of the analytic process is a substantive-level theory inductively developed from data derived around a specific problem, condition, or population of people. This theory can then be subjected to further empirical study. For example, the process of prostate cancer treatment decision making may be described for upper-middle class Caucasian men. However, this process may or may not be similar in men of other ethnicities. Thus further studies among men of various ethnicities may be conducted, now using the initial grounded theory as a point of departure.

3.2.5. Rigor

Rigor is established through verification processes. These occur throughout the research process. Within each analytic phase, the investigators ask questions about the interrelationships of the categories. The investigator then returns to the data to verify or refute the answers to these questions. By constantly returning to the data, closeness of the grounded theory to the data is ensured. After the theory is developed, the investigator then reviews published literature for supplemental verification. Finally, the investigator may have participants review the written report and comment on whether or not the process or theory is congruent with their experience.

3.3. Phenomenology: Understanding the "Lived Experience"

In the broadest sense, phenomenology is study of everyday lived experiences of human beings *(27)*. Using this qualitative method, investigators seek to understand the meaning underlying human experiences of phenomena *(28,29)*. For clinicians, the phenomena of interest are related to health issues. For example, an aim of a phenomenologic study may be to understand how men experience and live with incontinence after a radical prostatectomy for cancer and to provide insights into managing this symptom. Phenomenology is deeply rooted in the writings of the German philosopher, Edmund Husserl (1859–1938), with a second major branch following the philosophic thought of Martin Heidegger (1889–1976). There are many subtypes of phenomenology, but the common thread is concern with meaning of the lived experience.

3.3.1. Types of Research Issues Addressed

Phenomenology is most appropriate when the purpose of a study is to understand or describe the essential structures and meanings of a phenomenon as experienced by a group of people. Examples of questions are: What does it mean to participate in deciding on prostate cancer treatment? What are the underlying themes and contexts that account for patients' decisions to call their surgeons regarding surgical wound healing postdischarge?

3.3.2. Sampling

Participants must be individuals who have experienced the phenomenon and can articulate their experience in detail, reflecting on its personal meaning. Typically, up to 10 participants will provide in-depth data on the selected phenomenon.

3.3.3. Data Collection

After identifying individuals who have experience with the phenomenon, data are collected through intensive interviews, maybe requiring a participant being interviewed several times. To gain depth in understanding, the interviewer will use verbal probes to explore various aspects of the phenomenon with the participant until the underlying meaning of the participant's experience is visible. Before interviewing begins, the investigator must first set aside his or her own beliefs and experiences. This bracketing requires self-insight and allows the investigator to approach the study with minimal preconceived beliefs or theories. In this way, the data can speak for themselves.

3.3.4. Analysis

Interviews are audiotaped and transcribed verbatim for analysis. Analysis begins with several readings of the transcripts to gain an overall sense of what is being said. The investigator then writes a description of her or his experience of the phenomenon. Next, passages are sought that describe how participants experienced the phenomenon and then the investigator writes a description of what happened. Finally, the interviewer reflects again on her or his own description and constructs an overall description of the meaning or essence of the experience.

The outcome of the analytic process is an exhaustive description of the meaning of the phenomenon. For example, "loss" has certain, recognizable characteristics whether it is loss of a loved one, a cherished object, or a bodily function such as erectile function.

3.3.5. Rigor

Verification and standards are primarily related to the researcher's interpretation. The basic criterion is whether or not the final description provides an accurate picture of the common aspects and structural features of the phenomenon. This can be demonstrated by verifying results with study participants and asking outside reviewers to assess the data for similar patterns. Finally, a reader who has experienced the phenomena, but was not interviewed during data collection, reviews the logic of the analysis and reflects on whether it is consistent with her or his own experience.

3.4. Narrative: Understanding the Stories People Tell

Use of narrative approaches to research grew out of the postmodern era during the 1980s. This approach was born out of the reconceptualization of people as story tellers. Narrative focuses on how people tell their stories to reveal how meaning is made of events and situations. Individuals develop a narrative to make meaning of their lives and then story events and occurrences so that they fit that narrative. Cultures have predominant narratives that influence how individuals develop their personal narratives and the roles in which they place themselves.

3.4.1. Issues Addressed

Narrative approaches allow the investigator to explore how people impose order on the events of their life to make sense of them. A major illness or surgery is an event that an individual must somehow work into their personal narrative to make sense of it for their own life. Narrative approaches are useful when research issues revolve around identity and identity disruptions such as might occur in women having mastectomies or psychologic process such as coping.

Narrative research is descriptive and explanatory. It can be used to describe particular life episodes, conditions under which one type of story prevails over another, the relationship between individual stories and cultural stories, and the function that certain events serve for individuals. Narrative research can also be used to understand why something happened as it did.

3.4.2. Sampling

Narratives are ubiquitous because of the universal impulse of humans to tell stories. In selecting respondents, the investigator seeks out individuals who have experienced the event of interest and are able and willing to articulate their story about it within the context of their lives. For further details, the reader is referred to "Telling Stories: Narrative Approaches in Qualitative Research" *(30)*.

3.4.3. Data Collection

Data collection involves having respondents tell their stories verbally or in writing, uninterrupted by investigator questions or probes. In this way, the individual constructs the story in a way that places the event into his or her life in a meaningful way and chooses to emphasize those aspects that have meaning. The sequence and how this story is told can then be analyzed to provide description and explanation. Data can also include story line graphs on which an individual participant labels events of a lifeline and describes the meaning of those events for the investigator.

3.4.4. Analysis

Analysis of narratives uses techniques more often seen in literary critiques. Within the narratives collected, the investigator searches for emplotments, the ways in which individuals sort and order life events. The manner in which the individual characterizes himself or herself and those cast in key roles in the stories told can provide insights into how the person has fit the event into their life story. Also, placement of events within narratives allows insight into their meanings. Riessman *(31)* provides a succinct overview of the analytic process. Narrative analysis results in a "metastory." It is the investigator's representation of the respondents' stories. This can take the form of a book, a research report, or an article.

3.4.5. Rigor

The goals of narrative investigation are believability and enhancement of understanding. However, narratives can and should change over time as settings, perspectives, and underlying social discourses and power relationships change. Of relevance for narrative are the criteria of persuasiveness, correspondence (verification of the truthfulness of the representation by respondents), coherence of speaker's goal with linguistic devices used with common themes, and pragmatic use or the extent to which a particular study becomes the basis for other work.

4. AN EXAMPLE OF QUALITATIVE INQUIRY: EXPLORING PATIENT AND PHYSICIAN PERCEPTIONS OF PERSONAL AND MEDICAL FACTORS RELEVANT TO TREATMENT DECISION MAKING FOR LOCALIZED PROSTATE CANCER

No other disease condition with the high incidence of prostate cancer has so many alternatives with so few certainties. Localized prostate cancer (LPC) can be treated with one or more of several modalities including observation alone, surgery, cryosurgery, hormonal therapy, brachytherapy, or external beam radiation therapy. There are no completed, randomized studies of these treatment modalities. The notorious complications of prostate cancer treatment, sexual, bladder, and bowel dysfunction cannot be easily compared between modalities. There is a growing body of evidence that men with LPC conduct the decision-making process by considering their personal characteristics and factors which may be much more influential than any medical factor *(32–38)*. Physicians, notably surgeons, have been advised to approach counseling the man with LPC in a way that takes into account individual factors *(38, 39)*. Yet, few empiric data have been reported as to what those factors are and how physicians incorporate those factors into a discussion of treatment options. The purpose of this study was to explore physician perceptions of personal and medical factors relevant to treatment decision making for LPC, in general and for specific patients, and how patients with preferences for certain LPC treatments or outcomes may conceptually link these factors.

4.1. Designing the Study

A cross-sectional descriptive study of 12 physician/patient-paired transcripts using both qualitative and quantitative methods was employed. Data were collected in 1998–1999. University of Washington Human Subjects Division approval was in place at all times during the study and analyses from 1998 to present.

4.2. Sample and Procedure

In 31 individual, tape-recorded, and transcribed interviews with men diagnosed with LPC (within 6 mo) *(32)*, patients identified particular physicians as having had some influence on the treatment decision. Purposive sampling guided our choice of matched physicians to include surgeons, radiation oncologists, and medical oncologists. These 12 physicians were then individually interviewed regarding what factors they believed were important to discuss with men recently diagnosed with LPC during the presentation of treatment options and what medical and personal factors they recalled about the particular patient. A semistructured approach to interviewing was used, including identification of broad topics with minimal prompts. Each physician was informed before the interview as to which patient was to be discussed, allowing a preinterview review of records. Patients had given written permission for us to approach their physician for this component of the study.

4.3. Data Analysis

Each transcript was entered into NUD*IST 4, a code-based data analysis software package. Physician and patient transcripts were then paired. Text for each pair was selectively coded *(26)* using inductive analysis for important/influential personal factors, medical factors, and preferences and then quantitatively counted and matched for concordance between patient and physician.

4.4. Findings

Ten men and two women physicians were approached and all agreed to be interviewed. Nine urologists, two radiation oncologists, and one medical oncologist reported a mean of 20.25 (SD = 9.89) years of postgraduate practice, ranging from 6 from 42 yr.

4.4.1. Factors and Roles Identified by Physician

Physician participants described nine essential information topics: pathology, medical history, expected longevity, treatment options, research findings, side effects, physician bias, patient personal values, and patient fear. During the analysis, four distinct physician roles became evident: expert, educator, navigator, and partner. Eight of the physicians described adopting the partner role, incorporating all four roles. Table 1 lists definitions that were synthesized from the physicians' descriptions of their own verbal approaches to the "options talk" and the roles that the physicians adopted while helping the patient prepare for the treatment decision.

Table 1
Physician (MD) Role, Definitions, and Quotes Derived From the Physician Transcripts

MD Role	*Definition*	*Exemplar Quote*
Expert	MD described telling patients specific *specialized knowledge* (e.g., facts, data, survival rates, options)	"I just tell them those are the risks and these are the percentages and the have to know that very well before they accept surgery."
Educator	MD described explaining specialized knowledge to patients, placed in the patient's own *medical* context. This role extends expert role	"I would say the first thing I do is based on the facts of their specific ca decide what I think their options are. And if it's appropriate I'll go into more detail about each option or tell why I think one option might be be than another . . . . and I might tell them why I think in their particular ca the facts and the symptoms that this guy has would better he suited for option A rather than *option* B."
Navigator	MD described putting the specialized information in an applied personal *context of any man*/other men. Extends expert and educator role	"I tell most people that if they want to, 4 months [after] surgery they ca begin training for a marathon, if they want; if they [could] run a marath before the surgery. But, there is a risk of incontinence and that's a diffic management problem if it should occur . . . So, they have to be aware o that risk."
Partner	MD described putting the specialized information in the personal context of each particular patient's life context, including personal factors and values	"Sometimes you try and get a sense of how they feel about . . . I ask the if they're comfortable treating the prostate and leaving it in their body. I ask them how they feel about surgery or radiation. Some people really want to do anything but surgery, some people are afraid of radiation; th think it's toxic. I ask them . . . how *physically* active they are and what kind of work they do.

At least one intersection between text coded for a physician role and text coded for an essential topic occurred in all 12 physician transcripts. Intersections between essential tell topics and physician roles for all transcripts indicated that physicians who adopted navigator and partner roles were more likely to acknowledge patient personal values and patient fear as essential topics to discuss with men diagnosed with LPC.

Nearly all the physician participants addressed one specific challenge in the physician/patient discussion of treatment options. The first physician interviewed stated, "Frequently they will say well ... what [would] you do if it was your cancer?" Answers recalled by the physicians varied. Only one participant reported that he never answered the question, stating that he would never be in exactly the same situation as that patient. Others gave reasons why not to answer the question for particular patients: when the physician doesn't feel comfortable treating that patient and when the patient is resourceful and able to make an informed decision. Seven physicians described answering the patient's question as an opportunity to either: get closure on the decision ($n = 4$), give an honest answer to a sincere question ($n = 3$), or to explain the physician's own bias ($n = 1$).

4.5. Personal and Medical Factors Identified

Table 2 presents frequency data regarding the personal and medical factors identified by the 12 patients. Physicians spontaneously recalled a certain percent of these factors when prompted to recall any personal or medical factors about this particular patient. The concordance between the patient/physician pair was higher for the more commonly identified personal factors. The majority of patient participants cited personal factors as influencing the treatment decision. The following is a quote illustrating the link of a personal factor and treatment preference, plus a match with physician recall of one personal factor.

> Patient: But knowing that you can live a more normal life and enjoy what you have left, theoretically. Because at age 66, you don't know anyway how much you got left. So, I chose that [seed implants] as a better method to go. If I had 5 to 7 to 12 years left, I didn't want to be straddled with being a weak little kid and not being able to do a darn thing [due to incontinence].
>
> Personal factors = age, fear of incontinence; preference = seed implant
>
> Physician: . . . his overall health wasn't fantastic. Though ... his age is right in the range he was 67 or is now 67, so his age is right in the range for any treatment option, so that certainly didn't limit him.
>
> Match = age

4.6. Placing the Findings of the Qualitative Research Into Perspective

The majority of both personal and medical factors were linked to patient preferences for LPC treatment or outcome indicating that these factors are strongly influential for a personal treatment choice. Notably, the consulting physicians did not commonly recall the personal factors articulated as influential by the patients. This may suggest a lack of communication about relevant personal factors. It is difficult to discern to what degree physicians in this sample did or did not discuss these factors or help these patients to clarify their preferences. Other studies exploring patient preferences for cancer treatment or outcome have identified a number of the same personal and medical factors identified

Table 2
Personal and Medical Factors Identified as Important or Influential by 12 Patients and Those Recalled by the Respective 12 Physicians

Personal Factors	*Patient Identified N (% of 12)*	*MD Recalled N (% of Patient factors)*	*Concordance MD and Patient N*	*Personal Factor Influenced Decision N (% of Patient Factors)*
Common (>50% of Patients)				
Age	11 (92)	10 (83)	9	10 (91)
Cancer in family	10 (83)	1 (8)	1	10 (100)
Potent/sexually active	8 (67)	2 (17)	2	8 (100)
Fear of incontinence	8 (67)	1 (8)	1	8 (100)
Has family responsibilities	7 (58)	6 (50)	5	7 (100)
Desire for longevity	7 (58)	3 (25)	3	7 (100)
Being informed decision maker	7 (58)	4 (33)	4	7 (100)
Less Common (25–50% of Patients)				
Desire to be pain-free	4 (33)	1 (8)	1	4 (100)
Localized prostate cancer experience of peer	5 (42)	0 (0)	0	2 (40)
Diet concerns	3 (25)	1 (8)	1	3 (100)
General outlook on life	3 (25)	2 (17)	2	3 (100)
Being anxious related to cancer	3 (25)	2 (17)	1	3 (100)
Information seeker	3 (25)	3 (25)	2	3 (100)
Uncommon (0–24% of Patients)				
Physically active	1 (8)	0 (0)	0	1 (100)
Ethnicity	0	1 (8)	0	0 (0)
Social class	0	1 (8)	0	0 (0)
Alcohol use	0	1 (8)	0	0 (0)
Medical factors				
Comorbidity	8 (67)	4 (33)	2	8 (100)
Pathology	7 (58)	6 (50)	4	4 (57)
Prostate-specific antigen	6 (50)	4 (33)	2	1 (17)

in this study. Yan and colleagues, in a study of 1809 men diagnosed with LPC, reported that age, race, continence, potency, and comorbid conditions were associated with various treatments for LPC *(40)*. However, the level of patient involvement in the treatment choice and the interaction with the physicians were not clear in this retrospective survey study. In another qualitative method study of 102 men with LPC, Holmboe and Concato *(41)* documented that patients typically used information gained from a variety of sources combined with their own "patient—centered factors" to arrive at a treatment choice.

The generalizability of these findings is limited due to the exploratory sample size and convenience sample method of patient recruitment. In addition, physician participants were pre-identified as "influential," which may have bearing on the findings. Perhaps the match between perceptions would be less in a pairing of noninfluential physicians. Because each interview was conducted without using a highly structured format, the content contained in the transcripts may be constrained by the conversational cues present (or absent) in each interview. It may be that physicians did not reveal the extent to which they recalled personal and medical factors about the patient with which they were paired or that they simply could not remember. Similarly, it may be that patients did not reveal all personal or medical factors about themselves or the preferences with which these factors may be linked.

5. SOFTWARE FOR QUALITATIVE DATA ANALYSIS

Conducting a qualitative research study typically results in large amounts of richly detailed (often called "thick") descriptive data. Discovering the commonalties and unique themes in such data can be tedious and laborious as well as exciting. Software packages developed to handle textual data have the potential to automate many of the routine tasks related to data processing and analysis.

Although some traditional researchers have resisted and continue with paper, highlighters, and much floor space, most qualitative scholars agree that the use of computers for qualitative data analysis facilitates data management and makes possible analytic techniques that once took inordinate amounts of time *(42)*. It must be remembered that any software package is a tool for data processing and management. Approaches to analysis and interpretation of data remain the responsibility of researchers.

Several qualitative software packages are available. Those investigators who are considering which software package is most appropriate for a specific use are referred to a thoughtful publication by Barry *(43)* comparing two popular and widely marketed applications. Surgeons may find that their collaborators with experience in qualitative methods of analysis have trained and already become familiar with one or another application.

6. SUMMARY

Qualitative research methods are a group of strategies that are well-suited for discovering the personal aspects and meaning in the many surgically related conditions experienced by our patients. The rigorous analysis involved in a qualitative study can illustrate important variables and hypotheses within multiple designs and can be partnered with other methodological approaches. Clinicians in the surgical disciplines are encouraged to include, explore, and develop qualitative research approaches to clinical inquiry.

REFERENCES

1. Heidegger M. Being and time. New York: Harper & Row, 1962.
2. Merleau-Ponty M. What is phenomenology? Cross Currents 1956;6:59–70.
3. Glaser B, Strauss A. The discovery of grounded theory. New York: Aldine Publishing, 1967.
4. Denzin NK, Lincoln, Y. Strategies of qualitative inquiry. In: Denzin YL, ed. The handbook of 1ualitative research. Vol. 2. Thousand Oaks, CA: Sage Publications, Inc, 2000; 346.
5. Burns N, Grove S. The practice of nursing research: conduct, critique, & utilization. Philadelphia: W.B. Saunders Company, 2001; 840.
6. Miller WL, Crabtree BF. Clinical research. In: Denzin YL, ed. The handbook of qualitative research. Thousand Oaks, CA: Sage Publications, Inc, 2000; 607–631.
7. Risjord MW, Dunbar SB, Moloney MF. A new foundation for methodological triangulation. J Nurs Scholarsh 2002;34:269–275.
8. Benoliel J. Advancing nursing science: qualitative approaches. West J Nurs Res 1984;7:1–8.
9. Strang P. Qualitative research methods in palliative medicine and palliative oncology. Acta Oncot 2000;39:911–917.
10. Berry DL. Return-to-work experiences of people with cancer. Oncol Nurs Forum 1993;20:905–911.
11. Berry DL. Detection and diagnosis experiences of employed persons with urologic cancer. Urol Nurs 1994;14:52–56.

12. Maliski SL, Heilemann MV, McCorkle R. Mastery of postprostatectomy incontinence and impotence: his work, her work, our work. Oncol Nurs Forum 2001;28:985–992.
13. Aaronson NK, Ahmedzai S, Bergman B, et al. The European Organization for Research and Treatment of Cancer QLQ-C30: a quality-of-life instrument for use in international clinical trials in oncology. J Natl Cancer Inst 1993; 85:365–376.
14. Litwin MS, Hays RD, Fink A, et al. Quality-of-life outcomes in men treated for localized prostate cancer. JAMA 1995;273:129–135.
15. McCorkle R. The measurement of symptom distress. Semin Oncol Nurs 1987;3:248–256.
16. Holley SK. Evaluating patient distress from cancer-related fatigue: an instrument development study. Oncol Nurs Forum 2000;27:1425–1431.
17. Sofaer S. Qualitative research methods. Int J Qual Health Care 2002;14:329–336.
18. Brown JB, Sangster LM, Ostbye T, et al. Walk-in clinics: patient expectations and family physician availability. Fam Pract 2002;19:202–206.
19. Silverman D. Analyzing talk and text. In: Denzin YL, ed. The handbook of 1ualitative research. Thousand Oaks, CA: Sage Publications, Inc, 2000; 821–834.
20. Allen DG, Hardin PK. Discourse analysis and the epidemiology of meaning. Nurs Philosophy 2001;2:163–176.
21. Tedlock B. Ethnography and ethnographic representation. In: Lincoln NDY, ed. Handbook of qualitative research. Thousand Oaks, CA: Sage Publications, 2000;455–486.
22. Spradley J. The ethnographic interview. New York: Holt, Rinehart & Winston, 1979.
23. Spradley J. Participant observation. New York: Holt, Rinehart & Winston, 1980.
24. Glaser B. Theoretical sensitivity: advances in the methodology of grounded theory. Mill Valley, CA: Sociology Press, 1978.
25. Strauss A. Qualitative analysis for social scientists. New York: Cambridge University Press, 1987.
26. Strauss A, Corbin J. Basics of qualitative research: grounded theory procedures and techniques. Newbury Park, CA: Sage Publications, 1998.
27. Creswell JW. Qualitative inquiry and research design: choosing among five traditions. Thousand Oaks, CA: Sage Publications, 1998.
28. van Mannen M. Researching the lived experience. Albany, NY: State University of New York Press, 1990.
29. Benner P. Interpretive phenomenology. Thousand Oaks, CA: Sage Publications, 1994.
30. Sandelowski M. Telling stories: narrative approaches in qualitative research. Image J Nurs Sch 1991;23:161–166.
31. Riessman C. Narrative analysis. Thousand Oaks, CA: Sage Publications, 1993.
32. Berry DL, Ellis WJ, Schwien CC, Woods NF. Treatment decision making by men with localized prostate cancer: the influence of personal factors. Urol Oncol 2003;21:93–100.
33. Feldman-Stewart D, Brundage MD, Nickel JC, MacKillop WJ. The information required by patients with early-stage prostate cancer in choosing their treatment. BJU Int 2001;87:218–223.
34. Feldman-Stewart D, Brundage MD, Van Manen L. A decision aid for men with early stage prostate cancer: theoretical basis and a test by surrogate patients. Health Expect 2001;4:221–234.
35. Davison BJ, Gleave ME, Goldenberg SL, et al. Assessing information and decision preferences of men with prostate cancer and their partners. Cancer Nurs 2002;25:42–49.
36. Patel HR, Mirsadraee S, Emberton M. The patient's dilemma: prostate cancer treatment choices. J Urol 2003;169:828–833.
37. Steginga SK, Gardiner RA, Yaxley J, Heathcote P. Making decisions about treatment for localized prostate cancer. BJU Int 2002;89:255–260.
38. Diefenbach MA, Dorsey J, Uzzo RG, et al. Decision-making strategies for patients with localized prostate cancer. Semin Urol Oncol 2002;20:55–62.
39. Dawson NA, Fourcade RO, Newling D. The management of localized prostate cancer. Prostate Cancer Prostatic Dis 2002;5(Suppl. 2):S3–S7.
40. Yan Y, Carvalhal GF, Catalona WJ, Young JD. Primary treatment choices for men with clinically localized prostate carcinoma detected by screening. Cancer 2000;88:1122–1130.
41. Holmboe ES, Concato J. Treatment decisions for localized prostate cancer: asking men what's important. J Gen Intern Med 2000;15:694–701.
42. Bourdon S. The integration of 1ualitative data analysis software in research strategies: resistances and possibilities. Vol. 2003: Forum Qualitative Sozialforschung/Forum: Qualitative Social Research, 2002.
43. Barry C. Choosing qualitative data analysis software: Atlas/ti and NUD*IST compared. Sociol Res Online [Online serial] 1998;3.

18 Systematic Reviews and Meta-Analyses

Timothy J. Wilt, MD, MPH
and Howard A. Fink, MD, MPH

Contents

Surgical conditions span a broad spectrum of health states that result in morbidity and resource utilization. For patients and physicians, identifying the risks and benefits of interventions can be difficult. Selecting the "best treatment" for a particular clinical situation from the vast array of available options can be confusing. Health care providers, policy makers, and educators are focusing on "evidence-based health care," the integration of the best research evidence with clinical expertise and patient values. Because they represent the gold standard for testing new interventions, randomized or controlled clinical trials (RCT/CCTs) are the centerpiece of research evidence. Systematic reviews and quantitative meta-analyses have been suggested as an even higher level of evidence, because they provide scientifically rigorous synthesis of all the known evidence from RCT/CCTs or other best evidence.

To make appropriate health care decisions, patients, physicians, and health policy makers must have access to high quality information. The goals in this chapter are to: describe systematic reviews and meta-analyses and how they differ from traditional reviews; summarize methods used in conducting systematic reviews and meta-analyses; and provide an example of a completed systematic review/meta analysis to aid clinician investigators in conducting and interpreting these reviews.

1. WHAT ARE SYSTEMATIC REVIEWS AND META-ANALYSES?

Systematic literature reviews are a method of locating, appraising, and synthesizing evidence. Their primary goals are to answer specific questions, to reduce bias in the selection and inclusion of studies, to appraise the quality of the included studies, and to summarize them objectively. Systematic reviews are applicable to all types of research

From: *Clinical Research for Surgeons*
Edited by: D. F. Penson and J. T. Wei © Humana Press Inc., Totowa, NJ

Table 1
Differences Between Traditional Narrative Reviews and Systematic Reviews

Feature	*Narrative Review*	*Systematic Review*
Question	Often broad in scope	Often a focused clinical question
Sources and search	Not usually specified, potentially biased	Comprehensive sources and explicit search strategy
Selection	Not usually specified, potentially biased	Criterion-based selection, uniformly applied
Appraisal	Variable	Standardized critical appraisal
Synthesis	Often a qualitative summary	Quantitative summary if meta-analysis
Inferences	Sometimes evidence based	Evidence based

designs. They can evaluate treatment interventions, diagnostic or screening tests, and prognostic variables. Systematic reviews efficiently integrate otherwise unmanageable amounts of information to support evidence-based clinical decision making. These reviews identify and disseminate best evidence, evaluate the consistency of findings, explore differences, and help to resolve uncertainty. The products generated are vital in developing quality improvement projects, creating practice guidelines or policy initiatives, enhancing shared-decision making, and identifying gaps in knowledge that require future research.

Systematic reviews differ from traditional narrative reviews as summarized in Table 1. Findings from systematic reviews produce the most unbiased estimates of the clinical effect of an intervention, diagnostic test, or prognostic variable *(1)*. They have had a profound impact on researchers, clinicians, medical educators, patients, and policy makers. Guideline groups, including the US Preventive Services Task Force and the American Urological Association, conduct high-quality systematic reviews, to provide the highest evidence level. Without systematic reviews, researchers may miss promising investigative opportunities, be unaware of evidence, or embark on studies of questions that already have been answered. Policy makers and administrators use systematic reviews to develop clinical policies that optimize outcomes using available resources *(2)*. Clinicians and medical educators use systematic reviews in their daily practice.

Systematic reviews are a uniquely powerful mechanism for education. They offer teachers a new opportunity to model rational and effective use of information. Systematic reviews link clinical questions with research results that would otherwise be difficult to locate, read, and appraise *(3)*. Consumers use systematic reviews to help them make decide among diagnostic and treatment options *(4)*.

Meta-analysis is the systematic, quantitative approach to combining results from comparable individual studies for the purpose of synthesizing and integrating results. It typically yields pooled or weighted average estimates of intervention effects. Statistical pooling can estimate the effect of an intervention on a particular outcome with more precision than the individual studies and can suggest whether results vary according to patient subgroup or intervention. However, pooling in systematic reviews may not always be feasible or appropriate. Furthermore, as with any type of research, systematic reviews and meta-analyses have limitations *(1, 3, 4)*. The data generated from systematic reviews are limited to the quality and type of data reported from studies. The ability to combine and quantitatively analyze results does not mean that this is clinically or statistically valid. Inappropriate pooling may lead to erroneous conclusions. In these instances, a qualitative systematic review and summary of findings is valuable and may be more appropriate. However, the findings from well-conducted reviews are beneficial because they provide an unbiased, concise summary of the evidence.

Several publications have described the science of reviewing research, differences of narrative reviews vs systematic reviews and meta-analyses, as well as how to carry out, critically appraise, and apply meta-analyses in practice *(1, 3–6)*. Systematic reviews and meta-analyses should be as methodologically rigorous as well-designed and adequately powered RCT/CCT. Guidelines recommend: (1) development of a prospective protocol whereby the hypotheses are derived before data abstraction and analyses; (2) use of standardized definitions of key outcomes; (3) quality control of data; (4) inclusion of all patients from all studies in the final analysis; and (5) adhering to quantitative standards and the use of appropriate statistical monitoring guidelines to indicate when the results of the data of a meta-analysis are conclusive.

2. STEPS INVOLVED AND FORMAT OF SYSTEMATIC REVIEWS

The design, methods, and reporting of a systematic review should follow a standardized format that adheres to previously described quality standards. This enables the user to find the objectives and results quickly and assess their validity and applicability *(5–8)*. A summary of steps involved in conducting a systematic review and meta-analysis is provided in Table 2. The format used for systematic reviews similar to those conducted by the Agency for Healthcare Research and Quality's Evidence-based Practice Centers is shown in Table 3 and described in the following section.

2.1. Identify Research Topic

Initial plans are made regarding clinical and research questions to be addressed. A research protocol is developed and organized as follows: background explaining the topics being reviewed, including the biologic basis of the condition and clinical and economic importance; objective and key research questions, which is a precise statement of the primary objective of the review, including the invention(s) reviewed and the problem(s) addressed; and study/patient selection criteria.

2.2. Identify Studies for Specific Health Conditions

2.2.1. Search MEDLINE and Other Databases

A detailed plan for the literature search is developed often with the assistance of a literature search specialist. This plan describes data sources, search terms, and inclusion and exclusion criteria. We use a multifaceted approach to identify studies-e.g., use of standard electronic literature databases and reviews of key journals, reference lists of relevant articles, and Cochrane Collaboration resources. Relevant studies for each review are identified, screened, and retrieved using a standardized search strategy specific for each review (5, 6). We review the abstract and title of articles (and if necessary the full text) to determine eligibility. Translation of non-English language journals or correspondence with the authors may be necessary. RCT/CCT are not the best sources for evaluating adverse events, preferences, health utilities, and costs. Other sources include patient

Table 2
Steps Involved in conducting Systematic Reviews/Meta-Analyses

The process of writing a review adheres to established standards *(6–8)*.

- Identify research topic and questions
  - — Recruit Medical Advisory Panel members if needed
  - — Develop key clinical questions to be addressed
  - — Create and refine review protocol
- Identify randomized or controlled clinical trials (RCT/CCTs) or other best evidence and develop registry
  - — Select trials/evidence to be included
  - — Create and maintain registry of disease specific RCT/CCTs/reviews
- Determine inclusion/exclusion criteria
- Data collection
  - — Create provisional evidence tables
  - — Develop data abstraction forms
  - — Assess methodological quality of trials/reports
  - — Collect data
- Data analysis and synthesis (including pooling if feasible)
- Examine data for publication bias and heterogeneity
- Economic and decision analyses (if indicated)
- Prepare systematic review evidence report
  - — Summarize findings and conclusions
  - — Obtain outside peer review/criticism/comments
  - — Incorporate peer review comments and revise evidence report
- Disseminate final systematic review evidence report
- Maintain up-to-date-status of systematic review findings through approximately biannual reevaluation process

preference and health status surveys, postmarketing reports, product inserts and Food and Drug Administration Medwatch (www.fda.gov/medwatch) announcements.

2.2.2. Electronic Databases

We start with electronic searches of the MEDLINE and EMBASE databases; EMBASE has considerable overlap with MEDLINE, but for thoroughness it is instructive to search this database for at least background materials. Results are tagged and downloaded directly using reference management software (e.g., ProCite, Biblio-Link, Reference Manager) to facilitate formatting for inclusion in the report bibliography.

2.2.3. Especially Relevant Journals and Reference Lists

We determine whether relevant peer-reviewed journals are not indexed in these databases. If so, hand-searching of these journals is performed for as much of the time period specified as possible. We review the reference lists of critical articles or reviews to identify additional studies.

Table 3
Systematic Review Format Used in Technology Assessment Reports

- Cover sheet
- Structured abstract: context, objective, data sources, study selection, data extraction, synthesis, and conclusions
- Patient/consumer page: lay summary of key findings and impact statements
- Table of contents
- Executive summary
- Text
  - — Background
  - — Objectives and key questions: Description of the topic and questions examined; targeted populations, including subgroups; specification of the causal pathway to link the literature to key questions; disease epidemiology
  - — Search strategy and inclusion and exclusion criteria: Appendices document search strategies, electronic and other literature databases searched, time frame of the search, and inclusion and exclusion criteria
  - — Analysis: Summary description of the methodologic approach for the review. Criteria for grading the quality of the studies and strength of evidence; methods for analyzing and synthesizing the evidence
  - — Results: Narrative synthesis of findings; synthesis of information on special populations; presentation of supplemental analyses, such as meta-analysis of selected studies; tables or graphs to convey findings effectively
  - — Summary of individual studies including patient/disease characteristics, efficacy, and adverse effects data
  - — Quantitative synthesis (when statistically feasible and clinically appropriate)
  - — Economic and decision analysis (if indicated)
  - — Balance sheet of risks, benefits, and costs of different treatment options (including absolute and relative risk reduction and number needed to treat)
  - — Discussion and conclusions (including implications for clinical practice and future research)
  - — Evidence tables and figures for critical key questions, organized in some consistent way (with subsidiary evidence tables, if any placed in appendices)
  - — Characteristics of the included studies
  - — Specification of the interventions that were compared
  - — Results of the included studies
  - — Pooled synthesis of efficacy and adverse event data including a priori defined subgroup and sensitivity analyses
  - — List of excluded studies and reasons for exclusion
  - — References
  - — Appendices and acknowledgments; analytic framework; details on methods (e.g., data abstraction forms, quality grading scheme); evidence tables; list of excluded studies and reasons; bibliography (references cited plus studies excluded); and comments about when the report might be reviewed for updating

2.2.4. Cochrane Collaboration

The Cochrane database is the best source for identification of RCT/CCTs *(9)*. Our Prostate Cochrane Review Group (CRG) registry contains more than 2600 citations to prostate disease and urological cancer trials. We prospectively hand search major urologic journals; conference proceedings for American Urological Association (AUA) and European Urological Association (EUA) meetings for 1990 to present and existing trials registers (e.g., www.TrialsCentral.org) for ongoing trials.

Table 4
Sample of Provisional Inclusion Criteria for a Specific Topic: What is the Efficacy and Adverse Effects Associated With Treatments for Urinary Incontinence in the Elderly?

Category	*Inclusion Criteria*
Population	Humans; age > 65, both sexes, all ethnic and racial groups
Conditions	All diagnoses and causes relevant to urinary incontinence (e.g., previous hysterectomy, child birth, radical prostatectomy, benign prostatic hyperplasia).
Study settings	Inpatient and outpatient settings
Interventions	condition-specific, but will include interventions to treat urinary incontinence: Treatment: pads, medications, surgery, exercise
Outcomes	**Clinically relevant outcomes:** disease-specific aspects of morbidity (e.g., urinary tract infections, hospitalization), disease specific treatments/ procedures (pad, catheters, artificial sphincters), functioning, symptoms **Intermediate outcomes:** important surrogate outcomes or measures related to disease-specific conditions (# pads/day); laboratory tests.
Time period	1980 and later (depending on database)
Geographic site of study	North America, Europe, English-speaking Commonwealth countries; Scandinavia; Japan
Language	English (may include non-English language pending topic and advice of MAP)
Admissible evidence	North America, Europe, English-speaking Commonwealth countries; Studies with the following designs: RCTs (double and single blinded); non-RCTs (prospective and retrospective cohort studies; case-control studies) Exclusions: animal studies, studies not addressing key questions, commentaries, letters, editorials, case reports, case series
Sample sizes/ attrition rates/ duration	Ending sample sizes >10 subjects in all groups Attrition rates no greater than 30% and similar in all groups • 3 mo

2.3. Inclusion/Exclusion Criteria for Peer-Reviewed Literature

Developing inclusion/exclusion criteria involves a series of filters that progressively focuses the literature search. Provisional inclusion and exclusion criteria for a "standard" search of the peer-reviewed literature are described in Table 4.

We usually exclude studies in which attrition was greater than 30% or was significantly different between treatment or control arms. We often exclude studies for which the samples (treatment and controls) were fewer than 10.

2.3.1. Retrieve and Review Full Articles

Full articles are retrieved for those studies meeting review criteria. A data abstractor reviews each article, abstracts data, and enters information into evidence tables. A senior project leader reads the article and checks the evidence table for accuracy. When disagreements occur, the pair resolves differences through a re-review of the article. If differences arise from interpretation of the information, the study director or one of the senior researchers not otherwise involved adjudicates.

2.3.2. Track Excluded Studies

Reasons that articles, at the stage of full review, are not included in both the evidence tables and report is recorded. Abstractors note a reason for exclusion on the data collection form created for that article. We then record that code in the reference management software file, so that we can compile a listing of excluded articles and the reasons for such exclusion.

2.3.3. Review Titles and Abstracts

Abstractors review identified abstracts to determine eligibility for inclusion. This helps ensure that all appropriate abstracts are included in our literature synthesis. Abstracts determined to be ineligible after the initial review are reevaluated; if either reviewer still believes the article should be retained, we retain it. Generally speaking, we err on the side of inclusion rather than exclusion.

2.4. Data Collection

2.4.1. Data Collection

Provisional evidence tables and data extraction forms are developed with criteria for evaluating study quality and strength of the evidence. The descriptions that follow reflect our usual procedures.

2.4.2. Evidence Tables

Dummy evidence tables are created to guide development of abstraction forms or procedures for abstracting data directly into evidence tables. We create separate tables for each key question, outcome, by type of research design, and then alphabetically by study author. Generally we opt for more separate tables that are less complex. The first part describes the purpose of the research, its design, setting, populations, and outcome measures; the second part describes study outcomes, differences between groups, and other salient features, and provides a quality grade.

2.4.3. Abstraction Forms

Paper (or electronic) data abstraction forms are used for the reviewer to record the specified information about each study in a standardized fashion. Customized abstraction forms are created and piloted to ensure efficient/accurate data collection. They begin with an identification of the publication and some exclusion criteria at the start and go on with sections on which to record details on study design and outcomes and quality grade. This helps avoid extracting data on studies that would be excluded. Abstraction forms capture study characteristics. To promote efficiency, comprehensiveness, and consistency, we test our evidence tables at the same time.

2.4.4. Assess Methodologic Quality of Included Studies

Included studies are categorized as to the type of study (e.g., randomized controlled trial, case-case control, case series). However, even among RCT, results can be influenced by quality of the concealment of the randomized treatment allocation (i.e., studies with poor quality are more likely to report favorable effects). Grading the quality of individual studies takes place at the time of data abstraction. For RCT, the quality of concealment of treatment allocation is evaluated according to a scale developed by Schulz *(10)*, assigning 1 to poorest quality and 3 to best quality. We assess whether trial participants and investigators were blinded to treatment provided, whether trials used an intention-to-treat analysis, and the percentage of subjects who dropped out or were lost to follow-up. Sensitivity analyses are conducted by examining results from similar types of study design. Additionally, we assess results derived from analyses in which only RCT of best quality regarding treatment allocation concealment are considered.

2.4.5. Collect Data and Monitor Reviews for Bias, Consistency, Accuracy, and Quality

Several mechanisms assure quality of reports. Selection of technical experts from a variety of backgrounds provides multiple perspectives. Review questions are formulated using a standard format that requires the technical experts to address multiple aspects of questions. Trained reviewers are used to reduce error and bias. The abstracted data are reviewed for accuracy and consistency. Use of abstraction and rating instruments that have been pretested and revised as needed promotes consistency. Checks at the "second review" ensure accuracy and appropriate interpretation of study findings.

Because of the size, complexity, heterogeneity, and uneven quality of the literature, we subject a 10% random sample of articles to second review and abstraction in order to identify difficulties that might cause inconsistency across articles or abstractors. Experience indicates that some eligible reports may be published in non-English journals. To prevent a language bias, we often identify individuals experienced in data abstraction of non-English language articles.

Reviews using aggregated data from published studies provide similar results to reviews using individual patient data. Because of the cost and time involved in individual patient data analyses, our reviews rely on aggregated published reports or additional evidence obtained from authors/sponsors. Where available, data are abstracted according to predetermined subgroups including patient age, gender, race, disease severity, and treatment variations (e.g., dose, within-class agent, duration).

Reports are prepared in a standardized format that facilitate uniform critical appraisal across studies. Unbiased evidence tables are created by determining what the tables will include without reference to study results. For many of our larger "evidence reports" conducted for private or government agencies, we enlist external peer reviewers to provide an independent evaluation of our draft document. They receive a structured critique form to facilitate comprehensive refereeing.

Because the quality of systematic reviews is only as good as the quality of the primary evidence, we attempt to limit data synthesis to RCT/CCTs except for adverse effects, quality of life, patient preference data, and costs. When sufficient data from RCT/CCTs are not available to adequately address treatment efficacy questions we: describe these limitations and discuss possible inclusion of studies with lower methodologic quality (e.g., case-control studies, observational cohorts). The remainder of this chapter is devoted to the actual analyses of summary data obtained from systematic reviews.

2.5. Data Analysis and Synthesis

2.5.1. General Activities for Synthesis of Literature

In addition to providing a qualitative summary of the identified evidence we attempt to perform quantitative meta-analyses if possible and appropriate. Meta-analysis may be

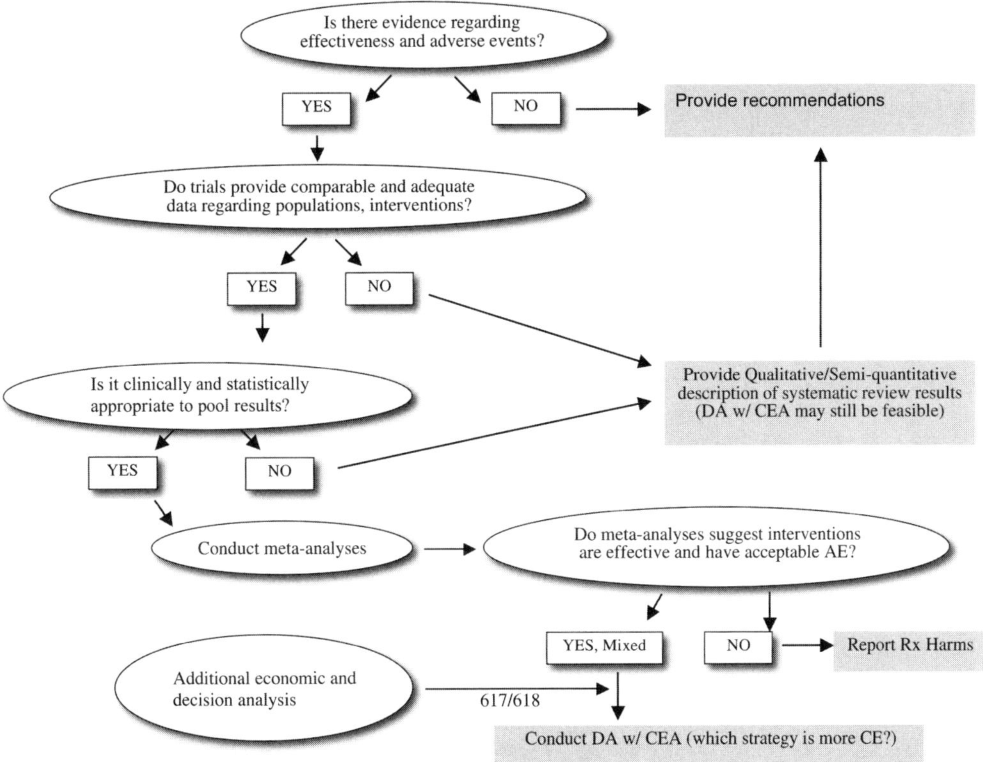

Figure 1: Flow diagram of systematic reviews with decision and cost-effectiveness analyses.

especially useful because much of the evidence may be comprised of many small, underpowered, or conflicting studies that do not provide definitive information (e.g., borderline significance, wide confidence intervals, conflicting efficacy results). Meta-analysis permits more precise estimates of possible benefit (or explore variations in effectiveness according to subgroups of patients and interventions) by quantitatively combining data from similar studies. However, it is possible that studies are so heterogeneous in clinical diagnoses, patient populations studied, therapeutic interventions, and outcomes measured that meta-analysis is significantly challenging or, indeed, improper and imprudent. Specific decisions and methods for meta-analysis are developed a priori. When it is not feasible or clinically appropriate to conduct a meta-analysis, our evidence report provides qualitative and semiquantitative summaries. Summaries are presented for individual studies as tables, figures, and text. Recommendations are made to improve and standardize the reporting of future research so that future results can be synthesized.

Figure 1 outlines the steps and products of data synthesis process that may also involve a cost-effectiveness or decision analysis (more detailed discussions regarding cost-effectiveness and decision analyses are provided in Chapter 16). As a first step, the evidence regarding the effectiveness and adverse events associated with each intervention is gathered and examined. In the absence of a body of evidence describing the effectiveness of an intervention, investigators develop recommendations for future research. If sufficient evidence is available, the adequacy and comparability of the data regarding populations, interventions, and outcomes will be reviewed, followed by an assessment of the clinical and statistical appropriateness of pooling the results. For studies that are not conducive to pooling, the review produces qualitative or semiquantitative summaries of the evidence. There are three possible outcomes when pooling data. Either results are consistent that the treatment is effective or ineffective, or there are mixed conclusions. In the case where studies agree that treatment is ineffective, no further analysis occurs and a qualitative summary of the therapeutic harms will be produced. When all studies agree that a treatment is effective or when the results are mixed, the question arises as to the magnitude of the effect.

A general description of standard meta-analytic methods follows: when pooling of data are clinically appropriate and statistically feasible, weighted risk ratios, risk differences, and their 95% confidence intervals are calculated. We typically use RevMan software *(11)* for categorical variables according primarily to the Peto method *(12)*. Other meta-analysis software programs are available and frequently useful for evaluation of data from studies other than RCT (www.metaanalysis.com). The number needed to treat for different outcomes along with their respective 95% confidence intervals will be calculated as 1/risk difference. For continuous variables, weighted mean differences (WMD) and their 95% confidence intervals are determined. Results are tested for heterogeneity at significance level of $p < 0.1$ according to the methods outlined by DerSimonian and Laird *(13)*. A p value < 0.1 is used because of the relatively low sensitivity for identifying heterogeneity. When analyses indicate heterogeneity exists (and pooling is still clinically appropriate), a random effects model is used or additional subgroup analyses are conducted to further explore reasons for heterogeneity and describe variations in effect. In general, prespecified subgroup analyses and meta-regression analyses may be conducted, as appropriate or relevant, to evaluate the consistency of effects and systematic variations in effect from differences in study design, patient characteristics, or intervention characteristics. An intention-to-treat or modified intention-to-treat analysis is used.

3. AN EXAMPLE OF A SYSTEMATIC REVIEW OF THE LITERATURE

The following example of a previously published report is provided to assist the reader in learning about the actual steps we used in developing, conducting, and presenting a systematic review and meta-analysis evaluating sildenafil (Viagra) for the treatment of male erectile dysfunction *(14)*.

3.1. Identification of Research Topic

Erectile dysfunction (ED), defined as the persistent "inability to achieve or maintain an erection sufficient for satisfactory sexual performance," *(15)*, is estimated to affect up to 30 million men in the United States *(15)* and may result in withdrawal from sexual intimacy and reduced quality of life (16). The prevalence of ED increases with age, diabetes, heart disease, hypertension, depression, and use of certain medications *(17, 18)*. ED also may be caused by spinal cord injury and prostate surgery.

Sildenafil was approved by the Food and Drug Administration for treatment of ED in March 1998. We were aware, in 2000, that many randomized controlled trials had evaluated sildenafil. We were unaware of any systematic review and quantitative meta-analysis that had examined the magnitude of treatment benefits and adverse effects associated with sildenafil treatment in men with ED, overall and according to age, comorbid conditions, and ED severity.

3.2. *Identification of Eligible Studies*

Trials were identified by searching the MEDLINE, HealthSTAR, Current Contents, and Cochrane Library computer databases between January 1995 and December 2000. The search strategy combined (impotence or erectile dysfunction) and (sildenafil or viagra or UK-92,480) and was limited by combining it with (clinical trial, controlled trial, randomized controlled trial, or multi-center study). In addition, bibliographies of retrieved trials and review articles were reviewed, and urology journals and national meeting abstracts were hand searched. All trials identified were written in English. Data for unpublished trials and supplemental data for published trials were obtained from the manufacturer and the Food and Drug Administration Internet web site.

Studies were considered eligible if they included men with ED, were randomized, compared sildenafil with placebo or active control, were at least 7 d in duration, and assessed clinical outcomes related to ED (e.g., success of sexual intercourse attempts, subject global assessment of treatment). Two reviewers independently assessed study eligibility. Differences were resolved by discussion.

3.3. *Data Collection and Outcomes*

Information on trial characteristics, patient demographics, inclusion and exclusion criteria, dropouts, treatment efficacy, and adverse events were extracted by two reviewers onto pretested data abstraction forms in a standardized fashion. The primary efficacy outcome was the percentage of all self-reported sexual intercourse attempts that were successful, defined as vaginal penetration that the subject found satisfactory. Additional outcomes included the percentage of subjects achieving successful intercourse at least once during treatment and the percentage of subjects reporting improvement in erectile function. For adverse effects, we examined the percentage of men reporting side effects and the percentage of men withdrawing from the trial. Missing or additional information was sought from authors/sponsors.

3.3.1. Assessment of Methodologic Quality

We assessed the quality of concealment of randomized treatment allocation according to a scale developed by Schulz *(39)*. We assessed whether participants and investigators were blinded to treatment provided, whether trials used an intention-to-treat analysis, and the percentage of subjects who dropped out or were lost to follow-up.

3.4. *Data Analysis*

For assessment of categorical treatment outcomes, we determined the percentage of men achieving each outcome according to treatment assignment. For measures of efficacy, we calculated weighted relative benefit increases and their 95% confidence intervals using RevMan software *(11)*. For adverse events and withdrawals, the percentage of men achieving each outcome according to treatment assignment, as well as the weighted relative risk increases and their 95% confidence intervals were determined. For assessment of continuous outcomes, we determined the mean value (e.g. percentage of successful attempts) for men within each treatment group and calculated WMD and 95% confidence intervals. Relative benefit increases, relative risk increases, and WMD were estimated using random effects meta-analyses.

Data from fixed dose studies suggested the presence of a meaningful dose-response effect for at least some treatment outcomes. Therefore, different fixed doses were not

Table 5
Baseline Characteristics of Subjects

Characteristic	*Sildenafil*	*Placebo*
Randomized subjects	4240	2707
Age (yr ± SD)	55 ± 10	54 ± 10
Ethnicity (%)		
White	71	68
Asian	21	21
Black	4	5
Other	4	7
ED duration (yr)	4.7	4.9
ED severity (%)		
Severe	47	47
Mild-moderate	46	44
None	2	3
ED etiology (%)		
Organic only	51	56
Psychogenic only	20	18
Mixed	29	26
Comorbid conditions (%)		
Hypertension	26	29
Diabetes	19	24
Ischemic heart disease	10	9
Depression	6	4
Spinal cord injury	4	7
Radical prostatectomy	3	4
Peripheral vascular disease	3	3

pooled in meta-analyses. A clinical decision was made to perform meta-analysis only between trials of similar design. Trials that employed a parallel group design, flexible dosing, and administration on an as-needed basis (PRN) were emphasized, primarily because this is the manner in which sildenafil is used in clinical practice. Efficacy data for specific subgroups also were derived from parallel, flexible-dose PRN studies.

3.5. Results

A summary of some of our findings is provided in Tables 5 and 6 and Figure 2. A table of baseline characteristics of subjects from the include studies (Table 5) provides a summary of relevant demographic and clinical information from the included studies (another table is often presented that describes characteristics of each of the included studies). Similar to reports from individual studies, this table assists the reader in evaluating baseline characteristics of the systematic review study population. The results indicated that 27 trials (6659 men) met inclusion criteria. Men were middle age, predominately white, and had ED for approx 5 yr duration. About half of the men had severe ED

Table 6
Efficacy Outcomes for Parallel Group, Flexible Dose Trials According to Subject Subgroup[a]

	Successful sexual intercourse, mean percentage of attempts per subject			Men with at least one successful sexual intercourse attempt during treatment			Men with self-reported improvement in erections		
	Sildenafil %	Placebo %	WMD (95%CI)(N)	Sildenafil %	Placebo %	RBI (95%CI)(N)	Sildenafil %	Placebo %	RBI [95%CI](N)
All subjects, primary method[b]	57	21	34 [29–38) (2283)	83	45	1.8 [1.7–1.9] (2205)	78	25	3.1 (2.7–3.5) (3535)
All subjects, alternate method[b]	66	25	39 [36–43] (2205)	NA	NA	NA NA	NA	NA	
Age ≥65	46	14	31 (24–38) (447)	74	36	2.0 (1.6–2.4) (426)	69	18	3.4 [2.7–4.2] (758)
Asian men	61	24	37 (31–42) (1220)	87	49	1.7 (1.6–1.9) (1170)	86	34	2.5 [2.2–2.8] (1363)
Black men	53	19	34 (16–51) (49)	78	31	2.3 (1.3–3.9) (47)	67	28	1.9 [1.3–2.8] (143)
Severe ED	47	11	34 (26–42) (844)	74	26	2.8 (2.1–3.7) (798)	67	15	4.2 [3.5–5.1] (1654)
HTN	50	16	33 (27–40) (628)	75	39	1.9 (1.6–2.2) (604)	68	21	3.1 [2.6–3.7] (1100)
Diabetes	44	16	27 (20–34) (551)	70	34	2.0 (1.6–2.3) (534)	63	19	3.0 [2.5–3.7] (1019)
Psychogenic	66	29	38 (32–44) (453)	91	61	1.4 (1.2–1.6) (440)	87	38	2.1 [1.7–2.5] (622)
IHD	42	14	24 (2–46) (202)	69	32	1.9 (1.3–2.7) (198)	63	20	2.6 [1.8–3.8] (376)
Depression	58	24	25 (4–47) (51)	86	43	1.8 (1.1–2.9) (49)	79	20	3.4 [2.4–4.7] (273)
PVD	57	13	39 (18–59) (49)	88	38	1.8 (0.9–3.6) (48)	70	14	3.0 [1.7–5.5] (107)
RP	25	3	24 (5–43) (42)	47	14	2.9 (1.1–7.3) (37)	48	10	3.8 [1.6–9.5] (116)
SCI*	53	8	45 (39–51) (332)	81	26	3.2 (2.4–4.2) (308)	83	12	7.2 [4.7–10.9] (345)

Abbreviations: SCI, spinal cord injury; WMD, weighted mean difference; N, number of men analyzed (except in SCI crossover data where N represents treatment arms); RBI, relative benefit increase; HTN, history of hypertension, IHD, ischemic heart disease, PVD, peripheral vascular disease, RP, history of radical prostatectomy

[a]No SCI data are available from parallel group, flexible dose trials; SCI data presented are derived from one crossover, flexible dose trial (n = 178 men).

[b]The primary method of analysis considered all sildenafil doses taken and intercourse attempts. The alternate method excluded from analyses intercourse attempts reported by the subject to have failed for reasons other than a sufficiently hard or long-lasting erection. Subgroup results were derived using the primary method.

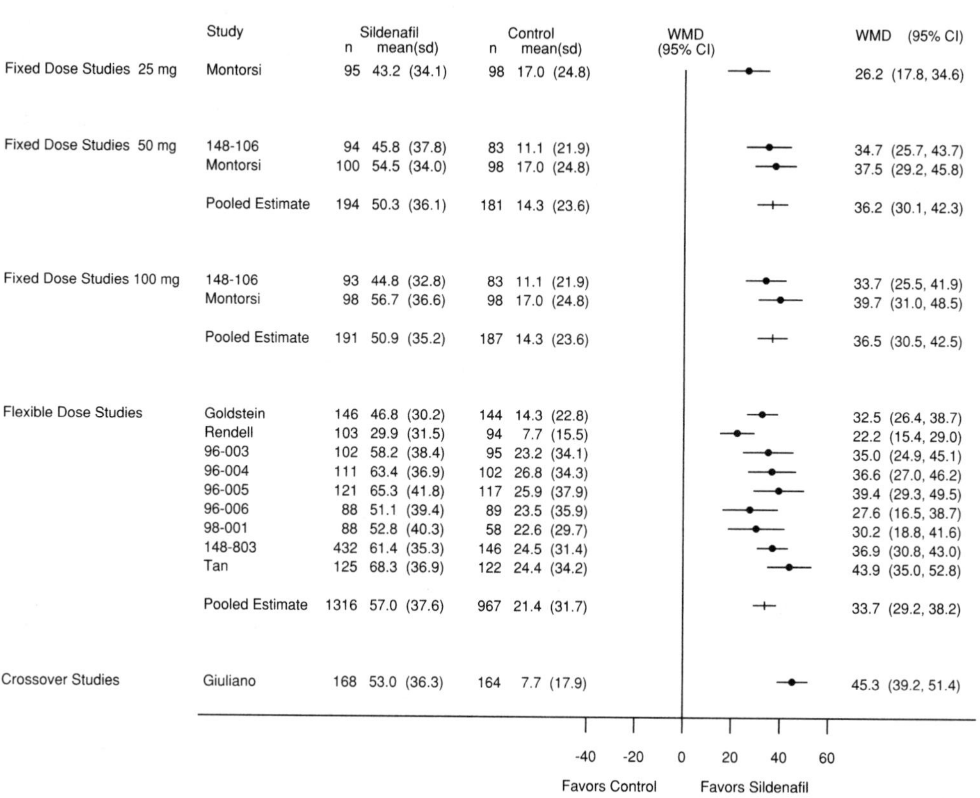

Figure 2: Weighted mean differences between sildenafil and placebo in the percentage of sexual intercourse attempts that were successful per participant according to specified study characteristics (e.g., fixed dose vs flexible dose vs crossover studies).

and similar percent were reported to have an "organic only" etiology. Clinically relevant comorbid conditions are listed.

Efficacy outcomes are reported for parallel group, flexible dose, PRN trials (Table 6). Results are reported for three specified outcomes for all subjects and according to prespecified subgroups. A typical meta-analytic "forest-plot" also shows the WMD between sildenafil and placebo in the percentage of sexual intercourse attempts that were successful per participant according to specified study characteristics (e.g., fixed dose vs flexible dose vs crossover studies) (Figure 2). In results pooled from 14 parallel-group, flexible as-needed dosing trials, sildenafil was more likely than placebo to lead to successful sexual intercourse, with a higher percentage of successful intercourse attempts (57% vs 21%); WMD = 33.7%; 95% CI, 29.2–38.2; 2283 men). In data pooled from six parallel-group, fixed-dose trials, efficacy appeared slightly greater at higher doses. Treatment response appeared to vary between patient subgroups, although relative to placebo, sildenafil significantly improved erectile function in all evaluated subgroups. Adverse effects were reported in the original manuscript and increased with higher doses of sildenafil.

We concluded that sildenafil improves erectile function and is generally well tolerated. Treatment response seems to vary between patient subgroups, although sildenafil has greater efficacy than placebo in all evaluated subgroups.

4. SUMMARY

Systematic literature reviews and quantitative meta-analyses are widely used as aids to evidence-based decision making. They serve as valuable resources for patients, clinicians, educators, health policy makers, and researchers by attempting to identify and summarize, in an unbiased fashion, the best evidence related to specific health care topics. As with any research endeavor, designing, conducting, and interpreting the results of these reviews requires rigorous adherence to quality standards. Information provided in this chapter should assist individuals interested in knowing more about this methodology.

REFERENCES

1. McQuay HJ, Moore RA. Using numerical results from systematic reviews in clinical practice. Ann Intern Med 1997;126(9):712–720.
2. Milne R, Hicks N. Evidence-based purchasing. Evidence-Based Med 1996;1:101–102.
3. Badgett RG, O'Keefe M, Henderson MC. Using systematic reviews in clinical education. Ann Intern Med 1997;126(11):886–891.
4. Bero LA, Jadad AR. How consumers and policymakers can use systematic reviews for decision making. Ann Intern Med 1997;127(1):37–42.
5. Moher D, Cook DJ, Eastwood S, et al. Improving the quality of reports of meta-analyses of randomized controlled trials: the QUOROM statement. Quality of Reporting of Meta-analyses. Lancet 1999; 354(9193):1896–1900.
6. Clarke M, Oxman AD, eds. Cochrane Reviewers handbook 4.1.6 [updated January 2003]. In: The Cochrane Library 2003(1). Oxford: Update software.
7. Systems to rate the strength of scientific evidence. File inventory, evidence report/technology assessment number 47. Rockville, MD: Agency for Healthcare Research and Quality, 2002. AHRQ Publication No. 02-E016.
8. Nixon J, Khan KS, Kleijnen J. Summarising economic evaluations in systematic reviews: a new approach. BMJ 2001;322(7302):1596–1598.
9. Dickersin K, Manheimer E, Wieland S, et al. Development of the Cochrane Collaboration's CENTRAL register of controlled clinical trials. Eval Health Prof 2002;25(1):38–64.
10. Schulz KF, Chalmers I, Hayes RJ, Altman DG. Empirical evidence of bias. Dimensions of methodological quality associated with estimates of treatment effects in controlled trials. JAMA 1995;273(5):408–412.
11. Review manager [computer program]: Version 4.1 for Windows. Oxford, UK: The Cochrane Collaboration, 2001.
12. Clarke M, Oxman AD, eds. Analysing and presenting results. Cochrane Reviewers Handbook 4.1.6 [updated January 2003], Section 8. In: The Cochrane Library 2003(1). Oxford: Update Software.
13. DerSimonian R, Laird N. Meta-analysis in clinical trials. Control Clin Trials 1986;7(3):177–188.
14. Fink HA, Mac Donald R, Rutks IR, Nelson DB, Wilt TJ. Sildenafil for male erectile dysfunction. Arch Intern Med 2002;162:1349–1360.
15. NIH Consensus Conference. Impotence. NIH Consensus Development Panel on Impotence. JAMA 1993;270:83–90.
16. Litwin MS, Nied RJ, Dhanani N. Health-related quality of life in men with erectile dysfunction. J Gen Intern Med 1998;13:159–166.
17. Panser LA, Rhodes T, Girman CJ, et al. Sexual function of men ages 40 to 79 years: the Olmsted County Study of Urinary Symptoms and Health Status Among Men. J Am Geriatr Soc 1995;43:1107–1111.
18. Martin-Morales A, Sanchez-Cruz JJ, Saenzd TI, et al. Prevalence and independent risk factors for erectile dysfunction in Spain: results of the Epidemiologia de la Disfuncion Erectil Masculina Study. J Urol 2001;166:569–574.

Index

About the Editors

David F. Penson, MD, MPH, is an associate professor of urology and preventive medicine at Keck School of Medicine, University of Southern California. He received his undergraduate degree from the University of Pennsylvania and his medical degree from Boston University School of Medicine. He served as a resident in surgery and urology at the University of California, Los Angeles. Upon completing his residency, Dr. Penson was simultaneously awarded an American Foundation for Urologic Disease Health Policy Research Scholarship and a Robert Wood Johnson Clinical Scholars Fellowship. As a fellow, he studied clinical epidemiology and health services research at Yale University, obtaining a master's degree in public health. Upon the completion of his fellowship, Dr. Penson joined the faculty of the University of Washington School of Medicine and was appointed affiliate investigator at the Fred Hutchinson Cancer Research Center. He was awarded a VA Career Development Award in 2001 and soon after obtained R-01 grant funding from the National Cancer Institute to study quality of life in prostate cancer survivors. He joined the faculty of the Keck School of Medicine at the University of Southern California in January 2004. Dr. Penson's specific research interests include the impact of prostate cancer and erectile dysfunction on patients' quality of life. He is currently funded by the National Cancer Institute and the Centers for Disease Control to study long-term quality of life and treatment decision making in localized prostate cancer. He has published on a wide range of topics, including prostate and bladder cancer, female urology, cryptorchidism, infertility, and erectile dysfunction. He served as the chairperson of the American College of Surgeons oncology group genitourinary organ site committee from 2002 to 2005 and was awarded the American Urological Association's prestigious Gold Cystoscope award in 2006.

John T. Wei, MD, MS, is an associate professor of urology at the University of Michigan Medical School and associate chairman for clinical research of its Department of Urology. Dr. Wei attended the honors program in medical education at Northwestern University, where he received his bachelor of science and doctor of medicine degrees in 1991. He completed urology residency at Cornell University Medical College in New York City. He then moved to the Robert Wood Johnson Clinical Scholars Program at the University of Michigan. During his 2 years there as a fellow, he also obtained a master's degree in clinical research design and statistical analysis from the School of Public Health, and his work has been recognized by the American Foundation for Urologic Diseases. Dr. Wei has since remained on the faculty of the University of Michigan and is currently the associate chairman for clinical research in the department of urology. Dr. Wei's research interests are in the area of urological health services research and include health-related quality-of-life assessments, disparity and patterns of care research, and evaluation of quality of care. Clinical topics for his research have included prostate cancer, urinary incontinence, and benign prostatic hyperplasia.